The Handbook of Environmental Chemistry

Founding Editor: Otto Hutzinger

Editors-in-Chief: Damià Barceló • Andrey G. Kostianoy

Volume 69

More information about this series at http://www.springer.com/series/698

Water Resources in Slovakia: Part I

Assessment and Development

Volume Editors: Abdelazim M. Negm · Martina Zeleňáková

With contributions by

P. Andráš · A. Arifjanov · M. Bálintová · D. Barloková ·
Š. Demčák · P. Dušek · M. Fendek · M. Fendeková · M. Gomboš ·
K. Halászová · M. Holub · J. Ilavský · Ľ. Jurík · T. Kaletová ·
B. Kandra · A. Luptáková · G. Markovič · A. M. Negm · D. Pavelková ·
J. Pokrývková · Š. Rehák · C. Siman · E. Singovszká · I. Škultétyová ·
M. Sokáč · Š. Stanko · A. Tall · Y. Velísková · M. Zeleňáková

 Springer

Editors
Abdelazim M. Negm
Faculty of Engineering
Zagazig University
Zagazig, Egypt

Martina Zeleňáková
Faculty of Civil Engineering, Department
of Environmental Engineering
Technical University of Košice
Kosice, Slovakia

ISSN 1867-979X   ISSN 1616-864X   (electronic)
The Handbook of Environmental Chemistry
ISBN 978-3-319-92852-4   ISBN 978-3-319-92853-1   (eBook)
https://doi.org/10.1007/978-3-319-92853-1

Library of Congress Control Number: 2018964914

This Springer imprint is published by the registered company Springer Nature Switzerland AG
The registered company address is: Gewerbestrasse 11, 6330 Cham, Switzerland

The Handbook of Environmental Chemistry
Also Available Electronically

The Handbook of Environmental Chemistry is included in Springer's eBook package *Earth and Environmental Science*. If a library does not opt for the whole package, the book series may be bought on a subscription basis.

For all customers who have a standing order to the print version of *The Handbook of Environmental Chemistry*, we offer free access to the electronic volumes of the Series published in the current year via SpringerLink. If you do not have access, you can still view the table of contents of each volume and the abstract of each article on SpringerLink (www.springerlink.com/content/110354/).

You will find information about the

– Editorial Board
– Aims and Scope
– Instructions for Authors
– Sample Contribution

at springer.com (www.springer.com/series/698).

All figures submitted in color are published in full color in the electronic version on SpringerLink.

Aims and Scope

Since 1980, *The Handbook of Environmental Chemistry* has provided sound and solid knowledge about environmental topics from a chemical perspective. Presenting a wide spectrum of viewpoints and approaches, the series now covers topics such as local and global changes of natural environment and climate; anthropogenic impact on the environment; water, air and soil pollution; remediation and waste characterization; environmental contaminants; biogeochemistry; geo-ecology; chemical reactions and processes; chemical and biological transformations as well as physical transport of chemicals in the environment; or environmental modeling. A particular focus of the series lies on methodological advances in environmental analytical chemistry.

Series Preface

With remarkable vision, Prof. Otto Hutzinger initiated *The Handbook of Environmental Chemistry* in 1980 and became the founding Editor-in-Chief. At that time, environmental chemistry was an emerging field, aiming at a complete description of the Earth's environment, encompassing the physical, chemical, biological, and geological transformations of chemical substances occurring on a local as well as a global scale. Environmental chemistry was intended to provide an account of the impact of man's activities on the natural environment by describing observed changes.

While a considerable amount of knowledge has been accumulated over the last three decades, as reflected in the more than 70 volumes of *The Handbook of Environmental Chemistry*, there are still many scientific and policy challenges ahead due to the complexity and interdisciplinary nature of the field. The series will therefore continue to provide compilations of current knowledge. Contributions are written by leading experts with practical experience in their fields. *The Handbook of Environmental Chemistry* grows with the increases in our scientific understanding, and provides a valuable source not only for scientists but also for environmental managers and decision-makers. Today, the series covers a broad range of environmental topics from a chemical perspective, including methodological advances in environmental analytical chemistry.

In recent years, there has been a growing tendency to include subject matter of societal relevance in the broad view of environmental chemistry. Topics include life cycle analysis, environmental management, sustainable development, and socio-economic, legal and even political problems, among others. While these topics are of great importance for the development and acceptance of *The Handbook of Environmental Chemistry*, the publisher and Editors-in-Chief have decided to keep the handbook essentially a source of information on "hard sciences" with a particular emphasis on chemistry, but also covering biology, geology, hydrology and engineering as applied to environmental sciences.

The volumes of the series are written at an advanced level, addressing the needs of both researchers and graduate students, as well as of people outside the field of "pure" chemistry, including those in industry, business, government, research establishments, and public interest groups. It would be very satisfying to see these volumes used as a basis for graduate courses in environmental chemistry. With its high standards of scientific quality and clarity, *The Handbook of Environmental Chemistry* provides a solid basis from which scientists can share their knowledge on the different aspects of environmental problems, presenting a wide spectrum of viewpoints and approaches.

The Handbook of Environmental Chemistry is available both in print and online via www.springerlink.com/content/110354/. Articles are published online as soon as they have been approved for publication. Authors, Volume Editors and Editors-in-Chief are rewarded by the broad acceptance of *The Handbook of Environmental Chemistry* by the scientific community, from whom suggestions for new topics to the Editors-in-Chief are always very welcome.

Damià Barceló
Andrey G. Kostianoy
Editors-in-Chief

Preface

The subject of water resources management is a very wide-ranging one, and only some of the most important aspects are covered in this volume. It soon became apparent that although a number of good books may be available on specific parts of the topic, no text covered the required breadth and depth of the subject, and thus the idea of water resources in Slovakia book came about. The book has been treated as the product of teamwork of 30 distinguished researchers and scientists from different institutions and academic and research centers with major concerns regarding water resources, agriculture, land and soil, rainwater harvesting, and water quality.

The book can serve as a reference for practitioners and experts of different kinds of organizations with responsibilities for the management of water, land, and other natural resources. Equally, we hope that researchers, designers, and workers in the field of water management and agriculture covered in the book will find the text of interest and a useful reference source. The landscape which is sustainably managed conserves water, lowers the rate and volume of runoff water from rain, snowmelt, and irrigation, and helps reduce the amount of pollutants reaching surface water.

Water is an important medium regarding the transport, decomposition, and accumulation of pollutants, whether of natural or anthropogenic origin, which in excessive amounts represent considerable risks for all kinds of living organisms, thus also for human beings. The step toward effective protection of water resources is to know their quality. Systematic investigation and evaluation of the occurrence of surface water and groundwater within the country is a basic responsibility of the state, as an indispensable requirement for ensuring the preconditions for permanent sustainable development as well as for maintaining standards of public administration and information. The basic requirement in this context is to optimize water quality monitoring and assessment and the implementation of necessary environmental measures.

This volume consists of 17 chapters in 6 parts. The first part is an introduction and contains two chapters. This part is prepared by Martina Zeleňáková from the Department of Environmental Engineering, Faculty of Civil Engineering, Technical University of Košice, and Mirka Fendeková and Marian Fendek from the Department of Hydrogeology, Faculty of Natural Sciences, Comenius University in Bratislava. The chapter "Key Facts About Water Resources in Slovakia" provides basic information about water resources in Slovakia which is a country rich in water resources. It presents climatic as well as hydrological conditions of the country focusing on river basins, surface water and groundwater of Slovakia, and general water quality. "Assessment of Water Resources in Slovakia" presents surface and groundwater bodies in more detail. It is also focused on geothermal water resources in Slovakia. The next chapter "Water Supply and Demand in Slovakia" pays attention to water abstraction, water demand, water consumption, and wastewater management in Slovakia.

The second part of the volume deals with water resources management in the agriculture sector in Slovakia. It is prepared by Ľuboš Jurík and coauthors from the Department of Water Resources and Environmental Engineering, Faculty of Horticulture and Landscape Engineering, the Slovak University of Agriculture in Nitra, and coauthor from Water Research Institute Bratislava. It consists of three chapters. The chapter "Irrigation of Arable Land in Slovakia: History and Perspective" addresses the development of irrigation worldwide from a historical point of view as well as in Slovakia. It is also devoted to organizations in Slovakia responsible for irrigation of arable soil and management of irrigation systems. The chapter "Quality of Water Required for Irrigation" deals with irrigation water quality parameters and indexes. It also deals with irrigation water quality regulation and irrigation water quality monitoring network in Slovakia. The last chapter of this part "Small Water Reservoirs: Source of Water for Irrigation" is oriented to surface water resources—small water basins and their function as storage of water for irrigation in agriculture.

The third part of the volume focuses on soil and water, groundwater. It was prepared by researchers from the Institute of Hydrology, Slovak Academy of Sciences. The first chapter "Interaction Between Groundwater and Surface Water of Channel Network at Žitný Ostrov Area" presents groundwater flow and methods of solving it. It also presents modeling of surface water and groundwater interaction and case study. The second chapter "Impact of Soil Texture and Position of Groundwater Level on Evaporation from the Soil Root Zone" quantifies the impact of soil texture on the involvement of groundwater in the evaporation process. The chapter presents the case study results obtained by numerical experiments.

The fourth part of the volume consisted of six chapters and is devoted to water quality. The chapters were written by Yvetta Velísková from the Institute of Hydrology, Slovak Academy of Sciences; Magdalena Bálintová and Eva Singovszká from the Institute of Environmental Engineering, Faculty of Civil Engineering, Technical University of Kosice; Alena Luptáková and P. Andráš from the Department of Mineral Biotechnology, Institute of Geotechnics, Slovak Academy of Sciences, Košice; and Danka Barloková and Ján Ilavský from the Department of Sanitary Engineering, Faculty of Civil Engineering, Slovak University of Technology, and their coauthors. The first chapter "Assessment of Water Pollutant Sources and Hydrodynamics of Pollution Spreading in Rivers" provides basic information from the EU and Slovak legislation regarding water resources protection. It is also devoted to water sources of pollution. The chapter also presents the case studies from the field measurements. The second chapter "Assessment of Heavy Metal Pollution of Water Resources in Eastern Slovakia" represents the pollution indices calculated for monitored sites. The heavy metals in sediments were monitored in six rivers in the eastern part of Slovakia. The third chapter "Influence of Acid Mine Drainage on Surface Water Quality" is devoted to the investigation of mine waters in Slovakia. It presents the case study of monitoring acid mine drainage in the Smolnik creek; results of analyses of water quality and sediments are presented in the text. The chapter titled "Formation of Acid Mine Drainage in Sulphide Ore Deposits" presents and discusses the creation of acid mine drainage (AMD), its influence on the environment, and its drainage. In the the chapter entitled "Groundwater: An Important Resource of Drinking Water in Slovakia," the authors discuss the groundwater quality, the occurrence of iron and manganese in water, and the methods of their removal from water. The results from experiments, from removal of iron and manganese, are also presented. The last chapter of this part "Influence of Mining Activities on Quality of Groundwater" deals with the occurrence of antimony in Slovakia. It presents the effects of antimony on the environment, and it is focused on methods of the removal of heavy metals from water.

The fifth part is focused on wastewater management. It was prepared by Štefan Stanko from the Department of Sanitary and Environmental Engineering, Faculty of Civil Engineering, Slovak University of Technology in Bratislava, and Gabriel Markovič from the Department of Architectural Engineering, Faculty of Civil Engineering, Technical University of Kosice, and coauthors. The first chapter of this part "Wastewater Management and Water Resources" is devoted to wastewater management in Slovakia with a focus on the assessment indicators. The second chapter "Possibilities of Alternative Water Sources in Slovakia" is about using rainwater, greywater, or other sources for the purposes of usable water in buildings.

The sixth part, written by editors, contains only "Update, Conclusions and Recommendations for Water Resources in Slovakia: Assessment and Development" which closes the book volume by the main conclusions and recommendations of the volume, in addition to an update of some finding which may be missed by the contributors of the volume.

Special thanks to all those who contributed in one way or another to make this high-quality volume a real source of knowledge and latest findings in the field of water resources of Slovakia. We would like to thank all the authors for their contributions. Without their patience and effort in writing and revising the different versions to satisfy the high-quality standards of Springer, it would not have been possible to produce this volume and make it a reality. Much appreciation and great thanks are also owed to the editors of the HEC book series at Springer for their constructive comments, advice, and the critical reviews. Acknowledgments must be extended to include all members of the Springer team who have worked long and hard to produce this volume and make it a reality for the researchers, graduate students, and scientists around the world.

The volume editor would be happy to receive any comments to improve future editions. Comments, feedback, suggestions for improvement, or new chapters for next editions are welcome and should be sent directly to the volume editors. The emails of the editors can be found inside the books at the footnote of their chapters.

We would like to close the preface with the statement of Heraclitus: "No man ever steps in the same river twice, for it is not the same river and he is not the same man."

Zagazig, Egypt   Abdelazim M. Negm
Kosice, Slovakia   Martina Zeleňáková
12 April 2018

Contents

Part I
Introduction to the Water Resources in Slovakia

Key Facts About Water Resources in Slovakia

M. Zeleňáková and M. Fendeková

Contents

Abstract Surface water and groundwater resources of Slovakia are rich enough to ensure current and prospective water needs. Surface water resources are bound to two different European river basins. The Danube River Basin covers 96% of the Slovak territory; Danube River flows towards the Black Sea. The Poprad and Dunajec River Basins cover 4% of the territory; both streams are tributaries of the Vistula River flowing towards the Baltic Sea. Surface water inflow into Slovakia through the Danube River amounts about 2,514 $m^3 s^{-1}$, representing 86% of the total surface water fund. The rest is amounting to app. 398 $m^3 s^{-1}$ rise in the Slovak territory, with the primary source in the precipitation. Natural groundwater amounts of Slovakia were estimated on 147 $m^3 s^{-1}$. Quaternary fluvial sediments and carbonate rocks (limestones and dolomites) create the most important groundwater aquifers. Usable groundwater amounts were estimated on 80 $m^3 s^{-1}$.

M. Zeleňáková (✉) and M. Fendeková
Department of Environmental Engineering, Faculty of Civil Engineering, Technical University of Košice, Košice, Slovakia
e-mail: martina.zelenakova@tuke.sk

A. M. Negm and M. Zeleňáková (eds.), *Water Resources in Slovakia:*
Part I - Assessment and Development, Hdb Env Chem (2019) 69: 3–20,
DOI 10.1007/698_2017_200, © Springer International Publishing AG 2018,
Published online: 30 December 2017

Keywords Groundwater, Surface streams, Usable amounts, Water fund, Water resources

1 Introduction

Slovakia is a country rich in water resources. Both the surface water and the groundwater resources ensure the present and also the prospective needs of the country. However, they are distributed unequally over the Slovak territory. The distribution depends on natural conditions – mostly on geomorphologic, geological, hydrogeological and climatic ones.

Slovakia is a landlocked central European country (16°–23°E, 47°–50°N) with the area of 49,034.9 km², bordered by Poland, Ukraine, Hungary, Austria and the Czech Republic (see Fig. 1).

Slovakia is a mountainous country with almost 80% of the territory over 720 m a.s.l altitude. Central and northern areas are of mountainous character belonging to the Western Carpathian Arch; lowlands are typical in the South and East. The highest point is the 2,655 m Gerlachovský Peak in the High Tatra Mts. (Vysoké Tatry) in the North of Slovakia. The lowest point is at 94 m near Streda nad Bodrogom village in the Eastern Slovakian lowland.

2 Climatic Conditions of Slovakia

Slovakia is located in middle latitudes experiencing four alternating seasons. The distance from the sea creates a transitional climate between maritime and continental. The western part of Slovakia is influenced by oceans, and in its eastern part the continental moderate climate prevails, although the climate in Slovakia is mostly determined by altitude. Detailed descriptions of climatic characteristics are available in the Climate Atlas of Slovakia [1].

Fig. 1 Slovakia and the neighbouring countries

The climate varies between temperate and continental climate zones with relatively warm summers and cold, cloudy and humid winters. According to updated Köppen-Geiger classification [2], the snow-type climate, fully humid with warm summers of the Dfb type, is typical for the major part of the country. The Cfb type, being characterized as warm temperate, fully humid with hot summers was identified in the south-western and southern lowland parts of the country. The highest elevations of the High Tatra Mts. belong to the Dfc type, being characterized as snow climate, fully humid with cool summer. Temperature and precipitation are altitude dependent.

2.1 Air Temperature

Temperature conditions of Slovakia are diverse and vary from place to place, thanks to a relatively large variability of altitude. The noticeable relationship between air temperature and the altitude is modified by the shape of terrain (mostly in minimum air temperature) when the lower air temperature is present in the basins and higher air temperature at the slopes. Temperature dependence on altitude is subject to the influence of air masses moving towards the Carpathian Mountains which causes south-western parts of Slovakia to be warmer than its northern parts. The Carpathian Mountains prevent infiltration of cold air masses into the territory of southern Slovakia. This part of our country benefits from a low frequency of northern wind flow. The eastern part of Slovakia is further away from the ocean, so the climate there resembles the continental climate which is manifested by colder winters in the long-term mean.

While evaluating the temperature conditions, it is necessary to keep in mind that the mean values of climatic characteristics are compiled from temperature conditions resulting from different synoptic situations. Variability of these conditions is a typical general feature of Slovak climate and so is the variability of air temperature. Another important fact is that the climate change has a strong impact not only on the mean air temperature values but also on other characteristics like the number of days with characteristic air temperature. To sum up, we can frame the basic temperature conditions of Slovakia [1]. The lowest average monthly temperature in January (the coldest month) varies between −11.2°C at Lomnický Peak (High Tatra Mts.) and −1.2°C at Ivánka pri Bratislave. The highest average monthly temperature in July (the warmest month) varies between 3.4°C at Lomnický Peak and 20.5°C at Ivánka pri Bratislave [3]. The lowest air temperature in the period 1961–2010 was as low as −36.0°C and took place on 28 February 1963 in Plaveč. The highest air temperature in the period 1961–2010 was as high as 40.3°C and took place on 20 July 2007 in Hurbanovo. This extreme value was also the absolute one. The lowest air temperature −41.0°C was recorded on 11 February 1929 in Vígľaš-Pstruša.

2.2 Precipitation

The territory of Slovakia belongs to the northern temperate climatic zone with a regular alternation of four seasons – spring, summer, autumn and winter, with a relatively even distribution of rainfall throughout the year.

In Slovakia, according to the Slovak Hydrometeorological Institute (SHMI), there is sometimes an average of less than 600 mm of precipitation per year. Rainfall in Slovakia, in general, is increasing with an altitude of approximately 50–60 mm at 100 m of the high. Mountains in Northwest and North of the country are richer in atmospheric precipitation than the mountains in the Central, Southern and Eastern regions. This is conditional on the higher exposition of these mountains to the prevailing northwest air masses movement. It is typical especially for the eastern Slovakia that the high atmospheric precipitation occurs at windy positions of southern mountain ranges under the southern cyclone situations. The rainiest month is usually June or July, and the less rainfall is typical for the January to March period. Frequent and sometimes prolonged periods of drought are caused by great precipitation variability mainly in the lowlands. Lowlands are typical by the smallest precipitations (even less than 500 mm per year) and a little rainfall in the summer. At the same time, these areas are also the warmest and relatively the windiest ones. As a result, highest values of the potential evaporation occur right there. Average annual precipitation ranges from 450 mm in the southern lowlands to over 2,000 mm in the northern High Tatra Mts. Area [3]. The highest daily rainfall was 231.9 mm measured in 1957. Storms occur relatively frequently at the whole territory of Slovakia during summer months. A large amount of precipitation falls during the storm events; almost every year somewhere in Slovakia exceed the daily rainfall the value of 100 mm. In winter, much of the precipitation falls, particularly in the middle and the high altitudes of mountain ranges, in the form of snow. The average duration of snow cover is less than 40 days in the southern Slovakia. In the Eastern plains, the duration is longer than 50 days per year. The snow cover duration in the intra-mountainous depressions has a length of 60–80 days in average; in the mountains it makes 80–120 days [1].

Precipitation distribution from both the temporal and spatial perspectives is described, e.g. in [4]. The annual average precipitation amount of 487 stations was 720.2 mm during the period 1981–2013 (Fig. 2).

The absolute highest precipitation total on an annual scale was in 2010 when 2,075 mm were recorded at the Jasná meteorological station (Nizke Tatry Mts.). In contrast, the lowest amount was recorded in 2011 at the Male Kosihy meteorological station located in the Podunajská (Danube) lowlands.

From the seasonal (monthly) point of view, highest mean monthly precipitation amounts are recorded in June (87.6 mm) followed by July (86.2 mm) and May (78.9 mm). On the other hand, the lowest precipitation amounts are in February and January, with 39.8 mm and 44.1 mm, respectively. More precisely though, this annual distribution has two maxima: in June as described above and a secondary one in November with an average value of 54.8 mm of precipitation (Fig. 3). Annual precipitation totals are spatially depicted in Fig. 4.

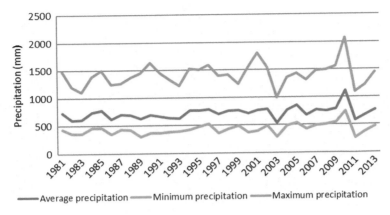

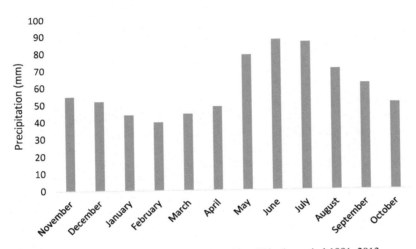

Fig. 2 Average, maximum, minimum precipitation in annual time scale

Fig. 3 Mean monthly precipitation totals over Slovakia within the period 1981–2013

The highest totals are recorded in the mountainous areas located in northern and north-western Slovakia (Western Carpathian Mountains) and in north-eastern Slovakia (Eastern Carpathian Mountains). On the other hand, the driest areas are located in the lowland (nížina) areas extending from the west to the south-east of the country: Borská nížina, Podunajská nížina, Juhoslovenská nížina and Východoslovenská nížina. These lowlands belong to the large Pannonian Basin (Carpathian Basin) geomorphological unit forming its northernmost borders. However, low precipitation totals are also recorded to the leeward of the Western Carpathian Mts.: Podtatranská kotlina Basin, Hornádska kotlina Basin, Šarišská vrchovina Highlands, Čergov Mts. and Ondavská vrchovina Highlands.

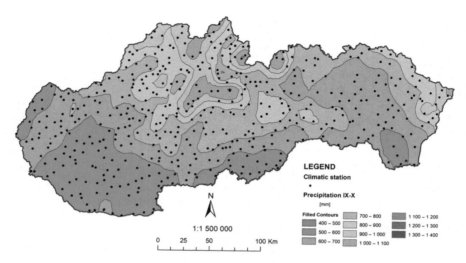

Fig. 4 Annual precipitation distribution over Slovakia in the period 1981–2013

2.3 Snow

The characteristics of snowfall and snow cover have been drawing more and more attention since the late twentieth century, due to their greater instability and disturbed annual regime. Hydrological aspects of this development are just as significant. The results of recent meteorological measurements and observations in Slovakia indicate that the vertical limit for stable occurrence of the snow cover is shifting to higher altitudes as it is in the neighbouring countries. Paradoxically, the snow cover is generally starting to rise in the positions from about 1,500 m a. s. l. This is related to warmer winters and higher precipitation totals during the winter seasons. Nevertheless, the temperature conditions in high mountain positions are still causing this increased precipitation to fall in solid form. However, the snow cover is generally on the decline in Slovakia. There are now more and more winters when the snow cover does not occur in the warmest regions of Slovakia. This is something that has been unimaginable in the Slovak territory in the past. The anomalies in the temporal and spatial distribution of snow are more frequent as well. For example, in the autumn of 2003, at the beginning of the last decade of October, the whole area of Podunajská nížina lowland was covered with a continuous layer of snow a few centimetres thick. Particularly in the early twenty-first century, at the end of winter and at the beginning of spring, drought related to the lack of snow occurred regularly. Avalanche risk is now acuter than in the past since an unusually thick layer of snow can be accumulated in a short time in the mountain positions during certain winters. Unstable temperature conditions and wind contribute to the formation of avalanches. There are also other practical reasons due to the increasingly unpredictable snow conditions for further analysis of future trends in the characteristics of snowfall and snow cover. One of the examples is motoring and expanding road infrastructure, which require systematic monitoring of snowfall and snow cover [1].

2.4 Evapotranspiration

The highest annual sums of the potential evapotranspiration occur in the Podunajská nížina lowland and in the southern part of Slovakia – more than 700 mm per year. The potential evapotranspiration decreases with an increasing altitude. In the High Tatra Mts., this vertical gradient reaches 18 mm per 100 m during the year. The annual course of the potential evapotranspiration is very similar to the annual course of the air temperature and reaches its maximum during the highest solar radiation balance (in July) and minimum in the winter (in December and in January) [1].

3 Hydrological Conditions of Slovakia

The total length of the river network is 49,775 km; average density is 0.88 km km^{-2}. The longest Slovakian river Váh is 367.2 km long. The surface water from the territory of Slovakia runs off towards two different seas. The majority of the Slovak territory with 96% (47,086 km^2) belongs to the Danube River Basin being drained to the Black Sea. The remaining 4% (1,593 km^2) is drained through the Poprad and Dunajec Rivers (the tributaries of the Vistula River) to the Baltic Sea. Therefore the Slovakia is often referred as the roof of Europe. The main European divide between the Black Sea and the Baltic Sea drainage areas follows the lower ridges and the flat landscape on the foothills of the High Tatra Mts. near Štrba and the Low Tatra Mts. (Nízke Tatry) at Šuňava villages. The view from the foothills of the Low Tatra Mts. at Šuňava towards the High Tatra Mts. is in Fig. 5.

A small monument is located at Šuňava village close to the Roman Catholic church of the St. Nicholas remembering on the position of Slovakia as the "roof of Europe" (see Fig. 6). The church itself lies exactly on the European watershed divide; the western part of the roof sends the rainwater to the Black Sea through the Vah and Danube Rivers and the eastern part of the Baltic Sea through the Poprad River.

Rivers also create quite a long part of the Slovak natural borders with the neighbouring countries: Morava with the Czech Republic, Danube with the Austria and Hungary, Ipeľ and Bodrog with the Hungary and Poprad and Dunajec with the Poland.

Three types of runoff regime are typical for Slovak rivers: (1) the temporary snow regime, (2) the snow-rain combined regime and (3) the rain-snow combined regime [3].

The temporary snow regime is typical for the high mountain area, involving Vysoké Tatry Mts. (High Tatra Mts.) and Nízke Tatry Mts. (Low Tatra Mts.) as the highest mountains of Slovakia. The accumulation period in this area starts already in October and last till March to April period. The high water-bearing period lasts from April until July to August. The maximum discharges are typical for May to June months and the minima for January to February period.

Fig. 5 View of High Tatra Mts. from the foothills of the Low Tatra Mts. near Šuňava village (Photo: Fendeková 2014)

The snow-rain combined regime occurs in the middle-mountain area, to which belong the Veľká Fatra, Malá Fatra, Slovenské Rudohorie and some other mountains. The accumulation period starts in November and lasts until March. The high water-bearing period lasts from April until June, with the maximum discharges in May and the minima in January to February period. The high water-bearing in lower altitudes is shortened to March to May period with the maximum discharges in April. The minima occur in the January to February period as in the previous runoff regime type. However, the second minima period could also occur in the autumn months of September and October.

The rain-snow combined runoff regime is typical for the rest of the country belonging to the upland-lowland areas. The accumulation period starts in November and lasts until February. The high water-bearing period lasts from March until April, with the maximum discharges in March and the minima in September. The high water-bearing in the lowest altitudes of Slovakia could start already in February and lasts until April, with the maximum discharges in March. The minima occur in September.

Most of the lakes are of glacial origin. More than 100 lakes created by glacier activity occur in the Tatras (Západné Tatry Mts., Vysoké Tatry Mts. and Belianske Tatry Mts.). The largest of them – the Hincovo pleso mountain lake – covers an area of 20.08 ha.

Fig. 6 Monument "Roof of two seas" (Strecha dvoch morí) at Šuňava village (Photo: Fendek 2017)

Except for lakes, many waterworks were constructed in Slovakia in the twentieth century. The most significant are the ones on the Danube (the Hrušovská zdrž reservoir and the Gabčíkovo) and a cascade of 19 reservoirs on the Váh River [1].

3.1 River Basins in Slovakia

The Danube River Basin is the largest one in Slovakia. The river network of the Danube River Basin is created by the Slovak part of the Danube and its most significant tributaries: Morava, Váh, Hron and Ipeľ. The Tisza River which mouths into the Danube River on the Hungarian territory takes its tributaries also from the Slovak territory. The most important of them are the eastern Slovakian rivers: Bodrog, Slaná and Hornád. The whole Danube River Basin is drained to the Black Sea. The Slovak tributaries of the Polish Vistula River are Poprad and Dunajec Rivers (see Fig. 7), being drained to the Baltic Sea.

The major part of Slovakia, as already mentioned, belongs to Danube River Basin. Danube River Basin is the second largest in Europe, covering the territory of 18 European states with the area of 801,463 km^2. Danube River is 2,780 km long

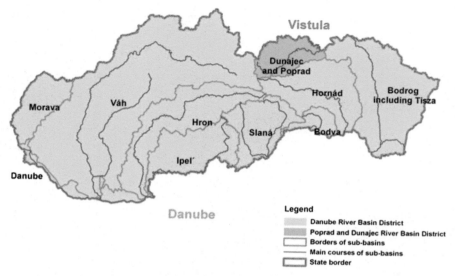

Fig. 7 River basin districts of the Slovak Republic and their sub-basins (Adapted from [5])

and flows roughly in the west-east direction, with a sliding path south to a long stretch between the Slovak Republic and Serbia.

The Morava River is the left tributary of the Danube. Its catchment area is 26,578 km² and lies in parts of the territory of the Czech Republic, Slovakia and Austria. Floods in Morava River Basin are common and occur due to various precipitation situations. The floods in Morava River Basin are caused mostly by precipitation in the upper parts of the basin, but floods can also occur in other parts of the basin.

The Váh River is a left tributary of the Danube that mouths into Danube River in rkm 1,766. The Váh River Basin flows in the North and West of Slovakia. The flood waves on the Váh River can be efficiently transformed, thanks to the water basin management system. The timing of the floods in the Danube and the Váh Rivers may have some influence on the Danube River flow conditions, but in those cases, the high stream flows in the Danube River influenced by backwaters, the lower section of the Váh River and its tributaries.

The Hron and Ipeľ Rivers do not have a significant impact on the course of the Danube floods, but floods occur fairly often in their own catchment areas. The Hron River sub-basin is bordered in the west and the north with the Váh River sub-basin; in the east with the Hornád River sub-basin; in the south-east with the sub-basins of Bodva, Slaná and Ipeľ Rivers; and in the south with the Danube River sub-basin. The Ipeľ River Basin neighbours with the Hron River sub-basin in the west and north and with the Ipeľ River sub-basin in the west, with the Hron River sub-basin in the northwest, with the Hornád River sub-basin in the north-east and with the Bodva River sub-basin in the east and south-east. The sub-basin of Bodva in the territory of the Slovak Republic is bordered in the northwest, north and east with the Hornád

River sub-basin and with the Slaná river sub-basin in the west side. The partial catchment areas of Hornád in the Slovak Republic are bordered in the west with the Váh River sub-basin. In the northwest Hornád catchment is bordered with the Dunajec and Poprad River sub-basins. In the north and east, Hornád catchment is bordered with the Bodrog River sub-basin. In the south it is bordered with the Bodva River sub-basin. And in the southwest it is bordered with the partial catchment areas of Slaná and Ipeľ Rivers. The sub-basin of the Bodrog River in the Slovak Republic is neighbouring in the northwest, in the section from the Slovak-Polish border to the Minčol peak, with the Poprad River sub-basin and in the west with the Hornád River sub-basin.

The largest sub-basin within the Danube River Basin is the Tisza River Basin covering an area of 157,186 km^2. The Tisza River Basin can be divided into three main sections:

- Mountainous area of the upper Tisza in Ukraine and Romania.
- Central Tisza in Hungary, which major tributaries Bodrog and Slaná collect water from the Slovak and Ukrainian Carpathian Mountains, as well as Samos, Kris and Mures River, the last one diverts water from Romanian's Transylvania.
- Lower Tisza is delineating the Hungarian-Serbian border, where it directly translates into Begej and other tributaries indirectly through a system of channels of the Danube-Tisa-Danube.

The Tisza is the longest tributary of the Danube River outside the Slovak territory with the length of 966 km. The largest flood protection system in Europe including channelization, construction of flood protection embankments and low walls, drain systems and pumping stations was gradually built up, later on complemented by another one retention reservoir.

The Vistula River Basin district is another one main river basin district in Slovakia besides the Danube River Basin district. Vistula (Polish Wisła, German Weichsel, Latin and English Vistula) is a 1,047-km-long river which is the longest Polish river, and, at the same time, it is the longest and after the Neva the second most dramatic influent flow of the Baltic Sea. The average discharge of the Vistula River amounts 1,080 m^3 s^{-1} at its mouth to the Baltic Sea. The area of the Vistula River Basin amounts 194,424 km^2.

The areas, long-term average precipitation and runoff (1961–2000) for the Slovak sub-basins of the Danube and Vistula River Basins [6] are shown in Table 1.

The most significant water courses draining the Slovak territory are in Table 2.

3.2 Surface Water and Groundwater Fund of Slovakia

The long-term average of about 2,514 m^3 s^{-1} of surface water inflows into Slovakia through the Danube River makes about 86% of the total surface water fund [8]. Approximately 398 m^3 s^{-1} rise in the Slovak territory within the main river basins of Morava, Danube (the Slovak part), Váh (together with Nitra), Hron, Ipeľ,

Table 1 Parameters of the main Slovak sub-basins of the Danube and Vistula River Basins (Adapted from [6])

River	Area (km^2)	Area (%)	Long-term average precipitation (mm)	Long-term average runoff (mm)
Morava	2,282	4.7	614	101
Danube	1,138	2.3	551	26
Váh (including Nitra)	18,769	38.3	822	307
Hron	5,465	11.2	790	293
Ipeľ	3,649	7.4	636	135
Slaná	3,217	6.6	713	200
Bodva	858	1.7	690	125
Hornád	4,414	9.0	701	203
Bodrog	7,272	14.8	718	223
Poprad and Dunajec	1,950	4.0	868	430

Table 2 Parameters of the main Slovak Rivers (Adapted from [7])

River	Length in Slovakia (km)	Average discharge (m^3 s^{-1})
Morava	114	120
Danube	172	2,025
Váh	367	152
Nitra	170	22.5
Hron	297	53.7
Ipeľ	248.2	21
Slaná	229	70
Bodva	110	5.2
Hornád	172.9	30.9
Bodrog	15.2	115
Poprad	144.2	3.31
Dunajec	17	84.3

Slaná, Bodva, Hornád, Bodrog, Dunajec and Poprad Rivers (see Fig. 7). It is representing about 14% of the surface water fund of Slovakia.

Sources of the surface water fund rising on the Slovak territory are precipitation, both rain and snow and groundwater.

The long-term water balance in Slovakia for the period 1961–2000 [6] can be written as

$$742 \text{ mm } (P) = 236 \text{ mm } (R) + 506 \text{ mm } (ETP) \tag{1}$$

where P is the precipitation, R is the runoff and ETP is the balance evapotranspiration.

The amount of the balanced evapotranspiration could be much higher at the lowest altitudes of the southern Slovakian lowlands reaching up to 94% of the total precipitation. On the contrary, the balance evapotranspiration reaches less than 25% in the highest altitudes of the high mountain areas.

What the year 2015 is concerned, the total atmospheric precipitations in the Slovak territory reached the value of 35,241 million m^3 (Fig. 8) in Slovakia [9].

A significant part of the Slovak surface water fund flows into Slovakia from the neighbouring states, and the usability of this fund is limited. Annual inflow to Slovakia (Figs. 9 and 10) in 2015 was 55,052 million m^3 which, compared to 2014, represents a decrease by 3,060 million m^3. Surface runoff from the Slovak territory (Fig. 11) has declined by 1,813 million m^3, compared to the previous year [9].

The surface streams are fed mostly by groundwater, which is recharged by precipitation. Therefore the geological structure and hydrogeological and climatic conditions are the main factors affecting the amount of surface water runoff. The important exception is the Danube River which recharges the groundwater in its

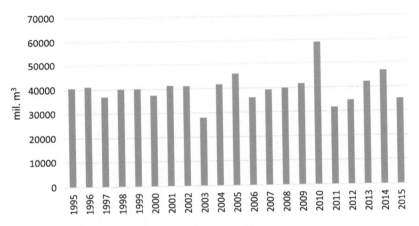

Fig. 8 Course of annual precipitation amounts in Slovakia

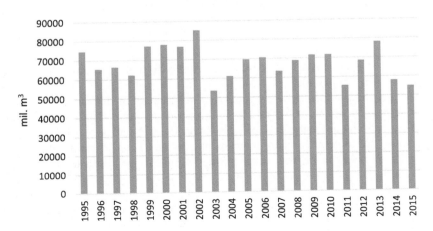

Fig. 9 Annual inflow to the Slovak Republic in Slovakia

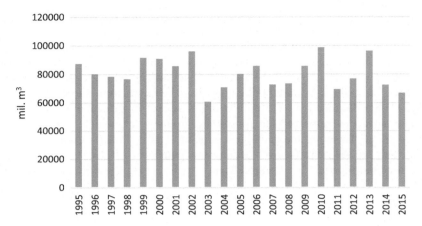

Fig. 10 Annual runoff in Slovakia

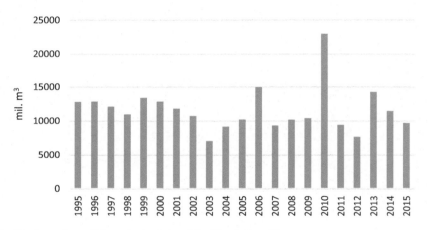

Fig. 11 Annual runoff from the territory of the Slovak Republic

alluvial plain forming the largest freshwater reservoir in the Central Europe called Žitný ostrov (Rye Island).

The groundwater sources are unevenly distributed in Slovakia. Favourable hydrogeological structures, which allow formation and accumulation of the significant groundwater amounts, are located in lowlands, in the alluvial river plains or in limestones and dolomites of the core mountains and karst plateaus. The biggest groundwater reservoir of fresh water in the Central Europe is situated in the Žitný ostrov area.

The total amount of groundwater accumulated in the rock environment (natural groundwater amounts) has the value of app. $147 \text{ m}^3 \text{ s}^{-1}$. The highest portion – app. $88 \text{ m}^3 \text{ s}^{-1}$ (60%) – is accumulated in the Quaternary alluvial sediments of rivers, from which app. $25 \text{ m}^3 \text{ s}^{-1}$ was estimated for the Žitný ostrov. Another 25% of natural groundwater amounts are accumulated in Mesozoic carbonate rocks (limestone and dolomite) of the Slovak core mountains. The rest comes from other types of geological environment. The amount of almost 80 m^3 s^{-1} was already verified by the results of hydrogeological research and represents usable groundwater amount.

According to Slovakia's Water act, groundwater is preferentially used for drinking water supply. Recalculation of usable groundwater amount into the available water amount for water consumption gives the number of 1,257 L per capita per day. This amount is app. 15 times higher than the present average water consumption per capita per day.

Almost 80% of inhabitants in Slovakia are supplied by groundwater; only 20% of drinking water supply refers to surface water sources. Surface water is used as drinking water source mostly in places, where the hydrogeological conditions are unfavourable for natural accumulation of groundwater. Such areas are located in the Northwest and Northeast of Slovakia in areas built of flysch sediments (sandstones, claystones), in some intra-mountainous depressions in the central part of Slovakia and in Central and Eastern Slovakian volcanic mountains. Unfavourable rock storage capacity is combined with the low precipitation amounts and high air temperatures in Southern and Eastern Slovakian lowlands, built of less permeable, mostly Neogene clayey sediments. The largest surface water reservoirs were built in groundwater-deficient areas, supplying the inhabitants with drinking water and also various economy sectors by drinking and technical water.

The oldest water supply reservoir in operation up to present days bears a date of construction in the year 1510. Natural conditions and economic demands of the society initiated the development of dam engineering in Slovakia already 500 years ago [1]. The world register of ICOLD (International Commission on Large Dams) comprises 48 dams in Slovakia, of which six are historical. By all of these dams, a total reservoir volume of about 1.84×10^9 m^3 is created. The largest group of small dams surpasses the number of 200.

3.3   Water Quality

Since 1994, volumes of discharged wastewater into surface water have been declining, despite year-to-year increments and reductions. In 2015, wastewater production declined by 51.6% compared to 1994, and by 43.2% compared to 2000 and by 1.3% compared to 2014. In 2015, volumes of organic pollution characterized by parameters of chemical oxide demands (COD_{Cr}), biochemical oxide demands (BOD) and insoluble substances (IS) are almost constant.

Quality indicators of surface water in 2015 were monitored according to the approved Water Monitoring Program for 2015. There were monitored 385 stations in basic and operating mode. The monitoring results were evaluated according to the Government Regulation no. 398/2012 Coll., amending the Government Regulation of the Slovak Republic no. 269/2010 Coll. and laying down requirements for the prospecting of good water status. The quality of surface water in 2015 in all monitored sites met the limits for selected general indicators and radioactivity indicators. The most significant exceedance of the limit values within the general indicators (Part A of the Government Regulation) was the nitrite nitrogen indicator in all sub-basins. The requirements for surface water quality for groups of synthetic and non-synthetic substances (parts B and C of the Government Regulation) were not met in the following: As, Cd, Hg, Zn, Cu, Ni, benzo (a) perylene + indeno (1,2,3-cd) pyrene, benzo (b) fluoranthene + benzo (k) fluoranthene, cyanides and PCBs. From the group of hydrobiological and microbiological indicators (Part E of the Government Regulation), the following indicators were not met: saprobic biosestone index, abundance of phytoplankton, chlorophyll, coliform bacteria, thermotolerant coliform bacteria, intestinal enterococci and cultivable microorganisms at 22°C [9].

4 Conclusions and Recommendations

Water is a vital component of the natural environment, but it is also a basic prerequisite for all human economic and social activities in general. Everybody carrying out some activity which may affect the state and relations of surface and underground waters has the obligation to make all necessary efforts for their preservation and protection. The first step towards effective protection of water resources is to know their size and distribution but also their quality. Systematic investigation and evaluation of the occurrence and condition of surface and underground waters within the European countries as well as worldwide are basic responsibility of the state, as an indispensable requirement for ensuring the preconditions for permanently sustainable development as well as for maintaining standards of public administration and information. Sustainable development of water management is based on the principle that water as a natural resource may be utilized only to that extent which ensures future generations sufficient usable supplies of water in the seas, rivers, lakes and reservoirs and that reserves contained in porous environments below the surface of the land remain preserved in the same quantity and quality. For this reason, it is necessary to devote all the more attention to the protection of water sources. The basic requirement in this context is to optimize their monitoring, the assessment of their quality and the implementation of necessary environmental measures.

References

1. Climate Atlas of Slovakia (2015) Klimatický Atlas Slovenska. Slovak Hydrometeorological Institute, Banská Bystrica, p 228
2. Kottek M, Grieser J, Beck C, Rudolf B, Rubel F (2006) World map of Köppen-Geiger climate classification updated. Meteorol Z 15(3):259–263
3. Landscape Atlas of the Slovak Republic, 1st edn (2002) Ministry of Environment of the Slovak Republic and Slovak Environmental Agency, Bratislava and Banská Bystrica, p 344

4. Zeleňáková M, Portela MM, Purcz P, Blišťan P, Silva AT, Hlavatá H, Santos JF (2015) Seasonal and spatial investigation of trends in precipitation in Slovakia. Eur Water 52:35–42
5. Water plan of the Slovak Republic, river basin management plan of the Danube River Basin District, actualisation (2015) Ministry of environment of the Slovak Republic, p 203
6. Majerčáková O, Škoda P, Šťastný P, Faško P (2004) The development of water balance components for the periods 1931–1980 and 1961–2000. J Hydrol Hydromech 52(4):355–364
7. Abaffy D, Lukáč M, Liška M (1995) Dams in Slovakia. T.R.T. Médium, Bratislava, p 103
8. Lieskovská Z (ed) (2016) Environment of the Slovak Republic in focus. Ministry of Environment of the SR and Slovak Environmental Agency, Bratislava and Banska Bystrica, p 64
9. Ministry of Environment of the SR, Bratislava, Slovak Environmental Agency, Banska Bystrica (1995–2015) Reports from environment of the Slovak Republic. http://enviroportal.sk/spravy/kat21

Assessment of Water Resources of Slovakia

M. Fendek

Contents

Abstract Water resources of Slovakia consist of surface and groundwater resources. The surface water ones are formed by surface water inflow to Slovakia and by surface water runoff rising at the Slovak territory. Groundwater resources formation is, among others, dependent on geological-tectonic conditions, hydrogeological parameters of the rock environment and climatic conditions. Surface and groundwater bodies were delineated on the Slovak territory according to the Water Framework Directive requirements. In total, 1,487 surface water bodies are on the list at present, 84 of them in the Vistula River Basin District, and 1,413 in the Danube River Basin District. The largest rivers of Slovakia besides Danube and Morava Rivers, which have their springs outside the Slovak territory, are Váh, Nitra, Hron, Ipeľ, Slaná, Hornád, Bodva, Bodrog, and Poprad with their tributaries. Groundwater bodies are divided into three levels – there are 16 Quaternary groundwater bodies, 59 pre-Quaternary, and 27 geothermal structures. Groundwater resources are evaluated on the annual base within the

M. Fendek (✉)
Department of Hydrogeology, Faculty of Natural Sciences, Comenius University in Bratislava, Bratislava, Slovakia
e-mail: hydrofen@hydrofen.sk

A. M. Negm and M. Zeleňáková (eds.), *Water Resources in Slovakia: Part I - Assessment and Development*, Hdb Env Chem (2019) 69: 21–62, DOI 10.1007/698_2017_201, © Springer International Publishing AG 2018, Published online: 17 April 2018

21

141 pre-Mesozoic, Mesozoic, Palaeogene, Neogene sedimentary, Neogene volcanic, and Quaternary regions.

Keywords Hydrogeological conditions, Surface and groundwater bodies, Surface and groundwater resources, Water Framework Directive, Water management sources protection

1 Introduction

Water resources of Slovakia have a manifold origin. Both surface and groundwater resources create the natural treasure of Slovakia, which is protected by law. Water protection comprises measures for the preservation of the quantitative and qualitative parameters of various kinds of water. The Slovak water legislation is fully compatible with the European legislation and in agreement with all international treaties concerning water protection. Surface and groundwater bodies were delineated on the Slovak territory according to the Water Framework Directive requirements [1].

Slovakia is a country rich in high-quality groundwater, which is formed in the rock environment of various ages, tectonic position, lithological composition and hydrogeological conditions. Groundwater is preferentially assigned for drinking water supply, as given by the Water Act [2]. Several types of protection measures are applied, among them protection zones of surface and groundwater management sources, protected water management areas and streams designated for the abstraction of water intended for human consumption.

2 Surface and Groundwater Bodies in Slovakia

Surface and groundwater bodies are the main units for evaluation of the status of water resources in Slovakia according to Water Framework Directive (WFD) requirements [1].

A surface water body is a discreet and significant element that is determined as a basic WFD element. It means that each WFD evaluation and activity (e.g. evaluation of water status, final identification of heavily modified water bodies, measures for status improvement, etc.) is related to the elemental unit of the water body.

Surface water bodies were delineated on the water courses with catchment area larger than 10 km^2. Water bodies on the water courses with catchment area smaller than 10 km^2 were not determined, and they are considered a part of the water body in the basin of which they are located.

Surface water typology has been developed for rivers and lakes using abiotic criteria. The ecoregion, altitude, catchment area, and geology were applied as descriptors as follows [2]:

- Ecoregion: Slovakia is a part of two ecoregions, the Carpathians and the Pannonian Lowland.
- Type according to the altitude: <200 m above sea level, 201–500 m above sea level, 501–800 m above sea level, and >800 m above sea level.
- Type according to the catchment area (small, 10–100 km^2; medium, 101–1,000 km^2; large, >1,000 km^2).
- Type according to geological composition: This descriptor is defined as a "mixed type" at present.

The ecoregion, altitude, depth, surface area, and geology were used as descriptors for lakes. The minimum size of lake water bodies for inclusion in the river basin management plans (RBMP) was set to 0.5 km^2, and the minimum catchment size of river water bodies was 10 km^2. None of the Slovak lakes exceed this threshold, and all reservoirs were assessed as heavily modified water bodies (HMWBs). Water bodies below the threshold size were not separately delineated and were considered to be a part of a water body in the catchment in which they are located.

A total of 1,487 surface water bodies are on the list at present within the territory of Slovakia, 84 of them in the Vistula River Basin District, and 1,413 in the Danube River Basin District [3].

Groundwater body is a basic territorial unit for all evaluations requested by WFD. Water bodies in Slovakia were delineated in three separate layers, based on the approach defined in the methodical instruction elaborated in the framework of CIS EC and basic data on groundwater:

- Groundwater bodies in Quaternary sediments
- Groundwater bodies in pre-Quaternary rocks
- Bodies of geothermal waters (geothermal structures) representing groundwater of deep circulation with the temperature above 15°C

A sum of 102 groundwater bodies are on the list at present. The numbers of groundwater bodies in respective River Basin Districts of Vistula and Danube are shown in Table 1. The example of groundwater water bodies delineated in Quaternary sediments of the Danube River Basin District is in Fig. 1.

Table 1 Numbers of groundwater bodies delineated in Slovakia according to WFD

	Groundwater bodies (GWB)					
	Quaternary GWB		Pre-Quaternary GWB		Geothermal structures	
River basin district	Number	Area (km^2)	Number	Area (km^2)	Number	Area (km^2)
Vistula	1	420.76	3	1,970.86	1	2,790.99
Danube	15	10,226.40	56	47,105.28	26	14,837.70
Total	16	10,647.16	59	49,076.14	27	17,628.69

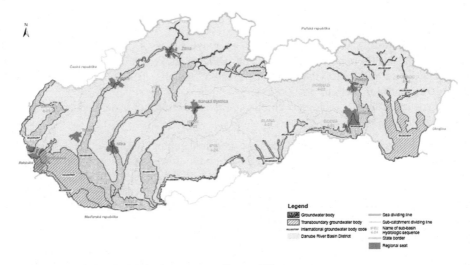

Fig. 1 Groundwater bodies in Quaternary sediments [3]

3 Surface Water Resources

The surface water resources take the second place with respect to abstracted and supplied amounts of water in Slovakia.

The surface water resources consist in the surface inflow to Slovakia through the Danube River, amounting in the long-term average approximately 2,514 m³ s^{-1} of surface water and of approximately 398 m³ s^{-1} of water rising on the Slovak territory [4], as already mentioned in Sect. 3.2 of first chapter in this volume.

3.1 Surface Inflow to Slovakia Through the Danube River

The Danube River is the second largest river in Europe, having its spring in the German mountain Schwarzwald (Black Forest). The total length of the Danube River is 2,857 km, and the basin area reaches 801,463 km². The Danube enters the Slovak territory at Devínska brána (Devín Gate) located on the 1,880.2 river km at Morava River mouth (border of Austria and Slovakia).

The main tributaries of the Danube River in the Slovak territory are the Morava, Váh, Hron, and Ipeľ Rivers. Among the Danube tributaries, only the Morava River does not rise on the Slovak territory. The spring of the Morava River is in the Czech Republic under Králický Sněžník Mountain at 1,380 m amsl. Morava River enters

Slovakia at the place of the confluence with Sudoměřický potok (brook) and forms the natural boundary between the Czech Republic and Slovakia along its whole course in the Slovak territory. The mouth of Morava into the Danube River is located close to Bratislava city under the Devín castle at 136 m amsl. Main Slovak tributaries of the Morava River are Myjava, Rudava, and Malina streams. The Danube leaves the Slovak territory at the 1,708.2 river km at Ipeľ River mouth in the Kováčovské kopce Mts. [5].

The maximum vertical dissection (energy of relief expressed as the difference between maximal and minimal altitudes) for the Slovak section of the Danube River has the value of 710 m. The highest altitude has the mount Čupec (Biele Karpaty Mts.) with 819 m amsl; the lowest one is located at the confluence of the Ipeľ River with the Danube River at 106 m amsl.

The total length of the Danube River on the Slovak territory is 172 km. Upstream Bratislava, the river flows in a concentrated channel with a relatively steep river bed. The river loses its slope after leaving the Malé Karpaty Mts. and flows over an alluvial cone creating a complicated river branches network up to Medveďov village. Danube River creates a natural boundary between Austria and Slovakia with the length of 7.5 km and between Slovakia and Hungary with the length of 142 km. The Danube River flows only on 22.5 km long course out of the total 172 km purely on the Slovak territory.

The Gabčíkovo Water Work – the biggest water work on the Slovak territory – was constructed on the Danube River between Čuňovo and Gabčíkovo villages. It was put into operation in 1992. The water work consists of Hrušov water reservoir (headwater installations) with an area of 40 km^2, the bypass canal (headwater canal and tailwater canal), and the series of locks on the bypass canal (hydropower plants and navigation locks). There were several main tasks for the Gabčíkovo Water Work construction. The first and the main one was the protection of the surrounding areas against floods. The water work should preserve the navigation also during the low-flow periods and stop deep erosion of the "Old Danube" River bottom. Production of the electric power using the energy of water was also one of the goals of the water work construction. The hydropower plants which are a part of the water work produce more than 30% of the total installed capacity of all hydropower plants in Slovakia. The Gabčíkovo Hydroelectric Power Station produces on average 2,200 GWh of electricity annually, making it the largest hydroelectric plant in Slovakia. The view on the Gabčíkovo Water Work at Gabčíkovo (end of the bypass canal) with the hydropower plant (installed capacity 720 MW) and two navigation locks is in Fig. 2.

There are 16 gauging stations measuring the water stage located at the Danube River within the Slovak territory. Discharges are calculated for seven stations: No. 5127 Bratislava – Devín, No. 5140 Bratislava, No. 5145 Medveďov, No. 5153 Dobrohošť, No. 6849 Komárno, No. 6860 Iža, and No. 6880 Štúrovo. The average and extreme discharges at the Bratislava – Devín and Štúrovo water stage gauging profiles – are shown in Table 2. None of the two gauging stations are influenced by the Danube River backwater which reaches up to No. 5140 Bratislava station.

Fig. 2 Gabčíkovo area of the water work – end of the bypass canal (http://www.vvb.sk/cms/)

Table 2 Average and extreme discharges of the Danube River at Bratislava – Devín and Štúrovo gauging stations (data source: SHMÚ [5])

Station no.	Station name	River (km)	$Q_{A\ 1961-2000}$ $(m^3\ s^{-1})$	Q_{min} $(m^3\ s^{-1})$	Date	Q_{max} $(m^3\ s^{-1})$	Date
5,127	Bratislava – Devín	1,879.8	2,061	754.9[a]	18.12.1991	10,390[a]	15.05.2002
6,880	Štúrovo	1,718.6	2,264	916.7[b]	06.01.2004	8,485[b]	04.04.2006

[a]Observation period 1926–2012
[b]Observation period 1934–2012

The Danube River with its branch Malý Dunaj forms the Žitný ostrov (Rye Island). Žitný ostrov is the biggest river island in Europe. Žitný ostrov is bordered in the South by the Danube River, in the North by Malý Dunaj River, and on a short course in the Northeast, also by the Váh River. Malý Dunaj River branches away from the Danube River at Bratislava-Vrakuňa city part and discharges itself into the Váh River at Kolárovo city. Geologically, the Žitný ostrov is a huge alluvial fan which was created by the Danube River by its cutting through the Malé Karpaty Mts. Žitný ostrov covers an area of 1,886 km^2 (728 mile^2), being approximately 84 km (52 miles) long and 15–30 km (9.3–18.6 miles) wide. The area is flat, with a very gentle slope and has a mild, moderately warm climate. Despite the fact that the precipitation amounts belong to the lowest ones at the Slovak territory, reaching only

500–550 mm annually [6], the area is the largest drinking water reservoir and, at the same time, one of the most fertile agricultural regions of Slovakia. This is enabled by the infiltration of the Danube River water into the alluvial sediments.

3.2 Surface Water Resources Rising at the Slovak Territory

The surface water resources rising at the Slovak territory comprise (1) the surface water flowing in the river network of all streams having its spring in Slovakia, creating the surface water runoff and (2) the surface water storage in reservoirs built on the Slovak territory.

3.2.1 Surface Water Runoff

The largest rivers of Slovakia besides Danube and Morava Rivers are Váh, Nitra, Hron, Ipeľ, Slaná, Hornád, Bodva, Bodrog, and Poprad Rivers with their tributaries. The amount of surface water rising on the Slovak territory is dependent on the precipitation amounts of the respective but also the previous years. As an example, the amount of precipitation and surface water runoff in the year 2015 is given in Table 3.

The surface runoff in the year 2015 reached only 84% of the long-term average despite the precipitation amount reaching 94% of the precipitation long-term average (Table 2). Therefore the year 2015 was characterised as dry according to runoff amount which comprised only 306.18 $m^3 s^{-1}$. The large part of the precipitation was depleted by high evapotranspiration, as the consequence of very warm summer and autumn 2015. Therefore the infiltration to groundwater was lowered substantially, causing the lowered recharging of the surface streams by groundwater runoff.

Some of the streams of Slovakia were designated by [8] for the abstraction of water intended for human consumption. A total of 102 streams, listed in Annex 2 of [8], were designated, majority of them in the headwater parts of mountainous catchments. They are located in all major river basins. Number of streams in the respective river basin with their total length is given in Table 4.

Table 3 Average amounts of precipitation and surface runoff on the Slovak territory in 2015 [7]		
Area (km^2)	49,014
Average precipitation total (mm)	719
Ratio of the long-term average (%)	94
Precipitation character of the year	Normal
Average surface runoff (mm)	197
Ratio of the long-term average (%)	84
Runoff character of the year	Dry

Table 4 Number and length of stream sections designated for water abstraction for human consumption in respective river basins of Slovakia (adapted from [8])

River basin	Number of streams	Total length (km)
Poprad	12	77.40
Váh	19	185.55
Nitra	2	24.15
Hron	12	110.75
Ipeľ	1	19.53
Bodrog	18	320.55
Slaná	7	52.05
Hornád	26	207.50
Bodva	5	39.60

The longest stream section designated for abstraction of water for human consumption is located on the Ondava River in the eastern part of Slovakia with the total length of 90.9 km.

3.2.2 Surface Water Reservoirs

There are 295 water reservoirs in Slovakia according to [9], 32 of them are regularly evaluated within the quantitative water management surface water balance [7]. More than 200 small water reservoirs serve mostly for irrigation purposes.

Eight water reservoirs, Nová Bystrica, Turček, Rozgrund, Hriňová, Málinec, Klenovec, Bukovec II, and Starina, were constructed for drinking water supply purposes.

Nová Bystrica water reservoir is located in the north-western Slovakia on the Bystrica stream (a tributary of the Kysuca River) belonging to the Váh River basin. The water reservoir was built during the period 1983–1989 and supplies by 210 L s^{-1} the cities of Čadca, Kysucké Nové Mesto, and Žilina, as well as many villages in the Kysuce region, as Zborov, Očšadnica, Ochodnica, Stará Bystrica, Nová Bystrica, and others [10]. The reservoir also serves for the transformation of the flood wave, coverage of minimum discharges of the Bystrica River during the low-flow periods, and electricity production using the hydroenergetic potential of water in small water power plants.

Turček water reservoir is located in the headwater part of the Turiec River (tributary of the Váh River) in the Turčianska kotlina basin (Central Slovakia), approximately 1.1 km above the Horný Turček village. The reservoir was built during the period 1992–1996. The water is diverted from the Turiec River basin into the Nitra River basin and used for drinking water supply of Prievidza, Handlová, and Žiar nad Hronom cities and districts [10]. Other functions of the reservoir comprise flood wave transformation, coverage of the minimum discharge during the low-flow period, and electricity production on small water power plant.

Rozgrund water reservoir is located in the Štiavnické vrchy Mts. (Central Slovakia) on the Vyhniansky potok stream (a tributary of the Hron River). Rozgrund is the oldest reservoir, the water of which is used for drinking water supply purposes. The reservoir was built in 1743–1744 [10]. The dam was heightened, later on reconstructed in 1749 and in the second half of the eighteenth century. In the past, the stored water was used for mining purposes. Since the beginning of the twentieth century, the stored water is used for water supply of the Banská Štiavnica city. Nowadays, the amount of 14 L s^{-1} is used for drinking water supply [10]. The reservoir also serves for the transformation of the flood wave and coverage of the minimum flow of 2.0 L s^{-1} during the low-flow period.

Hriňová water reservoir is located in the upper part of the Slatina River (a tributary of the Hron River) in the Poľana Mts., Central Slovakia. The reservoir was built during the period 1960–1965 [10]. Water is used for drinking water supply of the larger area connected by the water-main Hriňová-Lučenec-Fiľakovo, delivering the amount of 325.0 L s^{-1} (https://www.svp.sk/sk/uvodna-stranka/galeria/postery/). The reservoir serves also for the transformation of the flood wave, coverage of minimum discharges of the Slatina River amounting to 121.0 L s^{-1} during the low-flow periods, electricity production using the hydroenergetic potential of water in two small water power plants, and fish breeding (https://www.svp.sk/sk/uvodna-stranka/galeria/postery/).

Málinec water reservoir is located on the Ipeľ River between Málinec and Hámor villages in the southern part of Slovakia. The reservoir was built during the period 1989–1993 and supplies the Central Slovakian water-main by 450.0 L s^{-1} of high-quality drinking water (https://www.svp.sk/sk/uvodna-stranka/galeria/postery/). The maximum possible deliverable amount is 560.0 L s^{-1}. Other functions of the reservoir comprise flood wave transformation, coverage of the minimum discharge of 38.0 L s^{-1} during the low-flow period, electricity production on three small water power plants, and fish breeding (https://www.svp.sk/sk/uvodna-stranka/galeria/postery/).

Klenovec water reservoir is located on the Klenovecká Rimava River (a tributary of the Slaná River), approximately 1 km upstream of the Klenovec city (southern Slovakia). The reservoir was built during the period 1968–1974 [10] with the priority in water supply, amounting to 460.0 L s^{-1} at present (https://www.svp.sk/sk/uvodna-stranka/galeria/postery/). Besides this function, it also supplies the industry with the technical water amounting to 40.0 L s^{-1}, transforms flood wave, and covers the minimum flows of 150.0 L s^{-1} downstream during the low-flow period. Water is also used for electricity production in two small water power plants and for fish breeding (https://www.svp.sk/sk/uvodna-stranka/galeria/postery/).

Bukovec II water reservoir is located on the Ida stream, belonging to the Hornád River basin (the south-eastern part of Slovakia). It was built during the period 1968–1976 with the aim to supply the Košice city (the largest eastern Slovakian city) with drinking water [10]. Besides the surface water of the Ida stream, also discharges from the neighbouring catchment of the Myslavský potok are diverted into the Bukovec II reservoir. The amount of water used for drinking water supply reaches 120.0–140.0 L s^{-1}. The other functions of the water reservoir consist in flood wave transformation and coverage of the minimum discharge of the Ida stream during the low-flow period.

Starina water reservoir is located on the Cirocha River (the north-eastern part of Slovakia), a tributary of the Laborec River which belongs to the Bodrog River basin. The water reservoir was built during the period 1981–1988 with the aim to supply with drinking water the city agglomerations of Humenné, Strážske, Michalovce, Vranov nad Topľou, and Košice. The real abstraction reached 535.0 L s^{-1} in 2015 (http://www.vvb.sk/cms/); the maximum possible abstraction is 1,400.0 L s^{-1}. Other functions of the reservoir comprise flood wave transformation, coverage of the minimum discharge during the low-flow period amounting to 250.0 L s^{-1} in the period July 1 to October 31 and 140.0 L s^{-1} during the rest of the year [11], electricity production on small water power plants, and fish breeding. The length of the pipeline from Starina to Košice is more than 130 km with the altitude difference of 190 m.

The basic parameters of drinking water supply reservoirs of Slovakia are shown in Table 5.

There are several other water works in Slovakia storing surface water in large volumes. The Orava and Liptovská Mara belong to the largest ones. The main purpose of their operation from the water management point of view is the flood protection and balancing of discharges during the low-flow periods. Economically, electricity production is of major importance. Fish breeding and use for recreational purposes are also of great importance.

An interesting solution for the water use was chosen for the Čierny Váh Water Work with the pumped hydroelectric power plant, located in the northern Slovakia in the Liptov Basin. The water work consists of two water reservoirs. There was a compensation upstream reservoir with a storage volume of 3.7 million m^3 constructed on an irregular upland platform over the Čierny Váh River valley in an altitude of 1,150 m amsl. There is no surface inflow into the upstream reservoir. The function of the downstream reservoir is to accumulate water, which is pumped into the upstream reservoir with six pumps during the night (lower electric power prices), and to receive the head-flows from turbines in an amount of 180 m^3 s^{-1}. Three oblique armoured penstocks, each of them supplying two turbines or pumps,

Table 5 Basic parameters of drinking water supply reservoirs in Slovakia [12]

Name	Stream	Operated since (year)	Reservoir area (km^2)	Catchment area (km^2)	Retention volume (million m^3)	Storage volume (million m^3)	Maximum abstraction (L s^{-1})
Nová Bystrica	Bystrica	1989	1.920	59.3	3.19	32.83	700
Turček	Turiec	1996	0.540	29.5	0.40	9.96	700
Rozgrund	Vyhniansky potok	1744	0.057	2.6	0.03	0.41	14
Hriňová	Slatina	1965	0.480	71.0	0.10	7.15	325
Málinec	Ipeľ	1994	1.380	82.3	1.51	23.71	560
Klenovec	Klenovecká Rimava	1974	0.710	88.7	0.96	6.68	500
Bukovec II	Ida	1976	1.050	47.4	0.98	20.03	473
Starina	Cirocha	1988	2.810	131.0	8.17	45.03	1,400

and one underground communication tunnel provide the communication between the upstream and the downstream reservoirs. The aggregates of the pumped storage hydropower plant Čierny Váh may operate in three different operation modes: (1) turbine, (2) pumping, and (3) compensation. The first turbogenerator was put into operation in December 1980, the last one in August 1982. The whole project was realised in the territory which is geologically built by permeable limestone and dolomites of the Mesozoic ages; therefore the tightening of the upstream reservoir was quite complicated. The priority use of Čierny Váh Water Work is the production of the peak electric power of maximum 665 MW. The average annual amount of the produced electricity is 200 GWh.

4 Groundwater Resources

The amount of groundwater in the rock environment is closely connected to the lithological composition, geological-tectonic structure of the area, and hydrogeological conditions.

4.1 Geological Conditions of Slovakia

Geologically, the territory of Slovakia belongs to the Western Carpathians, which is the part of the Alpine-Himalayan mountain belt. The western boundary of the Western Carpathian belt is placed in the Danube valley to the depression west of the Hundsheim Hills (Austria) in the so-called Carnuntian Gate where the Eastern Alps meet the Western Carpathians [13]. The northern boundary is delineated by the edges of the Flysch belt nappes. The eastern boundary between the Western and the Eastern Carpathians is arbitrarily placed to the Uh River valley (Ukraine). The southern boundary is morphologically not visible, because of the huge cover of Neogene sediments of the Great Hungarian Lowland.

The Western Carpathians have a typical zonal structure. The Mesozoic and Tertiary formations, arrayed in a series of arcuate belts, have been tectonically transformed. The Western Carpathians themselves are divided into the outer belt and the inner belt [13]. The outer belt (Outer Carpathians) of the Western Carpathians is built of Neoalpine nappes, whereas the inner belt (Inner Carpathians) has Palaeoalpine, pre-Palaeogene structure. The two are separated by the Klippen belt (Fig. 3). The marginal units – Variscan consolidated Carpathian Foreland and Carpathian Tertiary Foredeep – do not crop out on the Slovak territory but, however, occur in the tectonic basement of the Inner and Outer Western Carpathians [13].

The Outer Carpathians (Flysch belt) are made of Tertiary series of rootless nappes – sedimentary sequences detached from their basement and thrusted over the North European Platform. The original basement of the sequences is not known [13]. The typical features of the series are flysch-like character of the Mesozoic and Palaeogene formations, a total absence of pre-Mesozoic formations, and a negligible

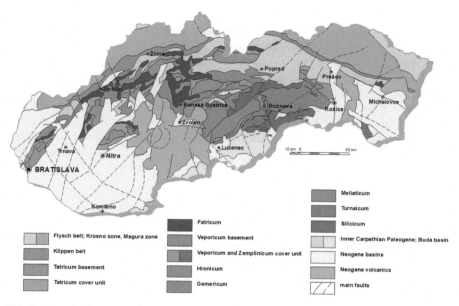

Fig. 3 Tectonic sketch of the Slovak Republic [13]

presence of the postnappe cover. The Flysch belt is composed of three groups of nappes: (1) marginal, (2) central, and (3) inner. The marginal group is totally absent in Slovakia; the central is scarcely extended in the north-western part of Slovakia (Krosno unit) but occurs in the north-eastern Slovakia as the Dukla unit. The inner – Magura group of nappes – predominates on the Slovak territory and consists of Palaeogene flysch sequences [13]. From the lithological point of view, claystones and sandstones alternation are typical for the Flysch belt; Cretaceous sediments crop out only rarely. The example of the flysch sequences in the Eastern Slovakia is in Fig. 4.

Klippen belt is the most complicated belt of the Western Carpathians, separating the Outer and the Inner Western Carpathians. Particular features of this unit are [13]:

- Absence of pre-Mesozoic rocks
- Facial variability during the Jurassic and Cretaceous period
- Flysch development during the Palaeogene period
- Characteristic Klippen-fashioned tectonic pattern represented by lenses of Jurassic-Early Cretaceous limestones penetrating the Late Cretaceous and Early Palaeogene marlstones and flysches [13]

The Inner Carpathians (Inner Carpathian belt) consists of following tectonic units [13]:

- Tatricum unit
- Veporicum unit
- Hronicum unit

Fig. 4 Flysch sequences of the Outer Carpathians – alternation of sandstones (pale colours) and claystones (dark grey colours) in the north-eastern Slovakia (Photo: M. Fendeková)

- Gemericum unit
- Meliaticum unit
- Turnaicum unit
- Silicicum unit
- Postnappe formations
- Neogene basins and grabens
- Neogene volcanics

The Tatricum and Veporicum units consist of crystalline core overlain by late Palaeozoic and Mesozoic cover. The crystalline core is made up mostly of medium- to high-grade metamorphic rocks (schistose gneisses, gneisses) and granitoids of the Palaeozoic age [13]. Besides the late Palaeozoic and Mesozoic cover, also the system of flat-lying nappes (Fatricum) made up of Mesozoic complexes composes the Veporicum unit. In the core mountain belt, the Fatricum often covers the Tatricum unit. The Hronicum unit is represented by a series of nappes created by rock sequences of the Permian to Lower Cretaceous ages overlaying either the Fatricum or the Tatricum unit. The composition of the Mesozoic cover of Tatricum, as well as Fatricum and Hronicum nappes, is similar, consisting of the:

- Early Triassic (Scythian – Werfenian) shales, sandstones and Lúžna sequence quartzites

- Middle to Late Triassic dolomites and limestones
- Late Triassic shales of the Carpathian Keuper
- Jurassic limestones of various types (sandy, crinoids, mottled, radiolarian, nodular, marly, cherty, and others) together with marlstones, marls, and shales of the Early to Late Jurassic and the Early Cretaceous sediments as sandstones, marlstones, and shales

There are some specific features typical for the Hronicum nappe, as the presence of Palaeovolcanic bodies (melaphyre) of the Ipoltica group (Permian) and Lunz layers of the flyschoid character (Late Triassic). The Werfenian and the Carpathian Keuper shales often contain layers of evaporites, mostly gypsum. The example of the crystalline core in the Vysoké Tatry Mts. is in Fig. 5. The forested area in front of the picture is built of Quaternary glacial moraine sediments.

Gemericum unit is exposed in the Volovské vrchy Mts. composed mostly of Hercynian metamorphosed assemblages with different grades of metamorphosis in the Northern and Southern Gemericum. The age of sequences is from the ?Late Cambrium up to ?Early Carboniferous, mostly Silurian to Devonian. The prevailing rock types are phyllites, metasandstones, paraconglomerates, and flysch with rhyodacite-andesite volcaniclastics, less basalts, lydites, and carbonates [13].

Meliaticum, Turnaicum, and Silicicum units occur mostly in the area of Slovenský kras karst (but also in some other areas as Muránska planina Plateau, Slovenský Raj Mts., and others) being composed of Mesozoic rocks. Clayey shales, radiolarites, sandstones, olistostromes, and limestones are the typical rock types for

Fig. 5 Crystalline core and glacial foreland of the Vysoké Tatry Mts., view on the Mlynická dolina valley (Photo: M. Fendek)

these three units [13]. The example of the folded dolomites of the Hronicum unit in the Čierny Váh River valley is in Fig. 6, of the limestones of Meliaticum unit is in Fig. 7.

Postnappe formations of the Inner Carpathians are of the Late Cretaceous to Palaeogene ages. The typical Inner Carpathian Palaeogene is composed of basal conglomerates, claystones, and a flysch formation (alternation of claystones and sandstones), occurring in the intra-mountainous depressions of the Northern Slovakia (Liptov Basin, Poprad Basin, and others) [14, 15]. The example of the Inner Carpathian Palaeogene, represented by the Podtatranská group, outcropping in the river bed of the Studený potok brook (Veľká Lomnica, Popradská kotlina basin) is in Fig. 8.

Basins and grabens are distinct morphostructural features connected to the development of the Carpathian arc at the close of Palaeogene and during the Neogene period [13]. After the Middle Miocene times, the pre-arc basins, such as the Vienna basin; the inter-arc basins, as the intra-mountainous depressions of the Eastern Slovakia; and the back-arc basins, as the Danube basin, were formed. They consist of huge thicknesses of mostly siliciclastic sediments (clays, sands, gravels), sometimes containing some coal or evaporites [14, 15].

The Neogene volcanics crop out mainly in the Central and Eastern Slovakia, being a part of an extensive volcanic region of the Carpathian arc and the Pannonian Basin ageing from 16.5 to 0.1 Ma. Neogene volcanic mountains of Slovakia are

Fig. 6 Folded dolomites of the Hronicum unit in the Čierny Váh River valley (Photo: M. Fendek)

Fig. 7 Limestones of the Meliaticum unit in the Slovenský kras karstic plateau (Photo: M. Fendek)

Fig. 8 Alternation of sandstones and claystones of the Inner Carpathian Palaeogene in Studený potok river bed (Photo: M. Fendek)

arranged in two provinces – the Central Slovakian and the Eastern Slovakian ones (Fig. 9).

The whole scale of volcanic rocks (rhyolites, andesites, dacites, basalts, pyroclastics) can be found in volcanic mountains of Slovakia. The calc-alkali volcanics within the stratovolcanic structures are typical for Middle to Late Miocene volcanism 16.5–9 Ma). The alkali volcanics are typical for Pliocene to Quaternary volcanism (8–0.1 Ma) where the volcano-sedimentary complex is missing [14, 15]. The ore mineralisation is connected to volcanism, mostly in Štiavnické vrchy Mts. and Kremnické vrchy Mts. The example of the Neogene basalts outcrop at Šomoška (Cerová vrchovina Mts., Southern Slovakia) is in Fig. 10.

The pre-Quaternary units are covered by various types of Quaternary sedimentary cover. The glacial moraine sediments are typical for the highest mountains of Slovakia (Vysoké Tatry Mts., Nízke Tatry Mts.) being composed of loamy to sandy gravels and coarse gravels with boulders. The glaciations and movement of mountain glaciers formed typical U-shaped valleys in the Vysoké Tatry Mts. – see Fig. 5. The thickness of the moraine sediments is different; in some places, they reach more than 300 m. They are visible in the Vysoké Tatry Mts. foreland as the typical morphostructural form of a rampart between the mountain peaks and the Poprad River valley (Fig. 5). Moraine sediments at Gerlachov are shown in Fig. 11. Glacial sediments redeposited by rivers create the glacio-fluvial cover of the bedrock in the Liptovská and Popradská kotlina basin (northern Slovakia) and in some other places in the Orava region (north-western part of Slovakia).

The slope sediments create another typical type of the Quaternary sedimentary cover of the foothills around the whole country. Their composition depends on the

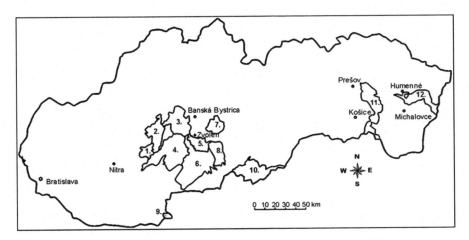

Fig. 9 Location of Neogene volcanic mountains in Slovakia. Central Slovakian Province: (1) Pohronský Inovec, (2) Vtáčnik, (3) Kremnické vrchy, (4) Štiavnické vrchy, (5) Javorie, (6) Krupinská planina, (7) Poľana, (8) Ostrôžky, (9) Kováčovské kopce, (10) Cerová vrchovina. Eastern Slovakian Province: (11) Slanské vrchy, (12) Vihorlat

Fig. 10 Outcrop of basalts at Šomoška, Cerová vrchovina (Photo: M. Fendeková)

Fig. 11 Moraine sediments at Gerlachov, foreland of the Vysoké Tatry Mts., Northern Slovakia (Photo: M. Fendek)

bedrock type. They consist of the rock debris (crystalline and neovolcanic complexes) or loam and sandy loam with rock debris.

The fluvial sediments are spatially the most widespread Quaternary sediments. They fill in the river valleys, being composed of floodplain humic loams, sandy loams with sands and gravels. The thickness can change from some tenths of centimetres in the headwater parts of the brook and river valleys through some first to 15 m in the middle parts and up to more than 500 m in the central part of the Podunajská nížina lowland (Gabčíkovo depression). The sediments of the river terraces are mostly composed of loamy sands and sandy and loamy gravels, often covered by loess or sandy loams.

The windblown sands are typical representatives of the aeolian sediments, widespread mostly in the Záhorská nížina lowland (south-western part of Slovakia). Loess, sandy loess, and calcareous loess are typical for the lowlands; the largest areas covered by loess are at the Trnavská sprašová tabuľa loess plateau in the eastern foreland of the Malé Karpaty Mts. (south-western part of Slovakia).

The fluvial organic and organic sediments are also developed on the territory of Slovakia. The swamp deposits are typical for the alluvial plains of large rivers, filling the buried oxbow lakes, e.g., along the Danube River and in the Danube River branch system. Peat, peat loams, and bog deposits are developed in the mountainous areas, e.g., in the Vysoké Tatry Mts., or in the Orava region.

The recent freshwater limestone – travertine – belongs to Quaternary chemical sediments. Travertine occurs in those places, where the mineral water flows out on the surface in the form of springs, mostly along the faults. The Bešeňová travertines (Liptov Basin, Northern Slovakia) are one of the most spectacular examples. The recent formation of travertine together with the old deposits at Gánovce (Poprad Basin, Northern Slovakia) is in Fig. 12.

4.2   Hydrogeological Conditions and Groundwater Resources

Hydrogeological conditions reflect the geological-tectonic structure of the territory of Slovakia. The amount of water stored in the rock environment depends on the lithological type of rocks, degree, type of rock disintegration, tectonic disturbance, hydraulic properties of the rock environment, and geomorphologic and climatic conditions. The chemical composition of groundwater depends on the primary groundwater sources, type of the rock environment, length and depth of the groundwater circulation, as well as human activities in the infiltration area.

The hydrogeological conditions of Slovakia enabled to delineate regions with similar hydrogeological conditions which were the base for hydrogeological regionalization of Slovakia. The whole territory of Slovakia is covered by 141 hydrogeological regions [16]. The evaluation of natural groundwater amounts and usable groundwater amounts is done on an annual basis at the Slovak Hydrometeorological Institute and published in the annual report on water management balance, part Groundwater [17].

Fig. 12 Recent formation of travertine at Gánovce (Photo: M. Fendek)

A hydrogeological region is a relatively closed balance area delineated according to geological, hydrogeological, and geomorphologic conditions. The region can be subdivided into partial regions and/or subregions. A partial region is a unit delineated within a region according to similar hydrogeological conditions; the subregion is a part of the region belonging to one main river basin (Morava, Danube, Váh, Nitra, Hron, Ipeľ, Slaná, Bodrog, Hornád, or Poprad).

Because the regions are delineated according to geological conditions, the following main types of regions were defined:

- Regions built of pre-Mesozoic complexes – mostly crystalline, labelled by G
- Regions built of Mesozoic complexes, labelled by M
- Regions built of Palaeogene complexes including Outer Flysch belt and postnappe formations of the Inner Carpathian Palaeogene, labelled by P
- Regions built of Neogene sedimentary complexes, labelled by N
- Regions built of Neogene volcanic formations, labelled by V
- Regions built of Quaternary sediments, labelled by Q

Some regions could be composed of formations of different stratigraphic units, and then the combination of labels is used for labelling of the region. As an example, the QG 139 region consists of crystalline complexes of the Vysoké Tatry Mts. and the Quaternary sediments of their foreland. Such a composition is based on the hydrogeological interconnection of the two parts – the Quaternary sediments of the Poprad basin are fed by groundwater runoff from the Vysoké Tatry Mts.

4.2.1 Hydrogeological Conditions of the Regions Built of Pre-Mesozoic Complexes

The crystalline complexes of the Palaeozoic ages form these regions. The regions represent the crystalline cores of the core mountains, as Malé Karpaty, Považský Inovec, Strážovské vrchy, Malá Fatra, Veľká Fatra, Nízke Tatry, Vysoké Tatry, and others, representing the crystalline of Tatricum basement (Fig. 3), Veporské vrchy and Stolické vrchy, representing the crystalline of the Veporicum basement (Fig. 3), and Volovské vrchy (Slovenské Rudohorie) representing the metamorphytes of the Gemericum unit (Fig. 3). The main lithological rock types are (1) granitic rocks, granites, granodiorites, biotite tonalities (locally porphyric), and hybrid granodiorites grading locally to migmatites, and (2) metamorphic rocks, phyllites, mica schists, para- and orthogneisses, banded gneisses, augen gneisses, and amphibolites.

The fissure permeability is the prevailing permeability type. The primary fissuring was formed during the magma cooling; the secondary one originates from tectonic disturbance. The laminar flow is typical for this type of rocks. The water-bearing properties are weak because of quite thin weathering zone in which the groundwater can circulate and can be stored. The storage capacity is higher in magmatic rocks than in the metamorphic ones. The spring yields are low – mostly less than $1.0 \, \text{L s}^{-1}$. As an example, spring of the Čierny Váh River under the Kráľova hoľa mount on the northern slopes of the Nízke Tatry Mts. is shown in Fig. 13.

The spring or well yield could be much higher under specific conditions, e.g., when the crossing of fault systems is drilled through, or the crystalline complexes are thrusted over the Mesozoic ones, or the Mesozoic complexes are infolded into the crystalline. The last possibility is represented by the Trangoška syncline in the Ďumbier Massif on the southern slopes of the Nízke Tatry Mts. (Central Slovakia). The Mesozoic sequences, consisting in Lúžna sequence quartzites, Middle and Late Triassic dolomites and limestones, as well as Keuper shales, are infolded into the banded and augen gneisses of the Ďumbier massif. As a result, the Trangoška spring yielding up to $490 \, \text{L s}^{-1}$ is located there. The view on the Ďumbier Massif is in Fig. 14; the side chamber from which the groundwater flows into the water storage reservoir is in Fig. 15.

Groundwater of the crystalline complexes is of the $Ca-HCO_3$ chemical type with the low mineralization (TDS), reaching mostly less than $150 \, \text{mg L}^{-1}$. The pH value is in the acid area, around 6.5. Only in the case of presence of sulphidic ore minerals, the chemical type changes into $Ca-SO_4$ one, the mineralisation increases to more than $1{,}000 \, \text{mg L}^{-1}$, and the pH value can decrease substantially. Generally, the groundwater is of good quality; springs are often tapped and used for local water supply. The water quality could be threatened by grassland farming, tourism, and mining activities.

Fig. 13 Spring of the Čierny Váh River, crystalline of the Nízke Tatry Mts. (Photo: M. Fendek)

4.2.2 Hydrogeological Conditions of the Regions Built of Mesozoic Complexes

The regions built of Mesozoic complexes are typical of very various rock types' occurrence. Typical aquifers are represented by the Middle and Late Triassic carbonates – limestones and dolomites; typical aquicludes (isolators) are represented by Early Triassic shales and quartzites, Late Triassic Lunz layers and Keuper shales, Middle Jurassic deep-sea marly limestones, or flyschoid sequences of the Early and Late Cretaceous. The rock permeability depends on the rock type. The fissure permeability is typical for quartzites and dolomites, the fissure-karst or karst-fissure permeability for limestones. The low porous permeability is developed in shales and flyschoid sediments.

The groundwater flow is laminar in fissured rocks, and it can be also turbulent in the karstified limestones. The karstification is typical mostly for the Middle and Late Triassic Guttenstein limestones of the Fatricum (e.g., a system of the Demänová

Fig. 14 The view on the Ďumbier Massif of the Nízke Tatry Mts. over the Trangoška spring area (Photo: M. Fendeková)

Fig. 15 The view into the side chamber with the groundwater flow from the rock massif (Photo: M. Fendeková)

caves, Nízke Tatry Mts.) or Wetterstein limestones of the Silicicum unit (e.g., Domica cave, Slovenský kras Mts.). The example of the underground karstic features of the Domica cave (Southern Slovakia) is in Fig. 16.

Fig. 16 Karstic phenomena in the Domica cave (Photo: M. Fendek)

There are three types of the carbonate rock structures:

• The tilted structures typical for Fatricum and Hronicum nappes, covering the crystalline cores of the core mountains and dipping under the Palaeogene or Neogene filling of the intra-mountainous depressions
• Horizontally placed structures of the Silicicum and Meliaticum units, forming the karstic plateaus (e.g. Slovenský kras or Muránska planina)
• Structures occurring in tectonic position with other units, e.g. infolded into crystalline rocks (e.g. the syncline of Trangoška)

The storage capacity of the Mesozoic aquifers is very high. It is higher but less stable in limestones and lesser but more stable in dolomites. The spring yields could reach up the first thousands of litres per second; more often they amount between 50 and 300 L s^{-1}. The highest spring yield is documented by the Pod hradom spring (Muránska planina plateau), which reached 6,380 L s^{-1} on 30 October 1990. Groundwater of the carbonatic complexes is of the Ca–HCO$_3$ chemical type with the average mineralization (TDS) of 300–600 mg L^{-1}. The pH value is neutral, with the values around 7.0.

Many karstic springs were tapped, and they are used for drinking water supply because of the excellent groundwater quality. As the example, the spring Pod starým mlynom at Horný Jelenec (Veľká Fatra Mts., Central Slovakia) is shown in Fig. 17. Groundwater is formed in the Krížna nappe Middle Triassic limestones. The spring

Fig. 17 Pod starým mlynom spring (Photo: M. Fendeková)

flows out from a huge deposit of travertines to the storage reservoir, from which it is conducted through the filter plant to the water-main system. The maximum yield of the spring reached 321.0 L s^{-1} on 7 April 2000; the minimum was 1.49 L s^{-1} on 22 January 2000. The average yield amounts to 34.4 L s^{-1}. The yield of the Pod starým mlynom spring is put into the Jergaly branch of the Pohronský water-main system which collects spring water from several large springs of the eastern part of the Veľká Fatra Mts. The Pohronský water-main supplies cities of Banská Bystrica, Zvolen, and Žiar nad Hronom and many villages in the surrounding area with the high-quality drinking water. Besides the Jergaly branch, the water-main is also supplied by the springs from the Harmanec syncline creating the Harmanec branch (Zalámaná, Veľké Cenovo, Malé Cenova, and Matanová springs) and by the groundwater wells in Podzámčok and Dobrá Niva groundwater sources.

4.2.3 Hydrogeological Conditions of the Regions Built of Palaeogene Complexes

The Palaeogene complexes occur mainly in the Outer Flysch belt and also in the Inner Carpathians intra-mountainous depressions. The flysch sediments of the Outer Flysch belt occur in the geomorphological units of Biele Karpaty, Javorníky,

Turzovská vrchovina, Moravsko-sliezske Beskydy, Kysucké Beskydy, Oravské Beskydy, and Kysucká vrchovina in the north-western Slovakia and in Ľubovnianska vrchovina, Čergov, Busov, Ondavská vrchovina, Laborecká vrchovina, and Bukovské vrchy in the north-eastern Slovakia. The Inner Carpathian Palaeogene sediments occur in the Skorušinské vrchy, Spišská Magura, Levočské vrchy, and Šarišská vrchovina highlands, and fill in the Podtatranská kotlina basin, consisting in Liptov, Poprad, and Levoča Basins. They can be found also in other intra-mountainous depressions of the Central Slovakia but in a lesser extent.

The prevailing rock types are claystones and sandstones, together with the conglomerates. The main difference between the Outer and Inner Carpathian Palaeogene sediments is in the type of tectonic disturbance. Whilst the Outer Flysch belt was folded into a system of nappes, the Inner Carpathian Palaeogene sediments are embedded horizontally and disturbed by a system of faults which caused the formation of horsts and grabens. This is typical, e.g., for the Liptov Basin (Northern Slovakia). The flysch sediments of the Outer Flysch belt could reach thicknesses of up to more than 8,000 m. The thickness of the Inner Carpathian Palaeogene depends on the tectonic disturbance of the area, reaching tens to hundreds of metres. In some areas, e.g., the Poprad Basin, thicknesses of more than 3,000 m were proven.

The prevailing type of permeability is the porous one. Whilst the sandstones represent aquifers storing groundwater and enabling the laminar flow, the claystones act as isolators. Because of the presence of isolators, there is not too much water available for water supply. Therefore, the majority of inhabitants are supplied by surface water reservoirs (Nová Bystrica, Starina) or by single small springs. The springs with the yields from 0.1 up to 3.0 L s^{-1} are tapped and used for the local water supply in the Kysuce and Orava regions (the north-western Slovakia) and in the Ondava and Laborec highlands (the north-eastern Slovakia). Larger groundwater amounts can only be found when the basal conglomerates are hydraulically interconnected with the underlying carbonatic rocks of the Hronicum or Fatricum units.

The areas built of Palaeogene sediments are rich in the occurrence of mineral water springs of low yields, mostly gaseous, containing the carbon dioxide (Fig. 18). The carbonatic mineral waters with the mineralisation of more than 1,000 mg L^{-1} could be formed when the tilted nappe carbonates from the surrounding mountains dipped below the Inner Carpathian Palaeogene sedimentary cover. Their temperatures reach up to 60°C. Such mineral waters are either used in natural healing spas, as Lúčky, Vyšné Ružbachy, or Bardejov, or they can be utilised in many aqua parks, as Bešeňová, Liptovská Kokava, Poprad, or Vrbov. The natural hydrocarbons (crude oil and gas) in small amounts are also present in the deeper parts of the Palaeogene sedimentary complexes.

The fresh groundwater is of the Ca–HCO$_3$ chemical type with the mineralisation of about 400–800 mg L^{-1}. The pH value is in the neutral area. Groundwater could contain the aggressive CO$_2$. The water quality is good, but it could be threatened by urban agglomerations, agriculture, and various industrial plants. Most of the small villages in the areas built of Palaeogene sediments have no sewerage systems, and water is discharged either into the rock environment or directly to the surface streams.

Fig. 18 Mineral water
spring flowing out from
travertine at Bešeňová
(Photo: M. Fendek)

4.2.4 Hydrogeological Conditions of the Regions Built of Neogene Sedimentary Complexes

The Neogene sedimentary complexes occur either in the intra-mountainous depressions as the Oravská, Žiarska, Zvolenská, Turčianska, Hornonitrianska, Prešovská, and Košická kotlina basins or in the southern marginal basins and lowlands as in the Podunajská nížina lowland, Borská nížina lowland, Juhoslovenská panva basin, and Východoslovenská nížina lowland. The Podunajská nížina lowland as the geomorphological unit corresponds to the Danube basin as the Neogene geological unit, and the Záhorská nížina lowland corresponds to the Vienna basin.

The main types of rocks are gravels, sands, clays, silts, but also conglomerates, siltstones, claystones, sandstones, tuffs, and epiclastic rocks. The lignite, coal seams, evaporites, and organodetritic limestones can be found in some areas. The thickness of the Neogene sedimentary complexes reaches from several tens of metres up to 8,000 m in the Gabčíkovo depression (central part of the Danube basin). The sediments are of marine to the brackish origin in the marginal parts and of the lacustrine to the fluvial origin in the intra-mountainous depressions. The multiple alterations of permeable and impermeable layers of various thicknesses and spatial extent are typical for the marginal basins, creating conditions for the occurrence of artesian horizons. Faults are the principal structural elements in the basins [14, 15].

The porous permeability and the slow laminar flow are typical for the sedimentary Neogene complexes. The groundwater storage properties are low. The amount of groundwater is influenced by the presence of impermeable silts and clays but also by very low recharging of the rock environment by precipitation and surface water. This is caused by the location of the lowlands in the areas with the lowest precipitation amounts (450–500 mm annually in average) and highest temperatures (around 10°C annually in average), resulting in the highest potential evapotranspiration on the Slovak territory. However, recharging of Neogene sediments could occur in the marginal parts of the basins at the contact with the mountains. The contact is mostly tectonic; the mountains are bordered by deep faults through which the groundwater from the Mesozoic sediments permeates into the Neogene sands and gravels. The well yields are around $3.0 \, \text{L s}^{-1}$, rarely up to $10.0 \, \text{L s}^{-1}$. The specific groundwater runoff amounts to 1.4–$2.5 \, \text{L s}^{-1} \, \text{km}^{-2}$ in the Borská nížina lowland, 0.5–$1.0 \, \text{L s}^{-1} \, \text{km}^{-2}$ in the Žitný ostrov, and not more than $2.2 \, \text{L s}^{-1} \, \text{km}^{-2}$ in other areas.

The geochemical zoning of groundwater chemistry is typical in the areas of the marginal basins and lowlands. It means that the chemical type of the fresh water in the first horizon is Ca–Mg–HCO_3 with the mineralisation of 0.4–$0.9 \, \text{g L}^{-1}$; in the deeper parts, it changes to Na–HCO_3 with the mineralisation up to $1 \, \text{g L}^{-1}$.

The groundwater quality is good; the possible threatening factor is the intense agricultural activity, which can result in increased nitrate contents. The increased concentrations of iron and manganese, which are naturally present in the Neogene sediments, could also cause problems.

The natural hydrocarbons (crude gas, oil) and lignite occur in the Vienna basin; the brown coal is mined in the Hornonitrianska kotlina basin. The occurrence of geothermal water is typical for the Neogene sedimentary complexes; they are widely used, e.g., in Dunajská Streda, Galanta, Nové Zámky, Veľký Meder, Poľný Kesov, Nesvady, Šaľa, and Sereď and in many other localities of the central depression of the Danube basin.

4.2.5   Hydrogeological Conditions of the Regions Built of Neogene Volcanic Complexes

The Neogene volcanic complexes (neovolcanics) cover about 10% of the Slovak territory, being concentrated in the central and eastern part of Slovakia. The full scale of volcanic rocks can be found in Neogene volcanic mountains – from effusive rocks as rhyolites, andesites, dacites, and basalts to sedimentary volcanic rocks as breccias, tuffs, conglomerates, and sandstones. The volcano-sedimentary complexes prevail in the ratio of 10:1 in comparison with the effusive rocks.

The fissure permeability is typical for effusive rocks, the porous one for the volcano-sedimentary complex. Laminar flow is typical for the Neogene volcanics. Three types of water-bearing structures can be found in neovolcanics: (1) weathered zone of the effusive rocks, (2) volcano-sedimentary complex, and (3) deep fault zones acting as drainage systems. The amounts of water stored in the weathering zone of the effusive rocks are low; the yields of natural springs range from several decilitres up to 1.5 L s^{-1}. The example of the fissure spring in the Štiavnické vrchy Mts. is in Fig. 19.

The amount of water stored in the volcano-sedimentary complex is a bit higher. The spring yields do not reach more than 1.5 L s^{-1}. However the well yields could amount up to 20.0 L s^{-1}. As the examples, wells in Plášťovce with 20.0 L s^{-1}, Litava with 11.0 L s^{-1}, or Dačov Lom with 5.0–10.0 L s^{-1} located in the Krupinská planina plateau can be mentioned.

The highest yields were found in the deep tectonic zones of the regional importance. One of them, called the Neresnica fault zone, follows the Neresnica brook valley on the contact of the Štiavnické vrchy Mts. and the Javorie Mts. to the south of the Zvolen city (see Fig. 10).

The spring with the high yields of more than 60.0 L s^{-1} (the amount unexpected in the Neogene volcanic area) occurred in the vicinity of the Podzámčok village, located to the south of the Zvolen city (Fig. 10). Therefore the hydrogeological investigation in the wider area started in the 1960s of the last century, and system of wells was drilled in the alluvial plain of the Neresnica brook close to Podzámčok village. The amount of water was proven by the hydrodynamic testing, and the water management source was put into operation. Later on, after reconstruction of wells in

Fig. 19 Handrlová spring at Podhorie, Štiavnické vrchy Mts. (Photo: M. Fendeková)

the early 1980s of the twentieth century, the water management source continued in operation with the amount of about 200 L s^{-1}. As the example, two exploitation wells are shown in Fig. 20. The band of trees behind the fence follows the Neresnica stream; the abandoned andesite quarry can be seen in the background on the left bank of the Neresnica brook.

The wells of the water management source are 150 m deep [18] and found water in the andesite lava flows. Another water management source was put into operation at Dobrá Niva village, located to the south of the Podzámčok water source. The Podzámčok water management source, comprising five exploitation and one observation wells, supplied the Pohronský water-main with up to 200 L s^{-1} of the high-quality groundwater in the early 1990s of the twentieth century [19]. However, this amount was too high and caused the over-exploitation of the hydrogeological structure. As a result, the groundwater heads decreased in more than 20 m in comparison with the uninfluenced stage. The Neresnica brook started to dry up during the low-flow period, and all springs in the adjacent area disappeared. The decrease in water demand in the mid-1990s of the twentieth century caused the lowering of abstraction to less than 80 L s^{-1}, and the groundwater heads started to increase [18, 19]. Nowadays, the groundwater level is on the stage prior to the start of the groundwater abstraction and the natural ecosystem functions without problems at the abstraction rate of about 70 L s^{-1}. One of the former exploitation wells is used for artificial lowering of the groundwater head at present (Fig. 21). The reason

Fig. 20 View on the part of the Podzámčok water management source (Photo: M. Fendeková)

Fig. 21 Artificial lowering of groundwater head by pumping of groundwater excess to the Neresnica brook (Photo: M. Fendeková)

is the flooding of the exploitation wells by groundwater due to the increase of groundwater head after lowering of abstraction amounts. There is a well head shown in Fig. 21 releasing the pumped water; the iron pipeline (on the right) conducts the groundwater into the Neresnica brook.

Groundwater in the Neogene volcanic effusive complexes is of the Ca–HCO$_3$ chemical type with similar low mineralisation and pH values in the acid area as the groundwater in the crystalline rocks. The presence of the silica acid is typical. The amount of total dissolved solids could increase substantially when the sulphidic ore minerals are present, and the oxidation processes go on. The mineralisation in the volcano-sedimentary complexes is higher, reaching up to 600 mg L^{-1}. The reason is in the presence of soluble calcareous cement and longer water-rock interaction due to slow groundwater flow in the porous media. The pH values are closer to the neutral band.

Generally, the groundwater is of good quality. The water quality could be threatened by grassland farming, tourism, and mining activities in the case of shallow groundwater flow in the weathered effusive rocks.

4.2.6 Hydrogeological Conditions of the Regions Built of Quaternary Sediments

Various types of Quaternary sediments form the cover of the pre-Quaternary bedrock. Among them, the alluvial sediments are the most important. Locally, the glacial sediments could be used for the drinking water supply, mostly in the Vysoké Tatry Mts. area and in its foreland. The well yields vary from 1.0 up to 20.0 L s^{-1}; exceptionally they could reach up to 80.0 L s^{-1}.

Generally, the porous permeability and the laminar flow are typical for Quaternary aquifers. The turbulent flow could occur in very coarse gravels at high flow velocities at the well filter.

The amounts of groundwater are dependent on hydraulic properties of sediments and on the recharge. Groundwater in the upper and middle sections of the surface streams alluvial plains are fed by groundwater inflow from the river valley slopes or from the upper valley sections (parallel groundwater flow with the stream). The well yield reaches 5.0–15.0 L s^{-1}. High well yields were documented in the upper part of the Torysa River and its tributary – Slavkovský potok brook basins. The well yields amounted 10.0–45.0 L s^{-1} [20] in Brezovica and Brezovička water management sources (to the west of the Prešov city).

The yields increase importantly in the downstream parts of the river valleys, where the groundwater can be recharged by the surface stream. This is the case of the Danube River, which feeds the groundwater in the Žitný ostrov area. The well yields amount tens of L s^{-1}. The volume of water pumped from wells in Rusovce water management source (close to Bratislava) amounts to 100.0 L s^{-1} from each out of ten wells of the water source. The highest amount of groundwater pumped during the pumping test was documented at Šamorín (to the south of Bratislava), amounting to more than 300.0 L s^{-1}. The well yields are reasonably high also in the downstream part of the Laborec River with the value of 10.0–20.0 L s^{-1} in the surrounding of Michalovce city (south-eastern part of Slovakia) but also in many other places.

The groundwater in Quaternary sediments is of the Ca–HCO$_3$ chemical type; the mineralisation value depends on the type of the sediment. Mineralisation is low in the case of windblown sands and glacial and slope sediments – up to 150 mg L^{-1} – and higher in alluvial sediments reaching up to 800 mg L^{-1}. The pH values vary around 7. Water is of high quality. However, the threat of pollution is quite high due to the intense utilisation of lowland by agriculture, daily life in large urban agglomerations, and various types of industrial plants. Therefore, in some places, the water from the second horizon is abstracted.

Comparison of areas covered by a respective type of hydrogeological region shows that 55.3% of the territory of Slovakia is covered by regions built of less permeable rock types – see Fig. 22. The crystalline complexes, Palaeogene and Neogene sediments, as well as Neogene volcanics belong to those regions. On the other hand, regions with well-permeable rocks, comprising Quaternary and Mesozoic regions, cover only 44.7% of the Slovak territory.

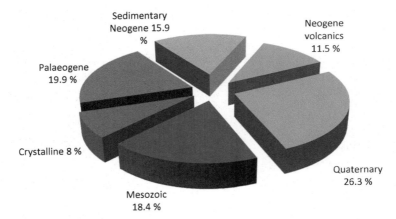

Fig. 22 Share of respective hydrogeological regions in the area of the Slovakia (based on data from [17])

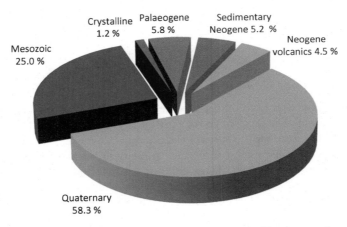

Fig. 23 Share of respective hydrogeological regions on the total usable amounts of groundwater in Slovakia (based on data from [17])

The figures are totally different when comparing the share of region type on the 80 m^3 s^{-1} of the total usable groundwater amounts. Figure 23 shows that in only two regions – Quaternary and Mesozoic – occur 83.3% of the total usable amounts in Slovakia, whilst in the rest of the regions, it makes only 16.7%.

There are areas in Slovakia which were declared by the law for protected water management areas [3]. Up to today, ten such areas were declared with the total area of 6,942 km^2 which makes 14.2% of the whole territory of Slovakia. The data on protected water management areas are shown in Table 6. Table shows that ground-water sources prevail among protected water management areas. The highest amount

Table 6 Data on protected water management areas of Slovakia (adapted according to [3])

No.	Name	Area (km²)	Area share on the Slovak territory (%)	Usable water amounts		
				Surface water ($m^3 s^{-1}$)	Groundwater ($m^3 s^{-1}$)	Total ($m^3 s^{-1}$)
1	Žitný ostrov	1,400	2.86	–	18.00	18.00
2	Strážovské vrchy Mts.	757	1.54	–	2.33	2.33
3	Beskydy-Javorníky Mts.	1,856	3.78	1.84	0.69	2.53
4	Veľká Fatra Mts.	646	1.31	0.97	2.98	3.95
5	Nízke Tatry Mts.					
	a. Western part	358	0.73	–	2.50	2.50
	b. Eastern part	805	1.64	2.33	2.43	4.76
6	Upper part of Ipeľ, Rimavica and Slatina Basins	375	0.76	1.09	0.11	1.20
7	Muránska planina	205	0.42	–	1.40	1.40
8	Upper part of the Hnilec Basin	108	0.20	0.16	0.10	0.26
9	Slovenský kras					
	a. Plešivecká planina	57	0.12	–	0.55	0.55
	b. Horný vrch	152	0.34	–	1.97	1.97
10	Vihorlat Mts.	225	0.46	0.08	0.43	0.51
Total		6,942	14.16	6.47	33.49	39.96

of protected groundwater resources is stored in the Žitný ostrov. The amount of groundwater usable amounts within the protected water management areas is more than five times higher than the surface water ones.

Another measure used for surface and groundwater resources protection is a delineation of protection zones. Three types of the zone can be distinguished. The first protection zone protects the proximate surrounding of the water source; the second protection zone protects the part or the whole infiltration area. If necessary, the third protection zone can be declared to protect the area against hazardous substances in case the second protection zone does not cover the whole infiltration area [21] (Table 7).

In this case, the total number of existing protection zones of groundwater management sources is more than ten times higher than the number of surface water sources. However, the total area of the protection zones belonging to surface water management sources is higher than that one of groundwater sources. The reason is that the drinking water supply reservoirs are included in the total. Drinking water supply reservoirs cover large areas, and therefore also the first and the second protection zones have large areas. On the contrary, the first protection zone of the single groundwater source has generally an area of 10×10 m.

Table 7 Numbers of existing protection zones of the water management sources [3]

Partial river basin district	Number of protection zones of water management sources		Area covered by protection zones of water management sources (km²)	
	Groundwater	Surface water	Groundwater	Surface water
Morava	31	0	138.65	0
Dunaj	29	0	60.30	0
Váh	447	14	2,116.71	194.36
Hron	173	7	569.17	95.42
Ipeľ	70	1	156.48	84.00
Slaná	76	6	137.89	137.62
Bodva	30	7	121.46	104.16
Hornád	124	18	193.24	726.93
Bodrog	230	17	70.82	3,394.59
Poprad and Dunajec	59	11	155.80	159.25
Total	1,269	81	3,720.52	4,896.33

4.3 Geothermal Water Resources

A systematic exploration and research of geothermal waters date for almost five decades in Slovakia. Geothermal aquifers can be found only in the Inner Western Carpathians due to favourable geological conditions. Geothermal aquifers are largely associated with Triassic dolomites and limestones of the Fatricum, Hronicum, and Silicikum nappes, less frequently with Neogene sands, sandstones, conglomerates, andesites, and related volcaniclastic sediments. Triassic carbonates can be hydraulically connected to Podtatranská skupina group represented by Middle–Late Eocene Borové basal formation composed of breccias and conglomerates that pass to detritic carbonates and rare organogene limestones, beneath top aquifuge recognised as Late Eocene–Oligocene Huty (claystones dominated), Zuberec (flysch dominated), or Biely Potok (sandstones dominated) formations [22]. The hydrogeological function of Neogene sequences and Quaternary cover varies regarding the drainage. The maximal thickness of the Hronicum sequences in the intra-mountainous depressions is up to 1,200 m. Fatricum sequence reaches the maximal thickness of 2,300 m.

Geothermal wells are located mostly in the intra-mountainous depressions or in lowlands bordering the Slovak territory in its southern part (Danube Basin central depression, Košice Basin, Vienna Basin). Up to today, 27 hydrogeothermal areas or structures have been identified in the Slovakian territory, comprising 34% of the whole territory (Fig. 24).

Geothermal waters were proven by 171 geothermal wells with the depth of 9–3,616 m [23]. The temperature on the well head ranges from 18 to 129°C; yields reach up to 70 L s^{-1}. Water is mostly of Na–HCO$_3$–Cl, Ca–Mg–HCO$_3$ and Na–Cl chemical type with the TDS value of 0.4–90.0 g/L. The total amount of 2,453 L s^{-1}

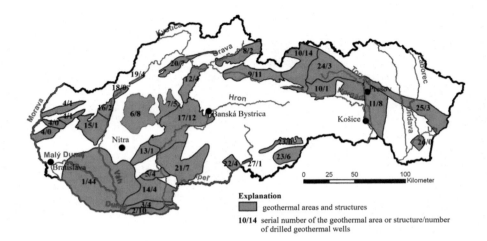

Fig. 24 Geothermal areas and structures in the territory of the Slovak Republic. *List of geothermal areas*: (1) Danube Basin central depression, (2) Komárno high block, (3) Komárno marginal block, (4) Vienna Basin, (5) Levice marginal block, (6) Bánovce Basin and Topoľčany embayment, (7) Horná Nitra Basin, (8) Skorušina Basin, (9) Liptov Basin, (10) Levoča Basin (W and S parts), (11) Košice Basin, (12) Turiec Basin, (13) Komjatice depression, (14) Dubník depression, (15) Trnava embayment, (16) Piešťany embayment, (17) Central Slovakian Neogene volcanics (NW part), (18) Trenčín Basin, (19) Ilava Basin, (20) Žilina Basin, (21) Central Slovakian Neogene volcanics (SE part), (22) Horné Strháre – Trenč graben, (23) Rimava Basin, (24) Levoča Basin (N part), (25) Humenné ridge, (26) Beša – Čičarovce structure, (27) Lučenec Basin

of geothermal water were documented by realised geothermal wells with a total installed thermal power capacity of 387.6 MW$_t$.

The effect of hydrogeochemical zoning is typical for Neogene hydrogeological structures where the chemical type of water changes with the increasing depth from $Ca–Mg–HCO_3$ throughout $Na–Ca–HCO_3–Cl$ and $Na–HCO_3–Cl$ up to $Na–Cl$ chemical type at the highest depths. The TDS increases in the same direction from less than 1 g L^{-1} up to 90 g L^{-1} at Marcelová (central depression of the Danube Basin). The geothermal water is enriched by dissolved gases, mostly by CO_2, CH_4, and H_2S.

The amended list of geothermal prospective areas/structures in the Slovak Republic, with the basic data on a number of wells, well depths, discharges, geothermal water temperatures, thermal potentials, and amounts of total dissolved solids, is shown in Table 8.

The total amount of thermal energy potential of geothermal waters in prospective areas (proven, predicted, and probable) represents 6,653.0 MW$_t$ (Table 9). This amount consists in 708 MW$_t$ of geothermal resources and 5,945 MW$_t$ of reserves.

The total amount of geothermal water utilised in the last period was 441 L s^{-1} in average per year. This utilisation makes only 18% of approved amounts of geothermal water. Geothermal waters are widely used for recreational purposes (45%), mostly in very popular aqua parks in many places of Slovakia. Space heating (18%), greenhouses and fish farming (16%), and heat pumps (21%) belong to other ways of geothermal water utilisation (Fig. 25).

Table 8 Basic characteristics of geothermal water in prospective areas/structures in Slovakia

Prospective area, structure	No. of wells	Well depth (m) Min-max	Discharge (L s⁻¹) Min-max	Total	Temperature (°C) Min-max	Thermal power (MW₁)	TDS (g L⁻¹) Min-max
Central depression of the Danube Basin	44	306–3,303	0.1–25.0	480.6	19–91	100.72	0.5–20.1
Komárno high block	10	160–1,021	5.5–70.0	268.7	20–40	18.86	0.7–0.8
Komárno marginal block	4	1,060–1,763	3.3–6.0	18.8	42–56	2.95	3.1–90.0
Vienna Basin	2	2,100–2,605	12.0–5.0	37.0	73–78	9.50	6.8–10.9
Levice block	4	1,470–1,900	28.0–53.0	181.0	69–80	47.94	17.6–19.6
Bánovce Basin and Topoľčany embayment	8	102–2,106	1.7–18.8	71.8	20–55	5.32	0.6–5.9
Horná Nitra Basin	5	150–1,851	2.5–22.0	57.9	19–59	7.05	0.4–1.9
Skorušina Basin	2	600–1,601	35.0–65.0	100.0	28–56	12.24	0.8–1.3
Liptov Basin	11	50–2,500	6.0–32.0	251.5	26–66	28.85	0.5–4.7
Levoča Basin (W and S parts)	14	9–3,616	0.0–61.2	251.5	20–62	34.92	0.6–4.0
Košice Basin	8	160–3,210	3.0–65.0	214.9	18–134	79.26	0.7–31.0
Turiec Basin	4	97–2,461	0.0–12.4	27.4	41–54	3.79	1.4–2.5
Konjatice depression	1	2,572	8.0	8.0	5.0	1.21	56.8
Dubník depression	4	350–1,927	1.5–15.0	36.0	18–75	3.70	1.6–30.0
Trnava embayment	1	118	14.5	14.5	24	0.55	2.52
Piešťany embayment	2	30–1,200	2.0–10.0	12.0	19–69	0.87	1.3–1.4
Central Slovakian Neogene volcanics (NW part)	12	92–2,500	0.0–30.0	115.2	27–57	13.95	0.4–5.0
Ilava Basin	4	100–1,761	0.0–10.0	18.1	22–40	1.43	1.7–2.6
Žilina Basin	7	105–2,258	0.0–22.0	66.4	24–42	3.63	0.4–0.8
Central Slovakian Neogene volcanics (SE part)	7	40–85	2.1–25.0	72.5	24–42	4.30	1.0–5.9
Horné Strháre – Trenč graben	4	320–625	1.5–10.5	16.0	21–38	1.04	0.4–3.1
Rimava Basin	6	158–1,100	0.0–45.0	61.4	18–33	1.76	1.7–31.5

(continued)

Table 8 (continued)

Prospective area, structure	No. of wells	Well depth (m) Min–max	Discharge (L s^{-1}) Min–max	Total	Temperature (°C) Min–max	Thermal power (MW$_t$)	TDS (g L^{-1}) Min–max
Levoča Basin (NE part)	3	3,500	4.0–5.0	13.5	51–65	2.29	8.7–12.3
Humenné ridge	3	250–823	0.7–4.0	6.9	23–33	0.44	4.4–11.9
Beša – Čičarovce structure	0	–	–	–	–	–	–
Lučenec Basin	1	1,501	11.2	11.2	37	1.04	12.1
Total	171	9–3,616	0.0–70.0	2,435.4	18–129	387.60	0.4–90.0

Table 9 Thermal energy potential of geothermal waters in the Slovak Republic

Resources (MW$_t$)			Reserves (MW$_t$)		
Proven	Predicted	Probable	Proven	Predicted	Probable
218	390	100	147	805	4,993
708			5,945		
Total amount: 6,653.0 MW$_t$					

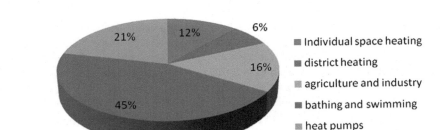

- Individual space heating
- district heating
- agriculture and industry
- bathing and swimming
- heat pumps

Fig. 25 Share of individual applications for annual energy use

5 Summary and Conclusions

Surface and groundwater resources create the water fund of Slovakia. According to requirements of the Water Framework Directive, the surface and groundwater resources are evaluated within the delineated surface and groundwater bodies. The groundwater bodies were delineated in three different layers – the Quaternary, pre-Quaternary, and geothermal groundwater bodies.

The surface water resources consist in the surface inflow to Slovakia through the Danube River and amounts of surface water rising on the Slovak territory. The surface inflow amounts to approximately 2,514 m^3 s^{-1} of surface water; the surface water rising on the Slovak territory comprises approximately 398 m^3 s^{-1}. Among the surface water resources, 102 stream sections designated for abstraction of water intended for human consumption and eight surface drinking water supply reservoirs are of the main importance. The stream sections designated for abstraction of water intended for human consumption are mostly located in the headwater parts of the river basins; the surface drinking water supply reservoirs were built in the areas with low groundwater resources.

According to the Slovakian Water Act, the groundwater is preferentially assigned for drinking water supply. Groundwater resources occur in various types of hydrogeological regions. Comparison of areas covered by a respective type of hydrogeological region shows that 55.3% of the territory of Slovakia is covered by regions built of less permeable rock types. The crystalline complexes, Palaeogene and Neogene sediments, as well as Neogene volcanics belong to those regions. On the other hand, regions with well-permeable rocks, comprising Quaternary and Mesozoic regions, cover only 44.7% of the Slovak territory. The figures differ when comparing the share of the region type on the estimated 80 m^3 s^{-1} of the total usable groundwater amounts. In only two regions – Quaternary and Mesozoic – occur 83.3% of the total usable amounts in Slovakia, whilst in the rest of the regions, it makes only 16.7%. There are ten areas in Slovakia, which were declared by the law for protected water management areas. Groundwater resources prevail in the majority of them except for Javorníky-Beskydy Mts. areas and upper parts of the Ipeľ, Rimavica, Slatina, and Hnilec Basins. The largest volume of usable groundwater resources is stored in the alluvial deposits of the Žitný ostrov area.

There are 27 bodies of geothermal waters in Slovakia delineated according to the Water Framework Directive requirements, occurring within the same boundaries as 27 prospective areas of geothermal water resources. Geothermal waters were proven by 171 geothermal wells with the depth of 9–3,616 m. The temperature on the well head ranges from 18 to 129°C; yields reach up to 70 L s^{-1}. Water is mostly of Na–HCO$_3$–Cl, Ca–Mg–HCO$_3$, and Na–Cl chemical type with the TDS value of 0.4–90.0 g/L. The total amount of 2,453 L s^{-1} of geothermal water was documented by realised geothermal wells with a total installed thermal power capacity of 387.6 MW$_t$.

Each surface and groundwater source used as drinking water supply source is protected by protection zones. Three degrees of protection zones exist according to the Slovakian legislation. The first protection zone (the protection zone of the first degree) protects the proximate surrounding of the water source; the second protection zone (the protection zone of the second degree) protects the part or the whole infiltration area. If necessary, the third protection zone (the protection zone of the third degree) can be declared to protect the area against hazardous substances in case the second protection zone does not cover the whole infiltration area. A total of 1,350 protection zones of the water management surface and groundwater sources were delineated up to present. No rules for geothermal water protection were legislatively adopted yet.

6 Recommendations

Slovak Republic is a country rich in geothermal waters. Geothermal energy is used in various ways. There are space heating facilities using the thermal energy of geothermal water. Geothermal water is also used in agriculture for greenhouse heating. Many wellness centres and aqua parks are already under operation around the country, and new business plans oriented on geothermal water use are being prepared. The continuously increasing activities connected to geothermal water use could lead to problems with the mutual influencing of geothermal wells located close to each other abstracting water from the same geothermal water aquifer. Therefore, more attention should be paid to preparation and regularisation of geothermal water protection measures and rules for utilisation of geothermal water resources.

References

1. Directive, Strategy Framework (2000) Directive 2000/60/EC of the European Parliament and of the Council of 23 October 2000 establishing a framework for community action in the field of water policy. Off J L 327
2. Act No. 364/2004 Coll. on Water and Amendments of Slovak National Council Act No. 372/1990 Coll. on Offences as Amended (Water Act)
3. Water Plan of the Slovak Republic (2011) Abbreviated version. Ministry of Environment of the Slovak Republic, Bratislava
4. Lieskovská Z (ed) (2016) Environment of the Slovak Republic in focus. Ministry of Environment of the SR and Slovak Environmental Agency, Bratislava and Banská Bystrica
5. Melová K, Šimor V, Lupták Ľ, Bucha B (2013) Basic information about the Danube River basin. In: Danube day 2013. Slovak Hydrometeorological Institute, Bratislava. http://www.shmu.sk/sk/?page=2177 (in Slovak)
6. Landscape Atlas of the Slovak Republic (2002) 1st edn. Ministry of Environment of the Slovak Republic and Slovak Environmental Agency, Bratislava and Banská Bystrica
7. Lovásová Ľ, Gápelová V, Podolinská J, Malovová J, Lupták Ľ, Melová K, Škoda P, Liová S, Síčová B, Staňová J, Fabišiková M, Pospíšilová I (2016) Water management balance of the Slovak Republic. Water management balance of surface water amount in 2015. SHMÚ, Bratislava
8. Resolution no. 211/2005 Coll. defining list of important water management and water supply streams
9. Slovak Water Management Enterprise (2015) Annual report. Banská Štiavnica
10. Abaffy D, Lukáč M (1991) Dams and water reservoirs in Slovakia. Alfa, Bratislava
11. Sabolová J, Dobrotka S (2014) 25 years of the water reservoir Starina. Vodohospodársky spravodajca 57:4–7
12. Kollár A, Fekete V (eds) (2002) Generel ochrany a racionálneho využívania vôd, 2nd edn. Ministry of Agriculture and Ministry of Environment of the Slovak Republic and Infopress, Bratislava
13. Biely A, Bezák V, Elečko M, Gross P, Kaličiak M, Konečný V, Lexa V, Mello J, Nemčok MJ, Potfaj M, Rakús M, Vass D, Vozár J, Vozárová A (1996) Explanation to geological map of Slovakia 1:500,000. Dionýz Štúr Publishers, Bratislava
14. Vozár J, Káčer Š, Bezák V, Elečko M, Gross P, Konečný V, Lexa J, Mello J, Polák M, Potfaj M, Rakús M, Vass D, Vozárová A (1998) Geologiccal map of the Slovak Republic 1:1,000,000. Ministry of the Environment of the Slovak Republic and Geological Survey of the Slovak Republic, Bratislava
15. Biely A, Bezák V, Elečko M, Kaličiak M, Konečný V, Lexa J, Mello J, Nemçok J, Potfaj M, Rakús M, Vass D, Vozár J, Vozárová A (1996) Tectonic sketch of the Slovak part of Western Carpathians. Geological Survey of the Slovak Republic, Bratislava
16. Šuba J, Bujalka P, Cibuľka Ľ, Frankovič J, Hanzel V (1984) Hydrogeological regionalization of Slovakia, 2nd edn. Slovak Hydrometeorological Institute, Bratislava
17. Belan M, Čaučík P, Lehotová D, Leitmann Š, Molnár Ľ, Možiešiková K, Slivová V (2016) Water management balance of the Slovak Republic. Water management balance of groundwater amount in 2015. SHMÚ, Bratislava
18. Fecek P (1996) Podzámčok – groundwater regime as reflected by its long-run use. Podzemná voda 2:47–51
19. Fendeková M, Ženišová Z, Némethy P, Fendek M, Makišová Z, Kupčová S, Fľaková R (2005) Environmental impacts of groundwater abstraction in Neresnica brook catchment (Slovak Republic). Environ Geol 48:1029–1039
20. Ženišová Z, Némethy P, Fľaková R, Fendeková M (2008) Quantitative and qualitative characteristics of water resources in upper part of Torysa catchment. Podzemná voda 14:118–128
21. Resolution no. 29/2005 Coll. on details of water management sources protection zones declaration, on measures for water protection and on technical improvements in water management sources protection zones

22. Franko O, Remšík Fendek M (1995) Atlas of geothermal energy of Slovakia. Dionýz Štúr Institute of Geology, Bratislava
23. Fendek M, Fendeková M (2015) Country update of the Slovak Republic. In: Proceedings, world geothermal congress 2015, Melbourne

Water Supply and Demand in Slovakia

M. Fendeková and M. Zeleňáková

Contents

Abstract The water supply in Slovakia is mostly assured from the public water supply systems based either on surface or on groundwater sources. Surface water sources are represented by water reservoirs or by direct water take-off from the surface streams. As groundwater sources, either wells or springs can be utilized. There are 295 water reservoirs in Slovakia, 32 of them are regularly evaluated within the quantitative water management surface water balance. Eight water reservoirs were constructed until now for drinking water supply purposes; more than 200 small water reservoirs serve mostly for irrigation. The amount of 247.5 millions of m^3, which makes 7.85 $m^3 s^{-1}$, was abstracted from the surface water sources in 2015. The main economic sectors using the surface water are industry, public drinking water supply and irrigation. The amount of water abstraction from the groundwater sources is generally higher than from the surface water. The total amount of 325.7 millions of m^3 (10.332 $m^3 s^{-1}$) was abstracted from groundwater sources in 2015. The number of inhabitants supplied with water from the public water supply sources has been increasing steadily, reaching the number of 4.7853 million (88.3%) of inhabitants of Slovakia. The number of supplied municipalities reached 2,380 with the share of 82.4% on the total number of municipalities of Slovakia. However,

M. Fendeková (✉) and M. Zeleňáková
Department of Hydrogeology, Faculty of Natural Sciences, Comenius University in Bratislava,
Bratislava, Slovakia
e-mail: miriam.fendekova@uniba.sk

A. M. Negm and M. Zeleňáková (eds.), *Water Resources in Slovakia:*
Part I - Assessment and Development, Hdb Env Chem (2019) 69: 63–78,
DOI 10.1007/698_2017_212, © Springer International Publishing AG 2018,
Published online: 27 April 2018

trend of the water consumption, both total and specific for private household, is declining in the long-term scale, from 195.5 L capita^{-1} day^{-1} in 1990 to 77.3 L capita^{-1} day^{-1} in 2015. The quality of drinking water from the public water supply systems has been showing a high level in the long-term period. Slovakia does not have any problem with a disease associated with the drinking water from public water supply systems. The development of public sewerage systems lags behind that of public water supplies. In 2015, totally 1,044 municipalities had the public sewerage system in place. This makes only 36.2% of the total number of 2,890 municipalities in Slovakia.

Keywords Connection to public sewerage systems, Surface and groundwater abstraction, Water demand, Water supply

1 Introduction

The first historical document on building the water supply systems on the Slovak territory comes from the year 1423 [1] when the first water conduit was built in Bardejov city (North-eastern Slovakia). The entry of the book of accounts from the year 1426 shows that the city paid for the water supply to a fortification ditch and for moss for pipe sealing. The first list of payers for water was created; the payment was called the "pipe" or "water fee." Although the household connections did not exist and water was taken only from the tanks located in the main square, each homeowner was obliged to pay the water fee [2].

The first water reservoir on the Slovak territory was built already in 1510 serving the Banská Štiavnica city (Central Slovakia) with the drinking water [3]. The water management system of the drinking water supply and mine dewatering in the wider Banska Štiavnica region was further developed mostly in the eighteenth century. This unique system of water management consisting of 27 artificial water reservoirs and connecting conduits was written into the UNESCO World Cultural Heritage in 1993.

The water supply system for mining activities was built in the fifteenth to sixteenth centuries in the surrounding of the Kremnica city (Grobňa and Turček water conduits). The Turček water supply system was built initially for delivery of technical water for the Kremnica mining city. The purpose of this original water engineering work was to transfer water from the Váh River Basin into the Hron River Basin. The conduit intercepted the surface water in 15 places. Its original length was 22.0 km. However its last 6 km long section was poor in water and demanding on maintenance; therefore its use was discontinued in 1859. In the nineteenth century, the conduit was 16.86 km long, realized along the contour line and surpassed an altitude difference of about 50.0 m. It was rebuilt in the twentieth century, and now it transfers water from the Turiec River Basin to the Nitra River Basin, both belonging to the Váh River Basin.

All streets of the Banská Bystrica city were supplied from water conduits bringing water through the wooden pipes from the springs in the surrounding mountains already in the end of the sixteenth century [2].

The largest development of the public water supply is dated to the nineteenth century and to the period after the World War II.

2 Water Supply

Nowadays, the water supply in Slovakia is assured from two different systems: (1) public water supply systems and (2) private water supply sources.

The public water supply sources are defined as those producing drinking water for at least 50 persons or abstracting volume of more than $10 \text{ m}^3 \text{ day}^{-1}$ in average. The public water supply systems provide only drinking water, which can be used for different purposes, mainly in the household, tertiary sphere, industry and agriculture. Water companies or other entities deliver water (see Figs. 1 and 2) through public water supply systems. Water delivered from drinking water supply sources must fulfil the qualitative criteria for water for human consumption according to [4]. The resolution also sets criteria for drinking water quality assessment.

The private water supply is mostly based on individual wells which are drilled for various legal entities (citizens, private companies), or on direct taking-off from surface streams. Water sources can be used as drinking water sources, but mostly, they produce technical water for various purposes, e.g. cooling water in power plants, water for fire extinguishing and water for irrigation or for recreational purposes. There are European and Slovak national standards for water quality assurance applied for different kinds of its utilization.

Fig. 1 Water companies in Slovakia

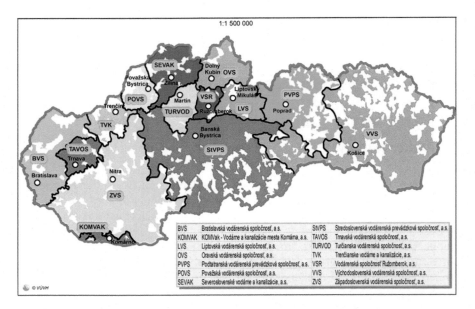

Fig. 2 Territorial distribution of water supply systems in Slovakia according to respective water companies

The Ministry of Environment of the Slovak Republic is the central body of the state administration responsible for water management, protection of water quality and quantity, flood protection and fishery, except aquaculture and sea fishing. There are several organizations founded by the Ministry of Environment, as Water Research Institute (VÚVH), Slovak Hydrometeorological Institute (SHMÚ), Slovak Water Management Enterprise (SVP), Water Management Construction Bratislava, Slovak Geological Institute of Dionýz Štúr (SGÚDŠ), Slovak Environmental Inspection, Slovak Environmental Agency and others. All the institutions together with regional and local environmental offices and municipalities share duties and responsibilities in various spheres of water resources administration, monitoring, assessment, research and protection.

The drinking water supply is based either on the abstraction of surface or on groundwater sources in Slovakia. Surface water sources are represented by water reservoirs or by direct water take-off from the surface streams. As groundwater sources, either wells or springs can be utilized.

2.1 Surface Water Abstraction

There are 295 water reservoirs in Slovakia according to [5], 32 of them are regularly evaluated within the quantitative water management surface water balance [6]. Only

Fig. 3 Water reservoir Bukovec II (photo: M. Zeleňáková)

eight water reservoirs (Rozgrund, Hriňová, Klenovec, Bukovec II (see Fig. 3), Starina, Nová Bystrica, Málinec and Turček) were constructed for drinking water supply purposes.

More than 200 small water reservoirs serve mostly for irrigation purposes. There was a substantial influence of water reservoirs on the hydrological situation in Slovakia in 2015. The maintenance of water amounts enabled to balance the unfavourable situation on surface streams below the reservoirs; in most cases the passive or strained stages were changed to the active one. That was the case of drinking water reservoirs of Starina (influencing the discharges positively in Cirocha River) or Bukovec II (influencing the Ida River positively). Some other large or smaller water reservoirs as Kozmálovce, Zemplínska Šírava, Palcmanská Maša and others helped to balance the low flows mostly in the summer-autumn period [6].

Besides of the water reservoirs, there are also streams, designated for the abstraction of water intended for human consumption. A total of 102 streams were designated, the majority of them in the headwater parts of mountainous catchments. The total length of these streams amounted 1,067 km in 2015 [6].

The amount of 247.5 millions of m^3, which makes 7.85 m^3 s^{-1}, was abstracted from the surface water sources in 2015 [6]. Surface water abstraction after 1995

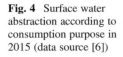 **Fig. 4** Surface water abstraction according to consumption purpose in 2015 (data source [6])

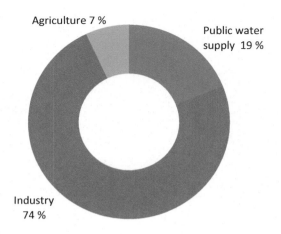

showed a significant decline despite minimal year-to-year increments and reductions. In 2015, volumes of abstracted surface water were 69.4% of abstracted volumes in 1995 and 66.4% of abstracted volumes in 2000. Between the years 2014 and 2015 abstracted volumes grew by 3.9%.

The main economic sector using the surface water was the industry with the share of 183.15 millions of m^3; the next was public water supply with 47.025 millions of m^3. The smallest share has the water abstraction for irrigation purposes with only 17.325 millions of m^3 in 2015. The share of surface water use in percent according to different categories in 2015 is given in Fig. 4.

The data from 1,133 users concerning surface water abstraction were processed in 2015. The water abstractions were related to 137 balance profiles on the surface streams. The total abstracted amounts increased from 17.729 m^3 s^{-1} in 2014 to 18.175 m^3 s^{-1} in 2015, at the same time the discharged amounts into the surface streams decreased from 19.088 to 18.846 m^3 s^{-1}.

The surface water abstraction development in individual economic sectors of Slovakia within the period 1995–2015 is depicted in Figs. 5, 6 and 7 [7].

In 2015, surface water abstractions increased to 247.581 millions of m^3, which is in 4% more than in the previous year. Abstractions for the industry in 2015 were at 183.29 millions of m^3, which represented only a small growth by 0.45 millions of m^3, i.e. 0.2%, compared to 2014. A slight growth was also recorded in surface water abstractions for water-supply networks, which, compared to the previous year, increased by 2.42 millions of m^3, that is, 5.4%. Surface water abstractions for irrigation grew and reached the value of 17.271 millions of m^3, that is, 63% [7].

2.2 Groundwater Abstraction

The amount of water abstraction from the groundwater sources is higher than from the surface water. The exploited groundwater sources are wells and springs. Wells are mostly located in Quaternary alluvial deposits of rivers; springs are tapped

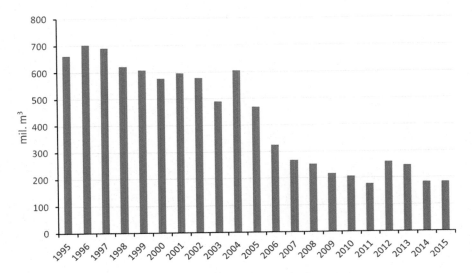

Fig. 5 Surface water abstraction by industry (data source [7])

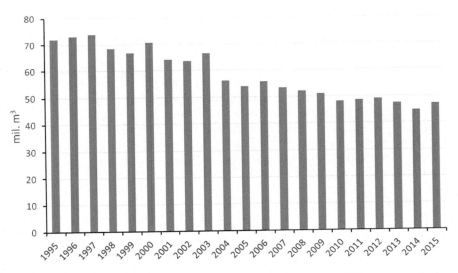

Fig. 6 Surface water abstraction for water-supply networks (data source: [7])

mostly in the fissure and fissure-karst rocks of the Mesozoic, Palaeogene and Neogene ages. Groundwater abstraction also declined after 1995; however, since 2000 its trend has been balanced, with very few increments and reductions. In 2015, volumes of abstracted groundwater were 43.5% of the abstracted volumes in 1995 and 27.4% of the abstracted volumes in 2000. Compared to 2014, abstraction grew by 1.4% [8]. The total amount of 325.7 millions of m^3 (10,332 m^3 s^{-1}) was abstracted from groundwater sources in 2015 [9].

M. Fendeková and M. Zeleňáková

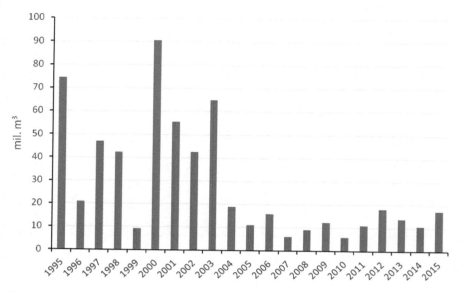

Fig. 7 Surface water abstraction for agriculture (data source: [7])

The diapason of groundwater users is published in more details comparing to surface water users. Among them, waterworks play the most important role, followed by industry (especially food industry), agriculture (livestock production and irrigation), social purposes and other uses. The greatest consumer of groundwater in 2015 was according to [6] the drinking water supply (waterworks) with 243 millions of m^3. The industry (as a whole) used 32.9 millions of m^3 and the agriculture 10.7 millions of m^3. The amount of 6.5 millions of m^3 was spent for social purposes and the rest of 32.6 millions of m^3 for other activities.

The share of groundwater use in percent according to different categories in 2015 is given in Fig. 8.

3 Water Demand and Water Consumption

The number of inhabitants supplied with water from the public water supply sources has been increasing steadily, reaching the number of 4.7853 million (88.3%) of inhabitants of Slovakia (https://www.enviroportal.sk/indicator/detail?id=441) in 2015, as it can be seen in Figs. 9 and 10. The number of supplied municipalities reached 2,380 with the share of 82.4% on the total number of municipalities of Slovakia. However, the trend of the water consumption, both total and specific for a private household, is declining in the long-term scale.

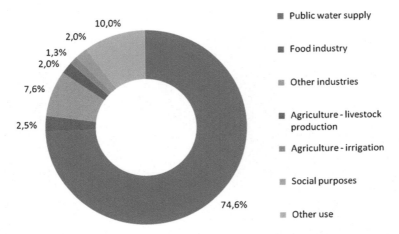

Fig. 8 Groundwater abstraction according to consumption purpose in 2015 (data source [9])

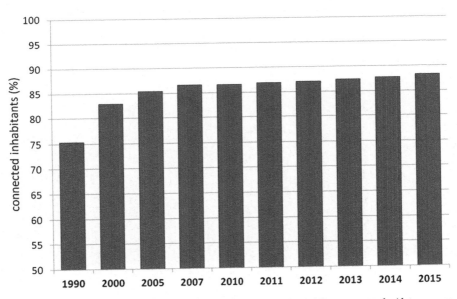

Fig. 9 Development in numbers of connected inhabitants to the public water supply (data source: Water Research Institute (VÚVH), https://www.enviroportal.sk/indicator/detail?id=441)

The public water supply in Slovakia, as already mentioned, is operated by water companies (Figs. 1 and 2). The length of water-supply pipelines is increasing with the increasing number of connected inhabitants. The development of water-supply pipelines length in kilometres is depicted in Fig. 11.

The water demand can be expressed by the total specific water demand (L capita^{-1} day^{-1}) which is the average volume of water produced by public water supply sources

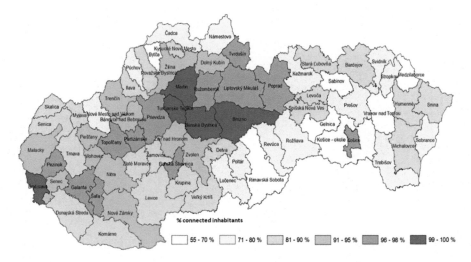

Fig. 10 Ratio of population connected to public water supply according to the districts (data source: [7])

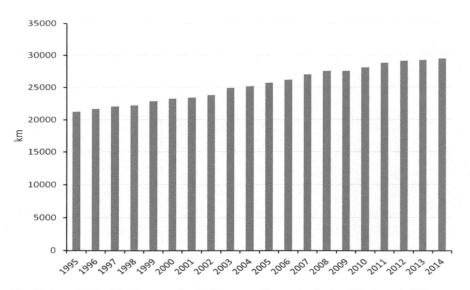

Fig. 11 Length of public water-supply pipelines according to the districts (data source: [7])

recalculated to each inhabitant and day of the year. The specific water consumption of the private households is measured by the volume of water which is withdrawn from the central public water supply network and supplied to households, under the consideration of the total population number within the respective regional or local supply area.

The amounts of almost 200 L capita^{-1} day^{-1} supplied to households (SC-H) were reached in Slovakia in 1990, and then the amounts decreased up to 100 L capita^{-1} day^{-1} until the 2007 and finally to less than 80 L capita^{-1} day^{-1} since 2013. The numbers on water consumption for selected years since 1990 are in Table 1.

More detailed numbers on the total specific water demand (TSD) and on the specific water consumption for households (SC-H) are shown in Fig. 12 (data source: https://www.enviroportal.sk/indicator/detail?id=1562).

The long-term decline of drinking water consumption became evident in all supplied municipalities of Slovakia. The decrease of specific water demand even below the lower limit of the hygienic minima (less than 70 L capita^{-1} day^{-1}) was registered in some villages. However, there is some stabilization around 80 L capita^{-1} day^{-1} notable since 2010 (see Fig. 12).

Table 1 Total specific water demand and household specific water consumption in L capita^{-1} day^{-1} in 1990–2015 (data source: Water Research Institute (VÚVH), https://www.enviroportal. sk/indicator/detail?id=1562)

Year	1990	1995	2000	2005	2010	2015
Total specific water demand (TSD)	433.2	321.5	273.4	204.7	180.8	164.9
Specific water consumption for household (SC-H)	195.5	142.5	123.6	95.1	83.4	77.3

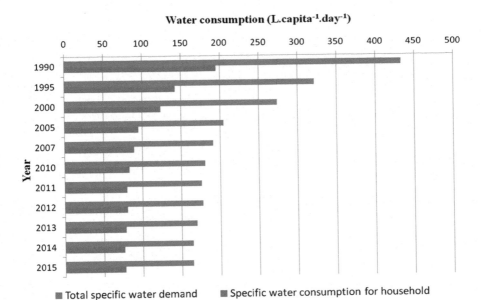

Fig. 12 Development of the total specific water demand (TSD) and the specific water consumption for household (SC-H) in L capita^{-1} day^{-1} (data source: Water Research Institute (VÚVH), https://www.enviroportal.sk/indicator/detail?id=1562)

The main reason for the decrease was the increased pricing of supplied drinking water and the collected wastewater. The first marked price increase was in 1993, amounting 130% of the previous price. After that the demand decreased the effort for more rational water management by saving, more precise consumption measurements and installation of devices with lower water consumption became evident. Another reason for the decreasing demand is that the high drinking water prices motivate the people to build their own drinking water sources whose drinking water quality might be below the sanitary standards.

The quality of drinking water from the public water supply systems has been showing a high level in the long-term period. The raw water (before the treatment) and also the treated water are sampled and analysed. Requirements on a number of samples of the raw water, treated water, water treated only by disinfection and treated water quality leaving the treatment plant are defined in [10]. The minimum required number of raw water samples according to [10] is given in Table 2.

The share of analyses meeting all limits of the drinking water quality parameters given by (https://www.enviroportal.sk/indicator/detail?id=441) was more than 94.54% of the total number of analysed samples. When omitting the free chlorine content, the share increased to 99.7%. The parameters exceeding the limits were mostly concentrations of iron, manganese, sulphates and opacity, in lesser extent nitrites and nitrates. The only organic matter found was dichlorobenzene, with the occurrence of over-limit values in 0.07% of analysed samples. The microbiological and biological parameters were over the limits in less than 2% of analyses. Slovakia does not have any problem with a disease associated with the drinking water from public water supply systems.

Table 2 Minimum number of raw water samples per year and type of chemical analysis within the operative drinking water quality control according to [10]

Volume of withdrawn water ($m^3\ day^{-1}$)	Number of supplied inhabitants	Minimum number of samples	
		Type of chemical analysis	
		Minimal	Complete
>10 ≤20	>50 ≤100	1 per 2 years	1 per 2 years
>20 ≤100	>100 ≤500	1 per 2 years	1 per 2 years
>100 ≤1,000	>500 ≤5,000	1	1
>1,000 ≤10,000	>5,000 ≤50,000	1 +1 for each 3,000 $m^3\ day^{-1a}$	1 +1 for each 5,000 $m^3\ day^{-1a}$
>10,000 ≤100,000	>50,000 ≤500,000	4 +2 for each 15,000 $m^3\ day^{-1a}$	1 +1 for each 30,000 $m^3\ day^{-1a}$
>100,000	>500,000	16 +1 for each 25,000 $m^3\ day^{-1a}$	4 +1 for each 50,000 $m^3\ day^{-1a}$

[a]Including each beginning limit volume

4   Wastewater Collection and Treatment

The development of public sewerage systems lags behind that of public water supplies. In 2015, totally 1,044 municipalities had the public sewerage system in place. This makes only 36.2% of the total number of 2,890 municipalities in Slovakia (https://www.enviroportal.sk/indicator/detail?id=1601).

The number of inhabitants connected to the public sewerage system reached 3.534 million in 2015. The number of connected inhabitants is increasing continuously, rising from 50.8% in 1990 through 54.7% in 2000 up to 65.2% in 2015 (Fig. 13). The spatial distribution of municipalities connected to the public sewerage system is not homogenous. The 14 districts with the low numbers of population connected to the sewerage system, where the ratio of connected municipalities reaches 31–50%, are located mostly in the Southern Slovakia. It can be seen in Fig. 14 that there are still two districts (Bytča and Košice-okolie) out of 89 with the ratio of only 25–30%.

The length of the sewerage system reached 12,833 km with the total number of 485,258 pieces of individual sewerage connections in 2015. The development in the length increase of water-supply pipelines in kilometres is depicted in Fig. 15.

There are also individual water treatment plants built in individual houses for water treatment in villages without the public sewerage systems. However, the data on them are not published.

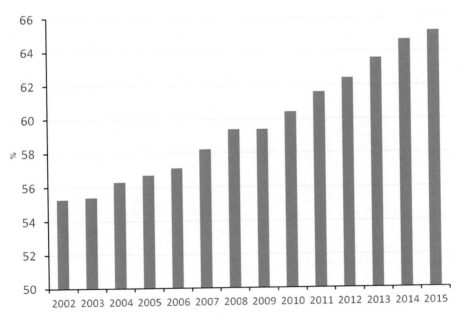

Fig. 13 Development in numbers of connected inhabitants to the sewerage network (data source: Slovakia [7])

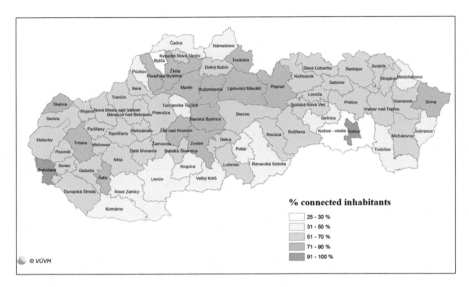

Fig. 14 Ratio of population connected to public sewerage systems according to the districts (data source: Water Research Institute (VÚVH), https://www.enviroportal.sk/indicator/detail?id=1601)

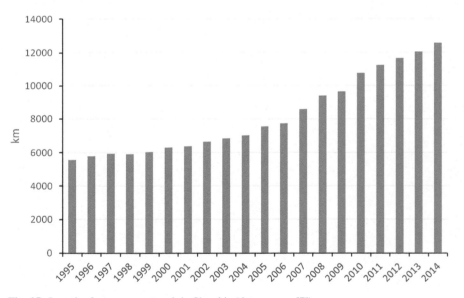

Fig. 15 Length of sewerage network in Slovakia (data source: [7])

5 Conclusions

The water supply in Slovakia has a long and successful history, going back to the fifteenth century. However, the largest development of the public water supply is dated to the nineteenth century and to the period after the World War II. The last published numbers on inhabitants supplied with water from the public water supply sources reached the number of 4.785 million (88.3%) of inhabitants of Slovakia. However, the numbers on inhabitants connected to public sewerage systems are much lower, reaching app. 3.534 million (65.2%) in 2015.

The system of water quality protection and checking is well organized; there were no water-related diseases recognized in Slovakia. The share of analyses meeting all limits of the drinking water quality parameters reached 99.7% when omitting the over-limit concentrations of the free chlorine content.

The only less favourable fact concerning the water demand in Slovakia is connected to the long-term decline of drinking water consumption which became evident in all supplied municipalities of Slovakia. The decrease of specific water demand even below the lower limit of the hygienic minima (less than 70 L capita^{-1} day^{-1}) was already registered in some villages.

6 Recommendations

The European Water Framework Directive put the strong accent on water quality in all parts of the hydrological cycle, but especially in surface and groundwater systems on which the natural ecosystems are dependent. The satisfactory natural water quality cannot be safeguarded without the ameliorated quality of water released from the sewerage systems to the surface streams.

Therefore the increased attention should be paid to the construction of highly effective sewage water treatment plants around the country. Especially, the municipalities in the southern districts suffer by deficient coverage by public waste water collection and treatment. Construction of the missing public sewerage systems and treatment plants should become one of the top items on the environmental protection agenda of the Slovak government.

References

1. Lieskovská Z (2016) Environment of the Slovak Republic in focus. Ministry of Environment of the SR, Slovak Environmental Agency, Bratislava, Banská Bystrica, p 64
2. Červeňan M (2016) Banskoštiavnická vodohospodárska sústava – najvýznamnejšia technická pamiatka Slovenska. Vodohospodársky spravodajca 43:7–8. (in Slovak)
3. Ružičková E (2000) Začiatky zásobovania obyvateľov Slovenska podzemnou vodou. Vodohospodársky spravodajca 43:8–9. (in Slovak)

 4. Resolution no. 496/2010 Coll. which amends Slovak Government Resolution No. 354/2006 Coll. defining requirements on water for human consumption and control of the water quality for the human consumption
 5. Slovak Water Management Enterprise (2015) Annual report. Banská Štiavnica, Slovakia
 6. Lovásová Ľ, Gápelová V, Podolinská J, Malovová J, Ľupták Ľ, Melová K, Škoda P, Liová S, Síčová B, Staňová J, Fabišiková M, Pospíšilová I (2016) Water management balance of the Slovak Republic. Water management balance of surface water amount in 2015. SHMÚ Bratislava, Bratislava
 7. Abaffy D, Lukáč M, Liška M (1995) Dams in Slovakia. T.R.T. Médium, Bratislava, p 103
 8. Ministry of Environment of the SR, Bratislava, Slovak Environmental Agency, Banska Bystrica (1995–2015) Reports on environment of the Slovak Republic. http://enviroportal.sk/spravy/kat21
 9. Belan M, Čaučík P, Lehotová D, Leitmann Š, Molnár Ľ, Možiešiková K, Slivová V (2016) Water management balance of the Slovak Republic. Water management balance of groundwater amount in 2015. SHMÚ Bratislava, Bratislava
10. Resolution of the Ministry of Environment of the Slovak Republic no 636/2004 Coll. on requirements on raw water quality and monitoring of water quality in public waterworks

Part II
Water in Agriculture

Irrigation of Arable Land in Slovakia: History and Perspective

Ľuboš Jurík, K. Halászová, J. Pokrývková, and Š. Rehák

Contents

Abstract The origin of irrigation in the world was based on the human's practical experiences in the past and gradually expanded with the development of advanced cultures. In the Middle Ages, irrigation structures did not develop and gradually ceased to exist. After World War I, alongside with industry, agriculture had been intensified in Europe. Irrigation once again became an important part of agriculture and countryside. In Czechoslovakia and Slovakia, conditions for designing, preparing implementation, and operation of irrigation were created. It was a long-term process, and only after 1960, organizations from designing to irrigation's operation were established. After 1990, the situation has changed, and nowadays, we are resolving the need for irrigation again. The following text analyses the development of organizations and management of irrigation in the past and looks at the future development. In Slovakia, the manager and organizer of irrigation constructions have become, and still are, the state.

L. Jurík (✉), K. Halászová, and J. Pokrývková
Department of Water Resources and Environmental Engineering, Faculty of Horticulture and Landscape Engineering, Slovak University of Agriculture in Nitra, Nitra, Slovakia
e-mail: lubos.jurik@uniag.sk

Š. Rehák
Water Research Institute (WRI), Bratislava, Slovakia

A. M. Negm and M. Zeleňáková (eds.), *Water Resources in Slovakia:*
Part I - Assessment and Development, Hdb Env Chem (2019) 69: 81–96,
DOI 10.1007/698_2018_284, © Springer International Publishing AG 2018,
Published online: 15 July 2018

Keywords Design of irrigation constructions, Irrigation water, Reservoir, Water management, Water sources

1 Introduction

Irrigation systems were developed in the era of the oldest civilizations thousands of years ago. In every history textbook, we read about irrigation channels in China or developed irrigation in India, China, and Mesopotamia. The extraordinary importance was devoted to irrigation in Egypt. Two important historical sites have been included on the World Heritage List: *the Aflaj Irrigation Systems Oman* – this property includes 3,000 of such systems still in use [1]. The origins of this system may date back to 500 AD. The second site is *Mount Qingcheng and the Dujiangyan Irrigation System in China* [2]. The origins of this system may date back to the third century BC to control the waters of the Min Jiang River and distribute it to the fertile farmland of the Chengdu Plain. Mount Qingcheng was the birthplace of Taoism, which is celebrated in a series of ancient temples [3].

Although the need for irrigation had been forgotten for centuries in many countries, in some Asian countries, amazing constructions of irrigation on terraces of slopes on the river banks have been preserved. Nowadays, the most beautiful structures are admired on the islands of Bali, the Philippines, or Yuanyang Province in China. Similar structures are also in the Moray-Urubamba area in Peru, anyway, as well as elsewhere in the world [2].

Europe in the Middle Ages had forgotten of irrigation for centuries. In Slovakia, and surrounding countries of central Europe, usually the natural rainfalls during the vegetation period or reserves from the winter months were enough for agriculture.

However, the twentieth century was a century of recurring climate fluctuations. In 1904, the summer months were down to autumn without precipitation, and almost all the water was lost in rivers. A similar situation had been repeated from the beginning of the next summer, and river levels had probably been the lowest throughout the century. After World War II, the range of built and functional irrigation facilities was very limited. During this time, it was necessary to secure food sufficiency, and drought, which occurred in 1947, gave the ground the need for irrigation. For example, in Hurbanovo, it did not rain for 83 days in that year. In nearby Vienna, it was possible to cross the river Danube. Moreover, the year 1953 was even worse because rainfall did not occur from summer to winter. Similarly, drought was repeated in 1983 and 1992. However, in between those tough times, there was a relatively appropriate climatic period for agriculture and the need for irrigation [3].

To sum up, years 1947 and 1953 were for the newly emerging states, created after World War II, which were extremely difficult. This had created a strong impetus for support and construction of irrigation in Central Europe [1].

2 Conditions for Establishment and Development of Irrigation in Slovakia

In Slovakia, this was anchored in the Water Management Plans, and thus funds were necessarily created for their implementation.

During the late nineteenth and early twentieth centuries, irrigations were built only on small areas and were operated by local organizations. The so-called ameliorative cooperatives were formed that operated the irrigation; however, the projects were prepared by experts – cultural engineers. Their task was to solve suitable water regime of the cultural landscape [4].

Until 1918 there were ten water cooperatives in the area of irrigation and melioration, near the river Danube and Východoslovenská nížina (Eastern Slovakian Lowland). To improve the cooperation, they created a joint organization – Melioračný zväz (Melioration Union). They worked together till 1950. Later, they were officially abolished by the new Water Act no. 11/1955 Coll. Hydromelioračné stavby (Hydromelioration constructions) and were part of the water management until the 1970s. At this time their separation started, and the management and operation were overtaken by the Ministry of Agriculture. From 1 January 1969, they created a new organization entitled Štátna melioračná správa (State Melioration Enterprise – SME). Over the years it had been an extremely important element for managing the development of all hydromeliorative constructions. Štátna melioračná správa (SME) also managed listed structures, their repairs, maintenance, and operation. By decision of the same Ministry, Štátna melioračná správa (SME) was abolished on 31 December 1991, and its property and ameliorative structures were overtaken by the Slovak Land Fund. It was only for 2 years, and then the property was overtaken again by another manager – Slovenský vodohospodársky podnik (Slovak Water Management Enterprise – SWME), and they subsequently handed it over to Hydromeliorácie, š.p., who still manages them up to today. Since 1991, following the abolition of Štátna melioračná správa (SME), the preparation of new constructions has not been realized, and as their operation is complex, the property is leased every year to interested parties for the operation. Since 1991 the irrigation area has been stagnating and slightly declining. Development of irrigation is realized only by changing irrigation devices for the water distribution. Nowadays, they are owned by farmers. To add, new devices allow higher water savings, better dosing, and quicker delivery of the required dose.

After 1950, the role of organization and management of irrigation was overtaken by the state in Slovakia. The state had to quickly address a number of questions in this period:

• To identify the appropriate type of prepared irrigation and their technological development through research facilities. Research institutes did not exist for this purpose, and it was inevitable to establish them.

- Workplaces for designing a larger irrigation range. Designers, solving already built irrigation systems, could not handle a range of irrigation structures.
- Ownership of irrigation. The built-up facilities were owned by landowners or joint cooperatives during the interwar period. After the process of land collectivization in the countryside, this method was not viable anymore, and it was decided that the main components of the irrigation structures would be financed and owned by the state. This allowed the preparation and operation of large-scale systems.
- Irrigation operators. The state had to set up a system of organizations to ensure the operation and maintenance of the facilities and, at the same time, the investment preparation for new constructions.
- To create water resources for irrigation, as irrigations must have accessible water for agricultural production during the lack of precipitation or small flow in rivers, such as it was in 1904.
- To provide knowledge and practical experience for farmers in growing crops under irrigation conditions.
- Prepare a financial plan for irrigation development.
- Prepare training of new staff for the design, construction, and operation of irrigation and other meliorative structures.

The biggest problem was that all these tasks must have been solved at the same time and with a small number of real professionals.

Between 1950 and 1960, the fundamental strategy for the development of hydromelioration was prepared. As first, suitable water resources for irrigation were identified. In Slovakia, the groundwater was designed to supply the population with drinking water. Thus, the surface water was allocated for the purposes of irrigation. Only small water reservoirs could have provided sufficient water in the period without precipitation. For the needs of irrigation, the programme for construction of a larger number of small water reservoirs and several large dams was prepared. Construction itself was planned for a long period between 1960 and 2000. Part of the water collection was realized directly from the rivers. Subsequent to the creation of resources, the irrigation had started. With regard to the organization of agriculture, several types of melioration systems have been prepared:

- Small irrigations with a range of up to 50 ha, mainly for growing fruit or vegetables.
- Medium-sized irrigations ranging from 50 to 500 ha, usually associated with a small water reservoir and production of vegetables, fruit, or field crops.
- Large irrigations which ranging from 500 to 2,000 ha had small water reservoirs and cultivated field crops, especially root crops or oil crops.
- Large-scale irrigations with a range of 2,000–30,000 ha, were the sources of production of all types of crops and dams were the sources of water.

Irrigation experts worked jointly with the climate scientists and pedologists on preparations. They identified areas with the lack of natural rainfall during the vegetation period for cultivated crops suitable for the development of irrigation, as well as areas with surplus water in spring period with the need of systematic drainage [5].

2.1 State Organizations for Hydromelioration Management

To manage the preparation and use of irrigation and other meliorative constructions, Melioračné družstvá (Meliorative Cooperatives) in Czechoslovakia was organized from the year 1955 [5].

Two types of design organizations have been created to support the preparation and design of constructions. Large-scale irrigations were primarily prepared by HYDROPROJEKT Bratislava. Other types of irrigations were designed, including small water reservoirs, in the company Pôdohospodársky projektový ústav (Agricultural Project Design Institute), later renamed to AGROCONS. A separate task was to secure construction capacities responsible for the project implementation as well as repairs and maintenance. The largest construction projects were realized by Vodohospodárske stavby (Hydraulic Structures) based in Bratislava but with branches in several locations in Slovakia [5].

Smaller constructions were carried out by companies Hydromeliorácie Nitra or AGROSTAV with workplaces throughout Slovakia.

The new training staff was ensured by creating specialized study programmes at the Slovak University of Agriculture in Nitra and at the Faculty of Civil Engineering of the Slovak Technical University [6].

At the same time, different types of legislation have been issued that have legislatively supported defined objectives, in particular, their timing.

Research of new irrigation technologies was realized at Výskumný ústav hydromeliorácií (Melioration Research Institute, MRI), whose offices were based in Prague and Bratislava. It should be remembered that tasks were solved centrally in Czechoslovakia, but in particular cases even in this period of time, the part of tasks was adjusted solely for the conditions in Slovakia.

Development of irrigation was coordinated with other water management projects, above all with the flow regulation, hydropower structures, and the development of drainage.

Historically interesting were irrigation research solutions. Probably the most important result of the irrigation technologies development was the development of the reel irrigation. The first prototype was created jointly by the staff of research institutes of meliorations in Bohemia and Slovakia. After successful tests, however, patent protection was not submitted, and therefore, they are now manufactured in countries around the world with patent protection out of Slovakia [5].

Not less interesting was an international solution of underground irrigation, a development of which due to the insufficient technology of plastics processing at that time did not expand in practice. Today, however, we turn again to the direct irrigation applied primarily to the root zone associated with the low energy consumption.

Výskumný ústav hydromeliorácií (Melioration Research Institute – MRI) also published the real values of water consumption for cultivated plants in the vegetation period. This has enabled to issue the ON 83 0635 standards "Irrigation water requirement for additional irrigation" used up to today, valid since 1 July 1974 [6] (Fig. 1).

Large-scale irrigations with a range of 2,000–30,000 ha (dams) were used for irrigating of all types of crops [7].

3   Development of Irrigation Constructions

The first irrigation constructions date back to 1890–1900, and their precise localization is not clear today. In 1935 the size of irrigation constructions was approximately 300 ha, opposed to the 1900 ha projected. Unfortunately, most of the projects were not implemented due to economic reasons [5].

In the postwar period, in 1953, the irrigated area was still very low– about 1,150 ha.

Since 1953, the State Water Management Plan of the Czechoslovak Republic (SVP 1953) had been the fundamental document for the development of irrigation. It

Fig. 1 Landscape without irrigation during drought (Photo: Jurík, 2007)

dealt with the need for waters in all sectors of the economy in the state. That is why it included a plan for irrigation on the whole territory of Czechoslovakia. Some water need plans were overestimated, and some were not included at all, so in 1967 there was a demand to redesign it [8].

For the construction of large-scale irrigation systems, a way to secure the necessary water was determined by the construction of dams. In Bohemia it was wider Polabie catchment, in Moravia Podyjí catchment, and in Slovakia Danube lowland and Východoslovenská nížina (Eastern Slovak Lowland). The necessary water for the planned irrigation should have been secured in Bohemia via the dams of the Vltava cascade; in Moravia, at least at the beginning, by Vír and Vranov dams; and in Slovakia, in particular, the rivers Danube and Váh and the Zemplínska šírava dam. In Slovakia, these original assumptions had been increased several times in the process of the irrigation implementation, and the need for water resources has increased by building small water reservoirs with a total number of 190 [9].

In 1960 the development of large-scale irrigation began, and the area of recorded irrigation was about 37,500 ha. After this period the method of irrigation changed. Mid-range to large-scale irrigation systems with one pumping station for the whole territory were built. Mid-range irrigations were built in the average area of 300 ha and large-scale up to 800 ha. Water was distributed to crops mainly by portable irrigation kits [10].

The task for the research institutes was to solve the design optimization for the large-scale pressure pipe networks. After 1970, large-scale and small-scale irrigation systems for fruit and vegetables have been preferred. As for the water distribution devices, new and powerful machines were used – reel irrigation machines and pivots or linear irrigation machines. From Russia, pivot irrigation machines FREGAT and VOLŽANKA or the East German machines Rollende Regneflügel were used [11].

It was necessary to solve the new concept of irrigation pumping stations and the hydraulic design of the pipe network for the new irrigation machines.

Large-scale systems needed plenty of water in the summer months. Therefore, small water reservoirs became part of them. For the fruit orchard needs, the theory and the practical use of antifreeze irrigation started to be solved [11].

4 Design Organizations for Irrigation

Slovakia was part of Czechoslovakia during resolving the irrigation constructions; however, the design institutes were addressed differently in Bohemia. In 1954, the government managed the establishment of institutes for design in agricultural practice – AGROPROJEKT – with branches in all regions of Bohemia. AGROPROJEKT exists in the Czech Republic with a couple of changes up to date.

In Slovakia in this period – and in the year 1954– Štátny oblastný ústav (State Regional Institute) was established in Bratislava and the regional counties of

Slovakia. Design organizations have undergone many changes till now and have almost disappeared, as today only reconstruction of constructed structures has been projected.

In the year 1967, Štátny oblastný ústav (State Regional Institute) was transformed into Pôdohospodársky projektový ústav (Agricultural Project Design Institute), which included Poľnohospodárska investičná správa (Agricultural Investment Enterprise), whose task was to resolve the financing of projected construction. Inclusion into the Poľnohospodárska investičná správa (Agricultural Investment Enterprise) did not last too long and from 1969 became independent again, as Štátna melioračná správa has been established in the Czech Republic and Slovakia. The impact of Štátna melioračná správa was very significant and important as it managed not only the preparation but also the hydromelioration devices. For Slovakia, it had its headquarters in Bratislava and investment units in the former counties, and the management of devices was managed through the offices in each district. The design institutes were in six towns altogether in Slovakia, and at the time of the highest progress, it had about 1,200 employees. The HYDROPROJEKT Bratislava collective of employees was allocated for the large-scale irrigations linked to dams [11].

5   The Melioration Programme of Slovakia

The melioration programme of Slovakia was developed to coordinate all the melioration arrangements that were dealt alongside with the irrigations. It identified the needs for all types of amelioration measures. In Slovakia, the estimated area for irrigation was set at 892,000 ha, and for soil drainage, it was planned on an area of 558,000 ha. Also, the necessary fertilization of light soils on the area of 62,000 ha and the regeneration of skeletal soils on the area of 206,000 ha were established. Erosion protection measures were the separate task to be implemented on an area of 325,000 ha. The melioration programme was gradually implemented between 1960 and 1990. However, this programme had almost stopped straight after the change in state management [11].

If you have to decide for an irrigation method of some space, a number of basic aspects have to be considered. Some of this belongs to them:

- The type of crops to be produced.
- The climate conditions.
- The amount of water required.
- Accessibility of the water source.
- The structure of the irrigated area (plane or hilly).
- The soil type (clay or sandy).
- The number of months in the year, when irrigation water is needed.
- The selection of the irrigation pumping stations.
- The consequences of irrigation failure for a certain period [12].

Based on consideration of the above aspects, a choice of irrigation method will be performed.

Legislative measures supported enforcement of hydromeliorative and so irrigation constructions. Irrigation and drainage itself were primarily solved in all versions of the Water Act but also in separate legal regulations, legislation about the quality of irrigation water.

In Slovakia, the major effort was concentrated on the development of large irrigation systems in 1980–1990. This primarily facilitated the development of irrigation water distribution systems. In the Czech Republic, a SIGMATIC irrigation machine was developed, and the reel irrigation machines were technically improved [13].

Again, it was necessary to adapt the irrigation pumping stations, to review solution for the irrigation pipe networks. The main reason was due to the operation of large irrigation machines with a huge inflow as well as the network flow or the pump requirements.

In 1990 the construction of new irrigation systems almost stopped, and only a few small-scale and medium-scale irrigations were built. The last major irrigation pumping station was completed in Východoslovenská nížina (Eastern Slovak Lowland) in 1992. Private companies have been investors in recent irrigation constructions.

The total balance of constructed irrigation constructions has stopped at about 320,000 ha. They are supported by about 500 pumping stations for which 258 km of irrigation channels have been built as a water supply. Today the range of the built irrigations remains; only the area of the functional one's changes. As the use of irrigation is constantly declining, some structures can no longer be used due to lack of maintenance.

The required power of the pumping station was based on a specific water flow to territory Q in l/s/ha, which moves within 16 h of operation within the limits of:

0.25–0.35 l/s/ha for cereals and technical plants

0.40–0.50 l/s/ha for fodder crops and root crops

0.50–0.80 l/s/ha vegetables + early potatoes + pumpkins

The operation of the pumping stations has been proposed as automatic, with a connection to the pressure or flow changes in the discharge pipeline or to the movement of the level in the storage tank.

Development of construction of irrigation structures was realized in different periods with different financing. A summary of the construction stages is shown in Table 1.

The resulting state of irrigation-related constructions in Slovakia today is:

- Irrigated area: 321,000 ha
- Irrigation channels and feeders: 275 km
- Pumping stations: Large: 40, Middle: 250, Small: 200

Land drainage was built on an area of 458,000 ha together with 6,282 km of drainage channels. Nowadays, mainly maintenance and repairs are realized on irrigation and the drainage. Some constructions have significantly exceeded their

Table 1 Construction stages of irrigation in Slovakia (according to the ŠMS data own processing) [14]

Period	Area of built irrigation in ha
1960–1965	48,915
1966–1970	57,685
1971–1975	81,176
1976–1980	46,592
1981–1985	70,884
1985–1991	15,620
Total Slovakia	320,872

estimated life span and are step by step slowly removed. In 2005, investment activity in the area of irrigation – the development of networks or service stations – was stopped.

There are five dams in Slovakia today, which have been primarily built for the large-scale irrigation. They are [13]:

Dam Zemplínska šírava – was built between 1961 and 1965 and now is the second largest water reservoir in Slovakia. The purpose of building was flood protection and irrigation. However, today it is mainly the recreation. Of the estimated irrigation range of about 15,000 ha, only about 7,500 ha was realized, and the water is transported through five pumping stations with the capacity of 3.2 m^3/s.

Dam Sĺňava – was built between 1956 and 1959 nearby the town Piešťany. Similar to the previous dam, the purpose of this one serves mainly for recreation, hydropower, and irrigation. Approximately 7,300 ha are irrigated directly from the dam. They are divided into six separate sections and operated by seven pumping stations. The water draw is up to 5.1 m^3/s.

Dam Horná Kráľová – was built on the river Váh in 1985. The dam was built for about 20,000 ha, but only about 10,000 ha of irrigation was realized. Allowed water draw is 5.7 m^3/s, which is the largest allowed draw in total. Water is transported to the first pumping station by gravity via two 2,200 mm diameter pipes.

Two joined water dams Teplý vrch a Ružiná (see Fig. 2) – were built in 1970 and are connected by pipeline. This facilitates to use their water volume together. Today they serve mainly for recreation. They were built for about 4,000 ha of irrigation with the water draw of about 2.0 m^3/s.

The next irrigation sources are three water reservoirs (WR) with large-scale irrigations up to 2000 ha. The first one is Budmerice WR with an irrigated area of 1,600 ha, the second Suchá nad Parnou WR with 1,850 ha of irrigated area, and third Lozorno WR with an irrigated area of 1,400 ha [14].

For irrigation purposes, a total of 192 small water reservoirs were at the territory of Slovakia built during these periods as a source of irrigation water. They provided plenty of water, especially during the summer months.

Large-scale irrigation is also supplied directly from the rivers. For example, from the river Morava, 15 pumping stations draw 4.5 m^3/s for about 13,000 ha. From the Little Danube and the Danube, 9 m^3/s is drawn for the irrigated area of about 30,000 ha using 19 pumping stations (see Fig. 3).

Fig. 2 Dam for irrigation water Ružiná (Photo: Jurík 2014)

Fig. 3 Small irrigation pumping station for 300 ha (Photo: Jurík, 2016)

In addition to these reservoirs, there are about 90 irrigation constructions with the irrigated area over 100 ha and about 100 with smaller irrigated areas [15].

Next 15 pumping stations draw 12.0 m^3/s for the area of 26,000 ha and use water from irrigation channels on Podunajská nížina [1].

A similar situation is also on the rivers Váh and Hron. Approximately 11.1 m^3/s is used from the river Váh using 21 pumping stations for 24,000 ha of irrigated area, and on the river Hron, it used 6.5 m^3/s with 11 pumping stations for about 15,500 ha of irrigation [15].

Unfortunately, some of these pumping stations are either unused or dysfunctional today.

6   Perspectives of Irrigation

The revival of irrigation has not occurred even during the serious drought in the last decade. Crop growers have somehow become accustomed to damages caused by drought, and therefore the use of irrigation to ensure production has significantly declined.

In 2016 and 2017, the drought was responsible for nearly the total destruction of sunflower and maize harvest.

In 2001–2002, the publication "Meliorations/What Next (Discussion on the proposal for the transformation of meliorations)" was created to reactivate irrigation, and later in 2015, the Ministry of Agriculture issued the strategy "Concept of revitalization for hydromeliorative systems in Slovakia." However, this document was withdrawn from the hearing and remained only in the archive of the Ministry.

In 2016, a call funded by EU – call from the Rural Development Programme of the SR 2014–2020 – was announced covering investments in construction, reconstruction, or modernization of irrigation systems, including infrastructure to increase production or its quality; nonetheless, funds were used for different purposes [16].

Climate change issues, such as hotter summers or persistent drought, have activated in many countries an interest for further development of irrigation and interest in the stabilization of production. In Slovakia, this interest is noticeable for entrepreneurs in agriculture located especially in southern Slovakia; however, the state support is not sufficient.

To clarify irrigation needs and to stabilize our own national production, several issues need to be addressed.

Firstly, the political conditions, which are supposed to give the framework and vision for the next decade, are fundamental. Then, there are tasks in the area of clarifying the needs of crop production in Slovakia. Finally, the water management strategy has to address the issue of the local resource creation in areas with a problematic balance of resources and water consumption [17].

Basic information about the use of irrigation is based on the quantity of consumed water. From the Slovak Environmental Agency – SAŽP statistics, we have prepared an overview of consumed surface and groundwater used for irrigation of crops (Table 2). Data were available until the year 2015.

As of 31 December 1987, 15.9% of the arable land was provided by irrigation systems in Slovakia. The water consumption for irrigation was 280 million in 1990 m^3 of water, and by 2005 this amount has fallen to only 3.6 million m^3. A critical intervention in the use of water regime treatment of agricultural land in Slovakia was the liquidation of the State Melioration Administration.

Irrigation and drainage systems are currently made up of major melioration facilities that have been built as part of the state investment melioration and hydromelioration details owned by land users (Fig. 4.) It consists of 2,935 water structures, consisting of 11,513 building objects. The state-owned company Hydromeliorácie, š.p., based in Bratislava is in present time the manager of the main melioration facilities owned by the state [13].

Managed irrigation systems owned by the state represent 320,872,000 ha of built-up irrigation area on farmland. Of the total number of 464 irrigation systems with 485 managed irrigation pumping stations, 194 irrigation systems are currently leased, which represents 169,025,000 ha (52.7%) of built-up irrigation area on agricultural land (Table 3 and Fig. 5).

7 Recommendations

The distribution of the climatic factors during the year, which is characterized by the uneven distribution of the precipitation and other elements of the water balance, significantly influences the amount of physiologically effective water in time and space and its dynamics. The interrelationship between the productive use of the landscape, the structure of the soil fund, the remediation measures, and the nonproductive functions of the landscape require that all measures be dealt complexly with functional technical, technological, and environmental continuity.

Table 2 An overview of the annual consumption of irrigation water in Slovakia, Source [18]

Year	mil. m^3	Year	mil. m^3	Year	mil. m^3	Year	mil. m^3
1980	80.9	1990	279.4	2000	108.74	2010	54.564
1981	121.6	1991	60.4	2001	70.94	2011	91.225
1982	162.2	1992	93	2002	77.26	2012	126.538
1983	218.4	1993	80.8	2003	445.91	2013	114.752
1984	172.5	1994	113.73	2004	84.105	2014	130.57
1985	121.9	1995	74.33	2005	106.076	2015	151.27
1986	280.1	1996	21.67	2006	110.814	2016	–
1987	190.7	1997	62.91	2007	152.286	2017	–
1988	265.2	1998	58.58	2008	76.653	2018	–
1989	223.5	1999	17.58	2009	106.119	2019	–

Fig. 4 Linear machine for irrigation at SUA farm Oponice (Photo Jurík 2008)

Table 3 Current usability of irrigation systems 2015 Source [18]

Irrigation region	No. of irrig. systems	Irrig. system used	Irrigation territory area (ha)	Area exploited (ha)
Záhorie	32	11	21,746	10,514
Podunajska	106	63	88,343	56,101
Dolné Považie	69	45	74,314	48,351
Horné Považie	64	36	41,735	33,761
Ponitrie	56	26	16,938	8,871
Pohronie-Poiplie	91	11	46,171	10,819
Bodrog-Hornád	46	3	31,625	608
Slovakia	464	195	320,872	169,025

Source: Hydromeliorácie, š.p., 2016

The regulation of the water using for irrigation and drainage and ensuring the balance in the protection and creation of the landscape, ensuring its biological, and technical functions has a special status in agricultural irrigation and drainage. They represent a set of measures which, by their effects, greatly influence the water regime

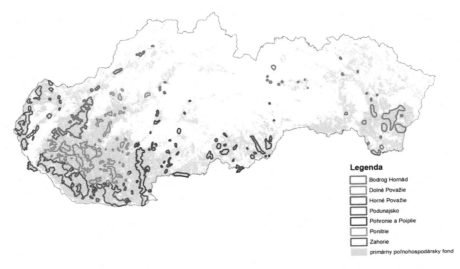

Legenda

☐ Bodrog Hornád
☐ Dolné Považie
☐ Horné Považie
☐ Podunajsko
☐ Pohronie a Poiplie
☐ Ponitrie
☐ Záhorie
▨ primárny poľnohospodársky fond

Fig. 5 Primary areas with best agricultural soils, suitable for irrigation. Source: VUPOP Bratislava 2016

of the landscape and the fertility of the soil. They are looking for a space for the people, plants, animals, and natural regeneration in a healthy landscape. The revitalization of hydromelioration systems requires and forms a new approach to the future of irrigation and drainage systems in Slovakia. Emphasis is placed on ensuring the necessary level of food security, using the potential of soil-climatic and the plant production conditions in Slovakia in the period of ongoing climate change.

Acknowledgements This chapter was supported by the Slovak grant APVV-16-0278 "Use of hydromelioration structures for mitigation of the negative extreme hydrological phenomena effects and their impacts on the quality of water bodies in agricultural landscapes."

References

1. Kenoyer JM (1998) Ancient cities of the Indus Valley Civilization. Oxford University Press, Karachi
2. Jensen ME, Rangeley WR, Dieleman PJ (1990) Irrigation trends in world agriculture. Irrigation of agricultural crops. Agronomy monograph, vol 30. American Society of Agronomy, Madison, WI, pp 31–67
3. Fukuda H (1976) Irrigation in the world. University of Tokyo Press, Tokyo
4. Hríbik J (2002) Experiences from realization and exploitation of irrigation projects in Slovak republic, 18th Congress on Irrigation and Drainage ICID. In: Seminár Lessons from failures in irrigation, drainage and flood control systems, Montreal
5. Rehák Š, Šanta M, Zápotočný V (2002) Irrigation water - an irreplaceable production and economic factor. Semisoft s.r.o., Bratislava, 120 s., ISBN 80-85755-11-4

6. Šoltész A, Baroková D, Červeňanská M, Janík A (2016) Water level regime changes in Danube lowland region. In: Colloquium on landscape management 2016: conference proceeding, Brno, 9 December 2016. 1. vyd. Brno: Mendel University in Brno, 103–110. ISBN 978-80-7509-458-2

7. Bárek V, Halaj P, Igaz D (2009) The influence of climate change on water demands for irrigation of special plants and vegetables in Slovakia. Bioclimatology and Natural Hazards. Springer, Dordrecht

8. Pierzgalski E, Jeznach J (1993) Stan i kierunki rozwoju mikronawodnień. In: Somorowski CZ (ed) Współczesne problemy melioracji, Wyd. SGGW, Warszawa, pp 35–42

9. Alena J, Takáč J (2007) The consequences of the climate change in irrigation management ČÚ 03, 2007 VÚMKI Bratislava

10. Heldi A (2004) Act on hydromeliorations and on amendment and supplement of some laws - proposal 2004. Hydromelioracie, Bratislava

11. Rehák Š, Bárek V, Jurík Ľ, Čistý M, Igaz D, Adam Š, Lapin M, Skalová J, Alena J, Fekete V, Šútor J, Jobbágy J (2015) Zavlažovanie poľných plodín, zeleniny a ovocných sadov. VEDA, Bratislava

12. Wriedt G, van der Velde M, Aloe A, Bouraoui F (2009) Estimating irrigation requirements in Europe. J Hydrol 373(3–4): 527–544. https://doi.org/10.1016/j.jhydrol.2009.05.018

13. Jurík Ľ (2013) Vodné stavby (Hydraulic structures). Slovenská poľnohospodárska univerzita, Nitra, 196 s. ISBN 978-80-552-0963-0

14. Kolektív: Analýza súčasného stavu správy, prevádzky a majetkovo – právneho usporiadania hydromelioračného majetku štátu, Hydromeliorácie, š.p. 2011

15. Lenárt R, Pokrývková J (2013) Správa a prevádzka hydromelioračných zariadení v podmienkach Slovenska v horizonte rokov 2003–2013 ENVIRO 2013–978–80-552-1101-5

16. Rehák Š (1998) Funkcia vody a jej regulácia v agroekosystéme krajiny [Water function and regulation in agroecosystem of landscape]. ENVIRO Nitra 1998. VŠP, Nitra, pp 239–242

17. Rehák Š (1999) Bases of irrigation management under conditions of water scarcity (Východiská riadenia závlah v podmienkach nedostatku vody.) In: 17th ICID congress, Granada, pp 22–39

18. MŽP (2000–2015) Správa o vodnom hospodárstve v Slovenskej republike v roku 2000–1026. Ministry of Environment, Bratislava

Quality of Water Required for Irrigation

T. Kaletová and Ľ. Jurík

Contents

Abstract Irrigation water can cause damage to irrigated crops and human and animal's health. Therefore, it is important to monitor the irrigation water regularly. There is a long-term tradition of irrigation water quality (IWQ) monitoring in Slovakia. A number of monitoring stations varied during years from more than 200 to 11 in recent years. The IWQ increased over the years. Advantages of the irrigation water monitoring helps in increasing quality of agriculture production, in a reduction of a risk of bacterial, respectively, and in reducing virus infection of humans and animals.

Keywords Irrigation water quality, IWQ monitoring, Water quality

T. Kaletová and Ľ. Jurík (✉)
Department of Water Resources and Environmental Engineering, Faculty of Horticulture and Landscape Engineering, Slovak University of Agriculture in Nitra, Nitra, Slovakia
e-mail: tatiana.kaletova@uniag.sk; lubos.jurik@uniag.sk

A. M. Negm and M. Zeleňáková (eds.), *Water Resources in Slovakia: Part I - Assessment and Development*, Hdb Env Chem (2019) 69: 97–114, DOI 10.1007/698_2017_214, © Springer International Publishing AG 2018, Published online: 25 May 2018

1   Introduction

The agricultural demands for water are still increasing as supplemental irrigation is adopted in regions where rainfall is not sufficient, to meet the optimal crop water requirements [1]. Irrigation can be understood as a controlling and stabilizing factor of the agricultural system, decreasing the influence of the casual elements. Estimate of irrigation water need for agricultural plants, its quantity, time reference for planning, design and operation purposes [2], as well as water quality is a concern to everyone who uses water [3].

It is a practical and financial challenge to manage water in specific situation, e.g. for irrigation. Plant and soil structure can be damaged by some irrigation waters [3].

Sustainability of water quality for a specific use is influenced by its physical, chemical and biological characteristics. Sustainability of water quality for a specific use can be understood as relationship between the water quality and the user's needs [4]. The main aim of the water quality monitoring is the verification of suitability of examined water quality for intended use or not [1].

A particular risk in the production of leafy vegetables that are eaten raw without cooking is the contribution of irrigation water. The contribution of irrigation water in the contamination of produce leading to subsequent outbreaks of foodborne illness can be found in the literature [5]. A pollution of irrigation water could be from point (e.g. untreated wastewater) or sheet source [6].

Dissolved salts and trace elements as results of natural weathering of the earth's surface occur almost in all waters [7]. According to Shalhevet and Kamburov [8], the total dissolved salts and its ionic composition are main parameters of irrigation water quality. Their concentration depends on the water source, location and time of water sampling.

The irrigation water quality primary depends on the water source [6]. The four main sources of the irrigation water used with pressurized irrigation techniques are [6]:

1. Dams and open reservoirs
2. Underground water (wells and boreholes)
3. Treated wastewater
4. Water from pipe networks [9]

The surface water collected in the dams and reservoirs, as well as usage of water directly from the rivers or open channels, is used in Slovakia.

The water used for irrigation should not negatively influence yield and quality of production, soil characteristics, human and animal health, as well as the quality of surface and subsurface water. There are different requirements for the irrigation water quality according to the region, soil and climate conditions, the type of irrigation technology and the type of crops [10]. Irrigation water quality may be evaluated according to three criteria:

1. Irrigation water cannot affect human and animal health, soil and yield and quality of surface and subsurface water.
2. Irrigation water cannot affect function of micro-irrigation elements, especially water distribution system, e.g. drippers and micro-sprinklers.
3. Irrigation water is evaluated according to the irrigated crops and their physiological requirements [11].

The water quality is naturally affected mainly by geological structures, through which water is flowing. The physical, chemical and biological characteristics of water are influenced by climate, hydrological and geological conditions, as well as anthropogenic activity. Therefore, those characteristics are changed during a year, and a regular monitoring of irrigation water quality is needed [6].

2   Irrigation Water Quality

2.1   Irrigation Water Quality Parameters

The water quality evaluation method, in brief, focuses on the essential parameters and criteria for more or less practical evaluation of the chemical, the physical and the biological quality of the water for irrigation with pressurized techniques as follows:

1. Chemical – salinity/toxicity hazards for the soil, the plants and the irrigation system such as it is pipe corrosion and emitter chemical clogging, especially calcium, iron, manganese, phosphorus and sulphide
2. Physical – emitter's blockage problems from suspended solid particles like sand, clay particles and other impurities content or organic one (algae, plankton, insects, snails, etc.)
3. Biological – problems from bacteria and other contents, harmful to human and animal health as well as for the soil, the plants and the irrigation systems [9, 11]

The temperature of water used for the irrigation should be higher or the same as the soil temperature [10].

The critical chemical constituents that affect the suitability of water for irrigation are the total concentration of dissolved salts, relative proportion of bicarbonate to calcium, magnesium and relative proportion of sodium to calcium. Water quality problems in irrigation include salinity and alkalinity [10, 12].

Parameters for evaluation of possible precipitation of salts, induction of salinity and sodicity due to irrigation practices are used to classify water for irrigation. However, a classification of water quality that considers the interaction of both salinity and soil sodicity with the toxicity risk is not available [13]. There is an increasing evidence of contamination of produce from irrigation water, but scarce information on the microbial quality of agricultural water is available [5].

Water quality problems, however, are often complex, and a combination of problems may affect crop production more severely than a single problem in isolation. It is easier to solve each factor individually than its combination, in case of the problems occur in combination. Therefore, the factors are evaluated for each problem and solution separately, such as:

- The type and concentration of salts causing the problem
- The soil-water-plant interactions that may cause the loss in crop yield
- The expected severity of the problem following the long-term use of the water
- The management options that are available to prevent, correct or delay the onset of the problem [4]

The kind and amount of salt determine the suitability of irrigation water [7]. As it is mentioned by Ayers and Westcot [14], there are primary groups for limitations which are associated with the quality of irrigation water:

1. Soluble salts total concentration (salinity hazard)
2. Sodium relative proportion to the other cations (sodium hazard)
3. pH values and concentrations of bicarbonate and nitrate (diverse effects)
4. Specific ions toxicity, such as chloride, sodium and trace elements

Except the toxicity also the amounts and combinations of these substances can define appropriateness of water for irrigation [7]. Chloride, sodium, boron, nitrates, bicarbonate, and an abnormal pH may create toxicity problems. Therefore, the evaluation of the water suitable for irrigation should include these and other parameters, as well as other factors influencing water quality [9].

Water pH, salts (electrical conductivity), manganese and iron are the most essential chemical tests for irrigation water quality for micro-sprinkler irrigation systems [3].

The salt concentration in most irrigation waters ranges from 200 to 4,000 mg L^{-1} total dissolved solids. The pH of the water ranges typically from 6.5 to 8.4 [3].

Dirt and suspended inorganic and organic matter compose the solid content in irrigation water. Inorganic matter consists of silt, sand, leaves, fine clay and rust dust, and organic substances are from the vegetative origin and living organisms and bacteria populations (algae, bacteria, protozoa). Its concentration in the irrigation water depends on the nature of water source and may vary in a wide range [9].

According to Act 364/2004, the water used for irrigation in Slovakia cannot negatively impact the human and animal health, soil and state of the surface and the subsurface waters. The qualitative aims of the surface waters used for irrigation are regulated by Regulation 269/2010.

2.2 Classification of Irrigation Water Quality

Several authors classified the irrigation water quality according to the values of chemical, the physical and the biological parameters [9].

It is usually difficult to understand a traditional, technical review of water quality data to political decision-makers, non-technical water managers and the general public [15]. The hard task that usually faces water managers is how to transfer their interpretation of complex water quality data into information that is understandable and useful to policymakers as well as the general public [16]. A number of indices have been developed to summarize water quality data in an easily expressible and easily understood format [15].

The engineers and farmers are able to understand the principal parameters for water classification (crop response to salinity, sodium hazard and toxicity); therefore, they are able to manage irrigation and follow-up purposes properly [9]. Such a quite simple and practical tool for the engineers and farmers are water quality indices [1].

Reviewing the literature has shown that different statistical analyses may be used for similar water quality objectives. This may create inconsistency in interpreting water quality data primarily for policymakers and the general public [16].

Consequently, a standard data analysis framework that meaningfully integrates water quality data sets and converts them into reliable information is an essential requirement. It ensures consistency especially when data may come from different origins and are analysed by different people [16].

The classification adopted by FAO in 1985 was proposed as an initial guide. It was proved to be a practical and useful guide in assessing water quality for on-farm water use [9].

Chemical characterization of groundwater for irrigation use has been attempted by adopting internationally accepted methods such as sodium adsorption ratio, Kelly's ratio, soluble sodium percentage, permeability index and residual sodium carbonate to define its suitability for irrigation purpose [17]. Manimaran [18] presents some other parameters to evaluate water suitable for irrigation – residual sodium carbonate, residual sodium bicarbonate, exchangeable sodium ratio, exchangeable sodium percentage, potential salinity or salt index.

Sodium hazard is usually expressed in terms of sodium adsorption ratio [4] and describes the relationship between soluble Na^+ and soluble divalent cations (calcium and magnesium) [19]. The higher the Na in relation to Ca and Mg is, the higher the SAR [20]. Soluble sodium percentage is also used to evaluate sodium hazard [21, 22]. Other methods for evaluation of sodium hazards are Kelly's ratio, Schoeller's index or Puri's salt index [17].

One way of classifying water quality is by means of indices, in which a series of parameters analysed are joined by a single value, facilitating the interpretation of extensive lists of variables or indicators, underlying the classification of water quality [23].

The water quality index (WQI) is a communication tool for transfer of water quality data [24]. The water quality index is a tool to summarize large amounts of water quality data into simple terms (e.g. good or fair) for reporting to the public and decision-makers consistently. Based on the results of the water quality index, we are able to evaluate and rank the water for various uses (irrigation, livestock water, recreation, habitat for aquatic life, etc.) [25]. It can be used in trend analyses,

graphical displays and tabular presentations. It is an excellent format for summarizing overall water quality conditions over space and time [26]. The WQIs cannot
replace a detailed analysis of environmental monitoring and modelling, as well as
they cannot be the universal tool for the water management [27].

Factor analysis, one of WQI, a multivariate statistical method, yields the general
relationship between measured chemical variables by showing multivariate patterns
that may help to classify the original data [28].

Cluster analysis helps in grouping objects (cases) into classes (clusters) on
the basis of similarities within a class and dissimilarities between different classes.
The class characteristics are not known in advance but may be determined from the
analysis [29].

The purpose of the principal component analysis is usually to determine a few
linear combinations of the original variables, which can be used for summarizing the
data with minimal loss of information. Principal component analysis as the multivariate analytical tool is used to reduce a set of original variables and to extract a
small number of latent factors (principal components) for analysing relationships
among the observed variables [30].

3   Irrigation Water Quality in Slovakia

3.1   Irrigation Water Quality Regulation in Slovakia

Nowadays, the irrigation water quality is evaluated according to the limits given by
the regulation 269/2010. Annex 2 contains the limits for the water used for irrigation
which are the same as limits in Slovak technical standard (STN) 75 7143 (the newest
version 1999) for the first class of water quality – water suitable for the irrigation. In
a case of values worse than limits in regulation, the evaluation process continues
according to the STN 75 7143 – water quality for irrigation. The standard evaluates
water in broader context according to its use in next two classes. The second class
is water condition suitable for irrigation so that water can be used for irrigation
providing that for each locality will be given special measurements according to the
level and character of water pollution, the local conditions and the type of irrigation.
The third class is water unsuitable for the irrigation. The standard does not allow
irrigating with the water, from which contamination could be transfer into a food
chain, e.g. microbiological pollution, heavy metals, phenols, etc.

Before issuing and applying the regulation 269/2010, the regulations 296/2005
and 491/2002 was working and before applying the STN 75 7143, the standard STN
83 0634 (1971) was used. The revision of regulation was required because of the
entrance of the Slovak Republic into the European Union. The changes in the
monitoring system that time were not so evident. The number of parameters
increases from the 40 in 491/2002 to 43 in 296/2005, and in the regulation
269/2010 are 42 parameters, for which the limits are given.

There were published the standards for irrigation with wastewater. The standard contained limits of ineligible substances (e.g. heavy metals, organic pollution and radiochemical pollution). Nowadays, the wastewater is not used for the irrigation in Slovakia.

The guidelines for evaluation of water quality for irrigation were published by FAO [4]. Those guidelines include laboratory determinations and calculations needed to use the guidelines. Comparison of limits in the regulation 269/2010 and limits mentioned by Ayers and Westcot [4] shows some differences. The regulation does not cover all the parameters, and some of them have different limits (Table 1). The regulation covers physical, chemical as well as biological parameters. The marginal limits for the irrigation water quality regard the qualitative aims of surface waters consequent upon the Water Framework Directive and at the same time regard the plant's requirements for the irrigation. Therefore, some limits of the parameters are different.

Table 1 Comparison of limits of irrigation water quality parameters

Water parameter	Symbol	Usual range in irrigation water	
		Ayers and Westcot [4]	269/2010
Salinity			
Salt content			
Electrical conductivity (or)	EC_w	0–3 dS m^{-1a}	–
Total dissolved solids	TDS	0–2,000 mg L^{-1}	800 mg L^{-1} (dry in 105°C)
Cations and anions			
Calcium	Ca^{2+}	0–20 me L^{-1}	100 mg L^{-1}
Magnesium	Mg^{2+}	0–5 me L^{-1}	200 mg L^{-1}
Sodium	Na^+	0–40 me L^{-1}	100 mg L^{-1}
Carbonate	CO_3^-	0–0.1 me L^{-1}	–
Bicarbonate	HCO_3^-	0–10 me L^{-1}	–
Chloride	Cl^-	0–30 me L^{-1}	300 mg L^{-1}
Sulphate	SO_4^{2-}	0–20 me L^{-1}	250 mg L^{-1}
Nutrients			
Nitrate-nitrogen	NO_3-N	0–10 mg L^{-1}	23 mg L^{-1}
Ammonium-nitrogen	NH_4-N	0–5 mg L^{-1}	–
Phosphate-phosphorus	PO_4-P	0–2 mg L^{-1}	–
Potassium	K^+	0–2 mg L^{-1}	–
Miscellaneous			
Boron	B	0–2 mg L^{-1}	0.5 mg L^{-1}
Acid/basicity	pH	6.0–8.5	5.0–8.5
Sodium adsorption ratio[b]	SAR	0–15 me L^{-1}	–

[a] dS m^{-1} = deciSiemen metre^{-1} in SI units (equivalent to 1 mmho cm^{-1} = 1 millimmho centimetre^{-1})
mg L^{-1} = milligramme per litre $\simeq$ parts per million (ppm)
me L^{-1} = milliequivalent per litre (mg L^{-1} ÷ equivalent weight = me L^{-1}); in SI units, 1 me L^{-1} = 1 millimol litre^{-1} adjusted for electron charge
[b] SAR is calculated from the Na, Ca and Mg reported in me L^{-1} [4]

3.2   Irrigation Water Quality Monitoring Network in Slovakia

The organisations for the irrigation water quality monitoring have been changed recently. The monitoring is managed by the Ministry of agriculture, and rural development of the Slovak Republic and the database is managed by the National agricultural and food centre – specifically the Soil Science and Conservation Research Institute (SSCRI) in Bratislava.

Irrigation water quality monitoring has a long tradition in Slovakia. As was mentioned, one of the oldest standards is from the year 1971. The systematic monitoring of irrigation water quality started in 1995, but the responsible person for the database in SSCRI confirmed that the monitoring started earlier. The data from that monitoring are not available.

The water samples for the evaluation of water suitable for the irrigation are taken once per a month during the main vegetation season of 1 year with a minimum of seven samples (usually from April to September). The general standards for the analysis are used. The irrigation water is evaluated according to the worst parameter. The classification of irrigation water is according to the characteristic value, which is calculated as an average of the three highest measured values of each parameter during the monitoring season (in the case of a number of samples less than 24). Each parameter is measured at the same level.

The number of monitoring station varied over the years. There were more than 200 monitoring stations before the year 2004, and from that year there was a rapid descent of irrigation water monitoring stations. Only 11 water resources were controlled in the last 4 years (see Table 2). A lower support for the monitoring system caused a decrease in the number of the monitoring stations. At the same time

Table 2 Number of monitoring stations, number of station in particular irrigation water quality classes

Year	Number of monitoring stations	First class	Second class	Third class
2002	273	83	116	74
2003	218	67	98	53
2004	60	7	48	5
2005	51	6	42	3
2006	80	47	28	5
2007	80	35	40	5
2008	80	23	47	10
2009	20	5	14	1
2010	35	7	23	5
2011	32	8	19	5
2012	17	7	8	2
2013	12	4	6	2
2014	11	3	8	0
2015	11	7	4	0
2016	11	3	8	0

started a monitoring of the drainage waters as a result of the requirement of Directive 91/676/EEC. The monitoring stations were situated mainly in the south part of Slovakia and covered all irrigation areas of Slovakia in 1995 (Fig. 1). In 2016, the official monitoring stations were situated only in south-west part of Slovakia (Fig. 2). The agronomists use water for irrigation also from other surface water resources in Slovakia, but those stations are not included in the state irrigation water monitoring network.

In regard to the limited financial support, it was necessary to select monitoring stations according to relevant criteria:

1. Station with permanent water demand with a high frequency for irrigation according to a long-term perspective
2. Station where water is used to irrigate crops with direct contact of irrigation water with the consumed vegetable or fruit or its part

The most of monitoring stations in 2002–2016 are in the second class of irrigation water quality according to STN 75 7143. This trend is not only for the presented years (Table 2 and Fig. 3) but also for the previous years (see [31]).

In 1995–1998 the most monitoring stations (average 46%) were in the second class of the irrigation water quality. In the third class were 21.3% of monitoring stations [10]. The results of monitoring in 1995–1998 showed the strong regional impact on the irrigation water quality. Relatively the clearest irrigation water was in Danube catchment, and the pollution is mostly created by high pH and dissolved

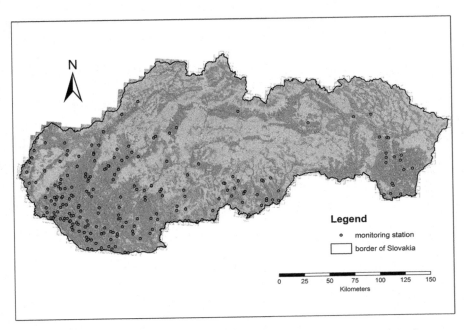

Fig. 1 Irrigation water quality monitoring station in 2003 (©GKU, adapted by Kaletová)

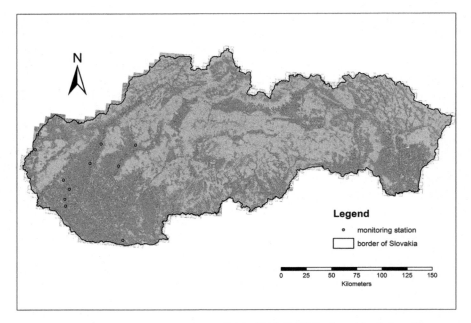

Fig. 2 Irrigation water quality monitoring station in 2016 (©GKU, adapted by Kaletová)

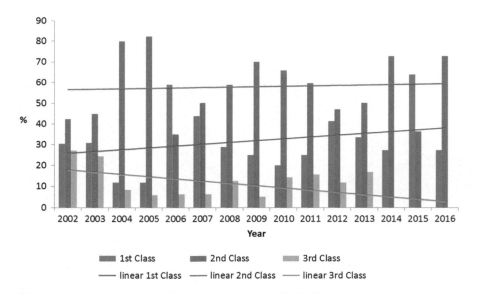

Fig. 3 Trend of each water quality class during the years 2002–2016

solids. The most polluted irrigation water was in Bodrog and Hornád catchment, and the cause of pollution was created by microbiological contamination. By the time, the irrigation water quality in the Danube and the Váh catchment increased. Slower increased of water quality was in the Hron catchment. The worst situation was in the Bodrog and the Hornád catchment where the number of monitoring stations of the third class increased (88.9%). Therefore, if there is a possibility to use for irrigation only polluted water, the specific measurements according to level and character of pollution should be applied. Those measurements can cause a change of irrigated crops, e.g. to technical crops which do not have a direct impact on the food chain. The microbiological pollution is a long-term problem in the Bodrog and Hornád catchment. A treatment of microbiological polluted water was technically possible, but it was also costly. The irrigation water was mainly polluted by untreated wastewater from households; therefore, the situation would be better after the construction of wastewater treatment plants [10, 11, 32].

The trend of irrigation water quality changed in the years 2000–2004, mainly in the Bodrog and the Hornád catchment. In 2001 it was for the first time when at least one monitoring station reached the irrigation water quality in the first class. There was the apparent increasing trend of irrigation water quality. The amount of monitoring station in the third class decreased, and the number of monitoring station in the first and second class increased comparing the previous years 1995–1999. As it was mentioned, the worse irrigation water quality was caused by the microbiological contamination. The build-up of wastewater treatment plants in the catchment and decrement of sources of pollution from the agriculture production (mainly decrement of number of animals) increase the irrigation water quality in the catchment [10]. Another reason that the monitoring IWQ in the station which was usually in the second or the third class especially in Eastern lowland was stopped of monitoring after the year 2011.

3.3   Irrigation Water Quality in Slovakia

The irrigation water quality decreased mainly in consequence of higher pH, dissolved solids (DS), calcium and microbiological pollution. The leading cause is microbiological pollution, created mainly by coli form bacteria and faecal coli form bacteria. The limits of those parameters were excessed almost every year; respectively, pH increased every year.

The values of pH were one of the most common excessed parameters in 2002–2016 (Table 3). The highest values were over 9.2, often more than 10.0. Problematic were water reservoirs in Lazany (2003, 10.3; 2006, 10.1; 2007, 9.98), Plavé Vozokany (2005, 10.3; 2008, 9.95), Sebechleby (2002, 10.3) and some others.

Dissolved solids were mainly excessed in the Danube catchment. The highest values were in sources Bohatá, Stračínsky creak or Vizaláš. The dissolved solids reached the highest value 1,336 mg L^{-1} in 2002, 1,590 mg L^{-1} in 2003 and 1,327 mg L^{-1} in 2004.

Table 3 Number of location with excessed parameters in particular years

Year	Number of monitoring stations	Parameter (number of monitoring stations with excessed values)
2002	273	pH (67), DS (1), SO_4^{2-} (11), Ca (30), NES[a,b], faecal coli form bacteria (165), *Enterococcus* (157), coli form bacteria (112), colofac virus (30), phytotoxicity (3), Pb (2), Fe (1), Al (11), Na (4)
2003	218	pH (49), DS (12), SO_4^{2-} (6), Ca (57), NES[b], chlorides (2), faecal coli form bacteria (117), *Enterococcus* (105), coli form bacteria (85), colofac virus (35), As (1), Fe (2), Mn (1), Al (12), K (1), Na (6)
2004	60	pH (6), DS (3), SO_4^{2-} (2), Ca (13), faecal coli form bacteria (10), *Enterococcus* (4), coli form bacteria (45), Al (2), Na (1)
2005	51	pH (13), DS (2), Ca (5), microbiological pollution – not specified (44)
2006	80	pH (13), DS (3), Ca (25), microbiological pollution – not specified (73)
2007	80	pH (23), DS (9), SO_4^{2-} (3), Ca (12), microbiological pollution – not specified (31)
2008	80	pH (25), DS (3), Ca (11), microbiological pollution – not specified (39)
2009	20	pH (8), DS (1), Ca (3), microbiological pollution – not specified[b]
2010	35	pH (9), DS (6), Ca (12), microbiological pollution – not specified[b]
2011	32	pH (11), DS (3), Ca (5), microbiological pollution – not specified[b]
2012	17	pH (8), DS (1), microbiological pollution – not specified[b]
2013	12	pH (5), microbiological pollution – not specified[b]
2014	11	pH (2)
2015	11	Ca (2), microbiological pollution – not specified[b]
2016	11	pH (2), SO_4^{2-} (1), Ca (1), microbiological pollution – not specified[b]

[a]Non-polar extraction substance
[b]Unknown number of locations

The sulphates were excessed in the same stations as dissolved solids in 2002–2004 and also in 2007 and 2016. The other years' sulphates were in the limits. The highest value was 789 mg L^{-1} in 2003; in the other years, they reached more than 400 mg L^{-1} (2002, 473 mg L^{-1}; 2004, 429 mg L^{-1}).

The microbiological pollution occurred almost all the years. The microbiological pollution with permanent character occurred in 2002, as well as in 2001.

No pollution with the heavy metals (Cd, Pb, Zn, Co, Ni, Cr, Cu, and Hg), as well as phytotoxicity (seed quality of *Brassica hirta Moench*), was measured.

The results of nitrates in 1995–2015 show that the highest amount of nitrates was mainly in gravel deposit and the lowest in water reservoirs. The highest amount of nitrates was 141 mg L^{-1} in water reservoir Vizaláš in 2003; over 100 mg L^{-1} were also exceeded in 1996 (107 mg L^{-1} and 104 mg L^{-1}, respectively). The level of 25 mg L^{-1} was not excessed in most of the results. Therefore, the irrigation water is only in small scale polluted by nitrates [6].

The higher values of pH were observed primarily in the water reservoirs with intensive eutrophication processes during the summer [33]. The development of eutrophication is strongly influenced by the amount of nutrients in the water, especially nitrogen and phosphorus. Nutrients together with appropriate temperature conditions cause an intensive development of phytoplankton which by photosynthetic activity invades carbonated balance in the water. Nutrients occur in the environment mainly from anthropogenic activities. Nutrients appear in water mainly from the soil erosion by the thoughtless use of the industrial fertilizers. Land use in agricultural stream basin exhibited high loads of sediments and surface runoff that affect stream water quality [34]. Most of the water reservoirs have not reclaimed surrounding; therefore, the soil with nutrients may travel directly to the reservoirs during the rain. Each water reservoir is affected by the eutrophication in the long-term perspective. The water reservoirs are a trap of nutrients. They collect more nutrients than they expend [33]. Also, Potužák and Duras [35] explain that there is a significant potential in the retention of nutrients from the different sources to naturally hidden in the water reservoirs. That is the process which will end only with the total silting-up of human extraction from sediments. More information about siltation of the water reservoirs may be found in, e.g. [36] or [37]. Hubačíková and Oppeltová [38] mentioned that the combination of sewage and use of manure as the fertilizer to increase the nutrient in the water reservoirs has a very adverse effect on the results of monitored indicators of water quality. If such water is used for irrigation, it can have an irreversible impact on the environment. Keesstra et al. [39] mentioned that vegetation in riparian zone slowing the flow rate, thus, reduces overall sediment yield from the catchment (nutrients including).

Except for the usage of fertilizers nutrients in the water, reservoirs are created from intensive fish production, mainly thoughtless feeding. Use of detergents with phosphorus compounds is not as frequent as it used to be in the past. Therefore, the sewage from the households also increased irrigation water quality.

Calcium is a permanent part of water. The higher values of calcium were measured in 2009–2011 (2009, 127 mg L^{-1}; 2010, 223 mg L^{-1}; 2011, 250 mg L^{-1}). The higher amount of calcium in the irrigation water does not have a massive negative impact on the soil and plants. Import of calcium by irrigation water into the carbonate soil is negligible. Irrigation water with calcium has a positive impact on the acid soils by increasing pH in soil. Calcium also has a positive impact on the soil structure and eliminates the negative impact of monobasic cations in the soil. Calcium has a negative impact on the technical facilities. It causes siltation of distributional system of irrigation network. Drip irrigation is very sensitive to the amount of calcium in the water; therefore, the amount of the calcium limit is stricter – 50 mg L^{-1}. Another negative impact may be caused by irrigation of container plants. In such case, calcium is accumulated in the small amount of soil [39].

4 Conclusion

In the past, economy of Slovakia was more agriculturally oriented than nowadays. Therefore, the monitoring of water quality and, namely, irrigation water quality has a long-term tradition in Slovakia. One of the first standards for the irrigation water quality was published in 1971. The valid regulation classifies irrigation water quality according to the limit value of several parameters. Changes which were applied in the past have a positive effect on the water quality. Reduction and modification of commonly used fertilizers, as well as treatment of wastewater from households, had main impact on it.

The last years were the monitoring network reduced to the minimum but on the other hand started the monitoring of drainage waters. Therefore, it is possible to analyse the impact and necessity of used fertilizers on the arable land. Increasing pressure on the plant production will also increase the demand on irrigation water in the landscape. It will focus on the enough quantity of water, as well as on the available irrigation water in appropriate quality. We can expect the extension of the monitoring network in the future.

The irrigation water quality is a matter not only of irrigation management. It is also an issue of food quality. Meeting the criteria of irrigation water quality requirements, we can assure healthy and safe food or livestock feeds. Charge of irrigation water quality also provides content of substances in the top soil horizon and groundwater protection. Therefore, the cost of monitoring the quality of irrigation water is reflected in the quality of agricultural (cultivated) crops, and human and animal health.

Agricultural production, food production, forestry and irrigation water itself is not possible without providing enough fresh and quality water in the landscape. Therefore, the need for water treatment and purification of the wastewater generated before discharge into the streams becomes a priority. Such an approach is also in line with the requirements of Directive 2000/60/EC of the European Parliament and of the Council of 23 October 2000 establishing a framework for community action in the field of water policy (Water Framework Directive) and its implementation. Slovakia has implemented Water Framework Directive by the Ministry of the Environment of the Slovak Republic.

5 Recommendations

Expected higher pressure on the plant production will increase the necessity of the irrigation water quality and quality. The sustainable quality can be achieved only in the case of sustainable use of landscape. Therefore, we recommend to:

- Continue with the monitoring in the current monitoring network.
- Increase the number of monitoring stations in the network.
- Respect the nitrate directive.
- Respect the crop requirements for the amount and time of used fertilizers.
- Respect the requirements arising from the implementation of the Water Framework Directive in the area of surface and groundwater quality.
- Support the research of water and nutrient transport in the porous media and within agricultural landscape.
- Support the research of connection surface, subsurface and ground water.

Acknowledgements This chapter was supported with the following grants and projects:
- APVV-16-0278 Use of hydromelioration structures for mitigation of the negative extreme hydrological phenomena effects and their impacts on the quality of water bodies in agricultural landscapes
- KEGA 028SPU-4/2017 Monitoring of Elements of Environment – practical course

References

1. Omran IM, Marwa FH (2015) Evaluation of drainage water quality for irrigation by integration between irrigation water quality index and GIS. Int J Technol Res Appl 3(4):24–32. https://documents.tips/engineering/evaluation-of-drainage-water-quality-for-irri

2. Bárek V, Halaj P, Igaz D (2009) The influence of climate change on water demands for irrigation of special plants and vegetables in Slovakia. In: Střelcová K et al (eds) Bioclimatology and natural hazards. Springer, Dordrecht, pp 271–282

3. Hopkins B, Horneck DA, Stevens R, Ellsworth JW, Sullivan D (2007) Managing irrigation water quality for crop production in the Pacific Northwest. Oregon State University, p 29. http://extension.oregonstate.edu/umatilla/mf/sites/default/files/pnw597-e.pdf

4. Ayers RS, Westcot DW (1994) Water quality for agriculture. Irrigation and drainage paper No. 29. FAO, Rome. http://www.redalyc.org/html/1802/180222945013/index.html

5. Allende A, Monaghan J (2015) Irrigation water quality for leafy crops: a perspective of risks and potential solutions. Int J Environ Res Public Health 12(7):7457–7477. http://www.mdpi.com/1660-4601/12/7/7457

6. Píš V, Nágel D, Hríbik J (2009) Kvalita závlahovej vody v rokoch 1995 - 2008 z hľadiska obsahu dusičnanov. Vodohospodársky spravodajca 52(1–2):14–16 (in Slovak)

7. Fipps G (2003) Irrigation water quality standards and salinity management, fact sheet B-1667. Texas Cooperative Extension, The Texas A&M University System, College Station. https://www.scribd.com/document/173645240/ASSESSMENT-OF-IRRIGATION-WATER-QUALITY

8. Shalhevet J, Kamburov J (1976) Irrigation and salinity: a world-wide survey. Caxton Press, New Delhi. http://www.redalyc.org/html/1802/180222945013/index.html

9. Phocaides A (2007) Handbook on pressurized irrigation techniques. Food and Agriculture Organization of the United Nations, Rome. http://www.fao.org/tempref/docrep/fao/010/a1336e/a1336e.pdf

10. Rehák Š, Bárek V, Jurík Ľ, Čistý M, Igaz D, Adam Š, Lapin M, Skalová J, Alena J, Fekete V, Šútor J, Jobbágy J (2015) Zavlažovanie poľných plodín, zeleniny a ovocných sadov. VEDA, Bratislava (in Slovak)

11. Bárek V (2008) Technické aspekty návrhu a prevádzky mikrozávlah v podmienkach Slovenska. Slovenská poľnohospodárska univerzita, Nitra (in Slovak)

12. Kumar M, Kumari K, Ramanathan A et al (2007) A comparative evaluation of groundwater suitability for irrigation and drinking purposes in two intensively cultivated districts of Punjab, India. Environ Geol 53(3):553–574. https://link.springer.com/content/pdf/10.1007%2Fs00254-007-0672-3.pdf

13. Meireles A, Andrade EM, Chaves L, Frischkorn H, Crisostomo LA (2010) A new proposal of the classification of irrigation water. Rev Ciênc Agron 41(3):349–357. http://www.scielo.br/scielo.php?script=sci_arttext&pid=S1806-69022010000300005

14. Ayers RS, Westcot DW (1985) Water quality for agriculture. Irrigation and drainage paper No. 29, Rev. I. U. N. FAO, Rome. http://www.fao.org/docrep/003/T0234E/T0234E00.htm

15. Couillard D, Lefebvre Y (1985) Analysis of water quality indices. J Environ Manag 21:161–179

16. Shaban M (2017) Statistical framework to assess water quality for irrigation and drainage canals. Irrig Drain 66:103–117. http://onlinelibrary.wiley.com/doi/10.1002/ird.2042/full

17. Bhokarkar S, Tiwari KC (2016) Chemical hydrogeological study of the Kim River Basin, South Gujarat, India: with special reference to impact of irrigation on groundwater regime. Int J Adv Res 4(3):1634–1645. http://www.journalijar.com/uploads/186_IJAR-9399.pdf

18. Manimaran M (2016) Assessment of quality contributing parameters of irrigation water by using standard formulae. Adv Appl Sci Res 7(3):155–157. http://www.imedpub.com/articles/assess ment-of-quality-contributing-parameters-of-irrigation-water-by-using-standard-formulae.pdf

19. Alrajhi A, Beecham S, Bolan NS, Hassanli A (2015) Evaluation of soil chemical properties irrigated with recycled wastewater under partial root-zone drying irrigation for sustainable tomato production. Agric Water Manag 161:127–135. https://www.sciencedirect.com/science/article/pii/S0378377415300597

20. Vyas A, Jethoo AS (2015) Diversification in measurement methods for determination of irrigation water quality parameters. Aquat Proc 4:1220–1226. https://www.sciencedirect.com/science/article/pii/S2214241X1500156X

21. Moyo LG, Vushe A, January MA, Mashauri DA (2015) Evaluation of suitability of Windhoek's wastewater effluent for re-use in vegetable irrigation: a case study of Gammams effluent. WIT Trans Ecol Environ 199:109–120

22. Shammi M, Karmakar B, Rahman M, Islam M, Rahaman R, Uddin K (2016) Assessment of salinity hazard of irrigation water quality in monsoon season of Batiaghata Upazila, Khulna District, Bangladesh and adaptation strategies. Pollution 2(2):183–197. https://jpoll.ut.ac.ir/article_56946_e301bbee372ccbc7794ac8ce6e75ae23.pdf

23. Maia C, Rodrigues K (2012) Proposal for an index to classify irrigation water quality: a case study in northeastern Brazil. Rev Bras Ciênc Solo 36(3):823–830. http://www.scielo.br/scielo.php?script=sci_arttext&pid=S0100-06832012000300013

24. Ball RO, Church RL (1980) Water quality indexing and scoring. J Environ Eng 106(4):757–771

25. Yousry MM, El Gammal HAA (2015) Factor analysis as a tool to identify water quality index parameters along the Nile River, Egypt. J Am Sci 11(2):36–44. http://free-journal.umm.ac.id/files/file/005_27939am110215_36_44.pdf

26. Stone J (1978) Water-quality indices for specific water uses. Geological Survey Circular 770. https://pubs.usgs.gov/circ/1978/0770/report.pdf

27. Jahad U (2014) Evaluation water quality index for irrigation in the North of Hilla city by using the Canadian and Bhargava methods. J Babylon Univ 2(22):346–353

28. Liu CW, Lin KH, Kuo YM (2003) Application of factor analysis in the assessment of groundwater quality in a blackfoot disease area in Taiwan. Sci Total Environ 313(1):77–89. http://www.sciencedirect.com/science/article/pii/S0048969702006836

29. Vega M, Pardo R, Barrado E, Deban L (1998) Assessment of seasonal and polluting effects on the quality of river water by exploratory data analysis. Water Resour 32:3581–3592. https://www.sciencedirect.com/science/article/pii/S0043135498001389

30. Mahmud R, Inoue N, Sen R (2007) Assessment of irrigation water quality by using principal component analysis in arsenic affected area of Bangladesh. J Soil and Nat 1(2):8–17. http://ggfjournals.com/assets/uploads/2.08-17_.pdf

31. Píš V, Nágel D, Tršťanská D (1995) Water quality for irrigation and its development during the last four years. In: Proceeding hydromelioration in Slovakia on the threshold of the 21 century, Bratislava

32. Zuzula I, Rehák Š, Minárik B (2001) Niektoré aspekty klimatickej zmeny z hľadiska závlah na Slovensku. In: Národný klimatický program SR (in Slovak)

33. Píš V, Hríbik J, Nágel D (2010) Monitoring kvality závlahovej vody v roku 2009. Vodohospodársky spravodajca 53(3–4):26–28 (in Slovak)

34. Halaj P, Halajová D, Bárek V, Rehák Š, Báreková A (2015) Recovery of natural resilience and resistance capacity of stream ecosystems in agricultural land in the context of climate change. International Multidisciplinary Scientific GeoConference Surveying Geology and Mining Ecology Management, SGEM 1(3):531–540

35. Potužák J, Duras J (2014) Jakou roli mohou hrát rybníky v zemědělské krajině? In: Vodárenská biologie 2014, Praha (in Slovak)

36. Kubinský D, Weis K, Fuska J, Lehotský M, Petrovič F (2015) Changes in retention characteristics of 9 historical artificial water reservoirs near Banská Štiavnica, Slovakia. Open Geosci 7:880–887
37. Fuska J, Bárek V, Pokrývková J, Halaj P (2014) Comparison of actual and presumed water capacity of fish pond in Lukáčovce. J Int Sci Publ Ecol Saf 8(1):409–414
38. Hubačíková V, Oppeltová P (2017) The impact of pond on water quality in the Čermná stream. J Ecol Eng 18(1):43–48
39. Keesstra SD, Kondrlova E, Czajka A, Seeger M, Maroulis J (2012) Assessing riparian zone impacts on water and sediment movement. Neth J Geosci 91(1–2):245–255

Small Water Reservoirs: Sources of Water for Irrigation

Ľ. Jurík, M. Zeleňáková, T. Kaletová, and A. Arifjanov

Contents

Abstract Small water reservoirs are part of the irrigation system in Slovakia, and the total volume of these reservoirs is about 56.404 million m^3. However, a volume of 158 million m^3 is required to meet the specified need. The construction of small water reservoirs in Slovakia was formerly undertaken by the State Amelioration

Ľ. Jurík (✉) and T. Kaletová
Department of Water Resources and Environmental Engineering, Faculty of Horticulture and Landscape Engineering, Slovak University of Agriculture in Nitra, Nitra, Slovakia
e-mail: lubos.jurik@uniag.sk; tatiana.kaletova@uniag.sk

M. Zeleňáková
Department of Environmental Engineering, Faculty of Civil Engineering, Technical University in Košice, Košice, Slovakia
e-mail: martina.zelenakova@tuke.sk

A. Arifjanov
Tashkent Institute of Irrigation and Agricultural Mechanization Engineers, Tashkent, Uzbekistan
e-mail: obi-life@mail.ru

A. M. Negm and M. Zeleňáková (eds.), *Water Resources in Slovakia: Part I - Assessment and Development*, Hdb Env Chem (2019) 69: 115–132, DOI 10.1007/698_2018_301, © Springer International Publishing AG 2018, Published online: 15 July 2018

Administration, but economic changes in the 1980s to 1990s have had a negative impact on these reservoirs. With reducing interest in irrigation, these reservoirs have lost the main purpose for which they were built.

Keywords Climate change, Disposable water resources, Irrigation, Water reservoir

1 Introduction

The history of small water reservoirs (SWRs) in Slovakia is simple and short. It is very similar to the history of irrigation in this country, because the reservoirs are part of the irrigation system. Slovakia is located in a region where the snow melts from March to May, depending on the weather. During this period, field work in agriculture is just beginning. The main vegetation season occurs at a time when water flows in the rivers, but at a time when groundwater reserves are decreasing significantly. We need plenty of water for crops to reach potential harvest, and since the quantity of water found in nature during this period is insufficient, we need to use other sources: either water that is available from another area, or water that is in the given area at a time other than the vegetation season.

Around 1955 Slovakia decided on the second option, i.e., to store water from the time period when it is not required for agriculture, by using water reservoirs in water management. These reservoirs are filled with water during the period when there is sufficient water and minimal water consumption. The water is subsequently used during periods of water scarcity and maximum consumption.

Water reservoirs are categorized throughout the world as SWRs and large dams, depending on the depth of water. Small water reservoirs have become the best solution for good-quality and sufficient-quantity water supply for irrigation in Slovakia, and their construction began very soon after the decision was made to use this method of water management. A plan was prepared for their locations and the necessary volume of water for optimum agricultural production was determined; i.e., the total water demand for irrigation in a dry year should be met by a volume of about 158 million m^3 water in the SWRs. Subsequently, sites were selected and a hydrological balance study was done to decide whether it was possible to keep enough water from the flows in small streams to fill each of the SWRs.

On the basis of agricultural production in particular parts of Slovakia, priority was given to the urgent construction of the SWRs, and to their distribution over the territory of Slovakia.

Completion of the construction of the total number of SWRs was scheduled for around 2020. However, the construction continued until 1990, when construction was interrupted and even stopped. The need for SWRs has changed owing to changes in irrigation use. Small water reservoirs associated with irrigation systems are shown in Fig. 1. The importance of the SWRs for irrigation is gradually decreasing, and their use has recently been focused on fish farming. Their importance has again been a subject of discussion, with problems of climate change and dry years in 2014 and 2015, but even this discussion did not provide enough momentum for the further construction of SWRs.

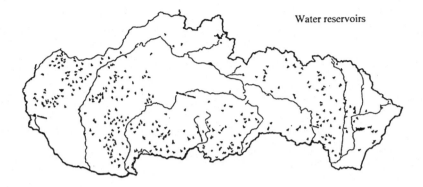

Water reservoirs

Fig. 1 Location of small water reservoirs in Slovakia

The construction of SWRs has secured a volume of 56.404 million m^3 of water; however, 158 million m^3 (two-thirds of the volume to meet the specified need) is still required. Management plans for the Danube and Vistula river catchments for the Slovak Republic have again emphasized the importance of SWR construction. So far, however, there has been no such construction.

2 Hydrological Conditions of Small Catchments

The collection of water in a reservoir or tank allows people to negotiate dry seasons. With time, water collections in Slovakia were classified for different sectors, and SWRs now provide water for various purposes to improve, support, and protect our life [1].

Slovakia is situated in the region of river distribution between the Black Sea and the Baltic Sea. Its natural characteristics create conditions in which most of the water from precipitation is subject to outflow. The water utilization situation in Slovakia also depends on the variable characteristics of water in time and space. Thus, one of the priorities (keeping the water in the landscape, accumulating it during the rainy seasons for use in the dry season) of the Slovak Water Management Enterprise (SWME) is determined by Nature.

Recent periods in Slovakia have been characterized by sequential increases in discharge, as well as the drying of springs. On the other hand, heavy rainfalls have caused flooding in both rural and urban areas. Also, the extensive rapid outflow of surface water causes soil erosion. Therefore, the volume of water accumulated in the catchment area and the total accumulation capacity of the landscape, as well as the storage of subsurface and groundwater decreases under these conditions.

The hydrological conditions of small catchments have been the subject of several analyses and research studies in the past, and statistical hydrological methods were mostly used to solve problems with these catchments. In recent years, climate characteristics, mostly in regard to changes in rain intensity, as well as in regard to the occurrence of particular rain events, have been highly dynamic, driven mainly by climate change. Precipitation has shown a relative balance over the year until recently, but currently we recognize more dry seasons followed by heavy rainfalls. These circumstances have led to distinctive changes in the hydrological balance of catchments, and we can see an increasing deficit of precipitation, particularly in certain seasons of the year.

A three-dimensional matrix of a catchment shows the area of water accumulation from rainfall, whereby the lowest point of the catchment represents the minimal level at which there is no more subsurface water outflow from the catchment. In this case, the minimal discharge can be zero.

The trend of the outflow depends on the water storage in the catchment, evapotranspiration from vegetation and evaporation from the catchment surface, and the actual precipitation. In long seasons without precipitation, the storage capacity of the accumulation area is exhausted and successive decrements of the subsurface water lead to the total withdrawal of the accumulated water.

The monitoring of hydrological balance components has a long tradition in Slovakia. The number of stations for monitoring the flow of water has varied and there has been an increasing trend in recent years (2001, 391 stations; 2016, 416 stations). These components are usually determined by the measurement of surface water levels with limnigraphs, the measurement of precipitation with ombrometers, and the measurement of subsurface water levels with probes.

3 Small Water Reservoirs

The SWRs are the most numerous group of water reservoirs in Slovakia. They have an important place in the area of Slovak water management, and are characterized by three criteria [2]:

- The maximum storage capacity at the controllable level in the reservoir is 2 million m^3,
- The maximum depth of the reservoir does not exceed 9 m (excluding greater depth locally at the site of the original riverbed),
- The maximum flow (Q_{100}) is up to 60 $m^3 \, s^{-1}$.

The main aim of building SWRs was for water accumulation to increase discharges for irrigation withdrawals. Therefore, it was better to build SWRs in areas where the conditions for agriculture production lacked sufficient water resources or had dry seasons. The determination of SWR volume depended on the water demand for irrigation, together with the effect of the natural conditions. The ultimate conditions for the building of an SWR were appropriate morphology of the surrounding area and the geological and hydrological conditions in the area. The parameters of the SWR were usually adapted to the natural conditions of the area rather than to the parameters for irrigation.

The SWRs were built as multipurpose entities. One of the purposes of the SWRs was to decrease flooding. The main calculation involved was the retention volume created by the elevation of the spill jet over a safety spillway in the case of Q_{100}. The design of spillways was problematic, because of missing discharge data from the profiles of future dams, or because the data were of the fourth class of reliability, meaning that the variance of data accuracy was ±60%.

Other purposes of the SWRs were only supplemental, mainly for sport fishing and recreation, and sometimes for industry. Therefore, conditions were created for intensive fish production by the build-up of appropriate areas. The banks were adjusted for the possibilities of active and passive recreation. The water surface in the landscape naturally creates the conditions for water fauna and flora.

The current state of SWR development in Slovakia, as well as future plans, is that most SWRs are to be used for irrigation and fish production. Therefore, attention is mainly focused on these aspects.

3.1   Water Management of Small Water Reservoirs in Slovakia

The construction of SWRs in Slovakia was formerly undertaken by the State Amelioration Administration. However, economic changes in this country in the 1980s to 1990s have had a negative impact on SWRs. With reducing interest in irrigation, the SWRs lost the main purpose for which they were built. At present, they are mainly used for fishing and recreational purposes, without any economic benefit for the managers of these waterworks, in the form of disposing of a quantity of surface water or realizing any hydroenergetic potential.

After the termination of the State Amelioration Administration, the SWRs were defined as water management structures for the SVP, š.p. Banská Štiavnica (SWME, a state enterprise). The new operator of these constructions was given other priority tasks especially aimed at ensuring flood protection measures. Therefore, some of the SWRs and hydro-melioration plants were brought under the administration of the new state enterprise – Hydromeliorace š.p. (Hydromelioracie, state enterprise). The bigger SWRs were again entrusted to the SWME.

The original purpose of using SWRs to store the volume of water required for irrigation has almost disappeared. Regardless of the economic aspect, the priority in the management of these water structures is their safety during operation. As soon as the SWRs were taken under SWME management, failures significantly affected their functionality. These failures were observed as bottom closures caused by improper operation, poor maintenance, and clogging (e.g., blockage).

3.2   Parts of Small Water Reservoirs

Each part of an SWR has a specific purpose. Not all of the parts can be found in each reservoir, and not all zones of storage water can be found in each SWR (see Fig. 2).

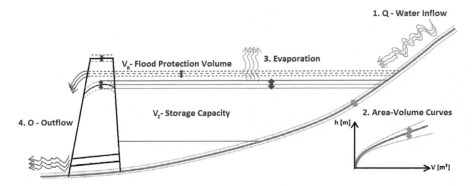

Fig. 2 Zones of storage in small water reservoirs (adapted from [3], with permission from MDPI AG, Basel, Switzerland)

An important aspect of SWRs is that the bottom of the discharge outflow is mostly located below the bottom of the valley, which is the level of the flow (stream, river) passing through the basin. Such a position is especially necessary for ponds that are emptied every year. Emptying is slow and causes higher outflow under the reservoir during autumn and winter, and this is particularly noticeable in small streams. The position of the discharge outflow at the bottom creates the so-called soil space, and its volume depends on the thickness of the soil layer, which is present after the small water basin is drained. The soil area is filled when the reservoir is again filled in the following year, and this soil area should also be considered for the water balance.

A space for constant retention (standing) is not used for normal operation. Quite often in irrigation SWRs a space is designed owing to the higher minimum operating level in relation to the location of the outflow devices, the preservation of a certain volume required as space for fish in winter, and the provision of the required water quality. That is the different reason as in case of large dams.

Storage capacity serves to increase flow rates and ensure water take-up from the SWR during periods of deficiency in the stream, and this capacity is essential in all accumulation reservoirs and is the major component of the total volume of the basin. Determination of the storage capacity volume is crucial during a critical shortage, when the flow in the stream is less than the required take-up.

In most SWRs, the flood protective volume (retention) is not controllable. It is defined by the highest operating level and the maximum level that is reached at the design of flood flow through the reservoir. The transformation effect of the retention volume on a flood wave is considered only if more accurate hydrological databases are available.

We quantified the volumes of the individual spaces and the volumes of their flood levels according to the characteristics of the SWRs.

3.3   Building of Small Water Reservoirs

The main advantages of SWRs are the simplicity of their building and the fact that there is no difficulty in regard to the water source. These factors allow them to be built in the upper parts of basins and wherever there are acceptable geological and morphological conditions and at least a small water source.

When deciding on the placement of larger reservoirs, we must carefully consider the placement from a safety point of view. We mean not only the dimensions of the dam itself and the objective of safe maximum flood inflow to the valley, but also, for example, the effect of catastrophic floods on the cascade of reservoirs and the particular case of side reservoirs, which close part of the floodplain area; in the event of a major flood there is a risk of flooding over the dams and water inundating the SWRs.

3.4   Water Utilization Planning of Small Water Reservoirs

The decisions to build SWRs depended on possible assurances of sufficient supplies of water for the volume created for the SWR. This question was solved by water utilization planning for the SWRs. The steps in this planning are defined in the Slovak technical standard [4], whereby the content and the reliability of water utilization planning is chosen according to the importance and purpose of the SWR design.

Water utilization planning of an SWR consists of a set of considerations, such as numerical (preferred nowadays) and graphical solutions dealing with the regulation of the outflow from the reservoir in terms of the quantitative water balance. These solutions lead to a water management plan for the SWR, which defines the method and safety conditions to ensure the water requirements and purposes of the SWR are met.

The term "water utilization planning" describes a set of calculations and graphs that are used for:

- determination of the accumulation and retention volumes of the SWR to fulfill the required functions and purposes,
- investigation of the optimal use of the SWR according to its volume,
- definition of the capacity of the operational parameters (safety spillway, water withdrawal, discharge),
- assurance of the required water management by appropriate manipulation of all the relevant items,
- determination of the impact of the SWR on the discharge process and other waterworks downstream.

The basis for water management in SWRs is the design of the water volume in the storage zone. This design is based on the simple balance of inflow and water consumption. The designs were usually made in steps of 1 month (in some cases 1 or 2 weeks). New models allow decreasing of the steps to 1 day, or even to 1 h. For the appropriate design of water utilization, planning of the following is necessary:

- requirements for the purpose of the SWR,
- SWR characteristics (line of flooded area, line of volume),
- hydrological data,
- water losses (evaporation, seepage, infiltration),
- other data.

Appropriate design of the storage function of a reservoir can be described as:

- determination of the needed storage volume for existing inflow and outflow,
- determination of possible withdrawal from the reservoir in terms of the particular storage volume and existing inflow into the reservoir.

In general, it is possible to express the water utilization balance, without considering the water losses, as:

$$V_t = V_t - \Delta t + \left(Q_p - Q_o \right) \Delta t \tag{1}$$

where V_t is the volume of water in the storage zone at the end of the particular time interval (m^3), Δt is the time interval (e.g., hour, day, week, month), $(V_t - \Delta t)$ is the volume of water in the storage zone at the beginning of the particular time interval (m^3), Q_p is the average value of inflow in the particular time interval (m^3), Q_o is the average value of outflow in the particular time interval (m^3).

The outflow consists of the guaranteed flow (Q_z) in the stream (river) downstream of the reservoir, the evaporation (E), and the sum of seepages (F), together with some other water losses specific for some types of SWRs. In general, in our calculations, we always include these losses, which can essentially influence the water utilization balance of the SWR. The water utilization balance can be expressed as:

$$V_t = V_t - \Delta t + Q_p \Delta t - \left(Q_o + Q_z + E + F \right) \Delta t \tag{2}$$

The guaranteed flow is not the real water loss, but it increases the discharge downstream of the reservoir. We can consider it as a loss in the case of water used for irrigation purposes. It is necessary to consider the loss of water by evaporation, which is possible to preliminarily calculate from the estimated average surface water level area of the future SWR. Loss by seepage through the dam has to be considered in cases of higher volume, as is the minimal discharge in the recipient.

To calculate the water storage volume, in this simplified solution, the series of average monthly discharges from April to October with the assurance of a total inflow of 90% or more (according to discharge repetition) is sufficient. The second important part of the water balance is the irrigation water need. We can determined this according to Branch technical standard ON 83 0635 [5] for a so-called standard dry year.

3.5   Water Losses from Small Water Reservoirs

Loss in the water utilization balance of the SWR is water that flows from the reservoir without our influence – that is, we have no effect on this parameter.

Losses in the water utilization balance of the SWR occur for the following reasons:

- evaporation,
- water that fills the pores in the bottom and the surroundings of the SWR,
- water leaking through the dam,
- water seepage through the bottom,
- water leakage through leakage of the closures.

As well as these losses, for the practical calculation of the water utilization balance of an SWR we also add a temporary loss of water by freezing of the water in the reservoir. The necessary flow under the SWR is not counted toward losses, but it is calculated for the overall water utilization balance of the SWR.

The loss of water by evaporation from the water surface depends on the temperature, the vapor tension in the air, the wind velocity, and the surface area of the SWR. The calculation of loss by evaporation for a given site is based on direct measurements or on data from the measurements of the Slovak Hydrometeorological Institute in Bratislava. Evaporation has different values for each month of the year. It is possible to calculate monthly evaporation in Slovakia according to the relevant percentage for a particular month (Table 1).

Water evaporation at the meteorological stations is usually measured from May to October. A technical standard was published in 1978, and since that time new measurements have been made and new research work has been performed. For example, for the area around Nitra, the percent evaporation is different from that published in the standard (Table 2). Even in the spring and winter months, there are clear differences from the original data for distribution, and the maximum evaporation values have also changed. Therefore, the technical standard values should be updated after revision.

In addition to surface evaporation, a large part of the water that evaporates from SWRs is in the form of evapotranspiration from aquatic and wetland plants. The

Table 1 Informative monthly evaporation (percentage) from free water surface during the year [2]

Month	I	II	III	IV	V	VI	VII	VIII	IX	X	XI	XII
Percentage of yearly evaporation	2	2	4	6	11	15	18	17	10	7	4	3

Table 2 Informative monthly evaporation (percentage) from free water surface during the year from measurements at Nitra station

Month	I	II	III	IV	V	VI	VII	VIII	IX	X	XI	XII
Percentage of yearly evaporation	0.2	0.9	5.1	10.1	15	17.2	18.1	15.4	10.3	5.5	1.9	0.3

Table 3 Approximate values for evapotranspiration of wetland plants in Slovakia [6]

Type of vegetation	Evapotranspiration (mm day^{-1})
Phragmites australis	3.2
Carex	2.2–4.5
Salix	2.4–4.8
Typha	3.2–5.7
Phragmites	1.4–6.9
Wetlands grasses	2.0–10.5
Phragmites	6.9–11.4

Table 4 Numbers and volumes of small water reservoirs in various regions of Slovakia

Region	Total need (million m^3)	Reality at 31.12.1985 (million m^3)	Number of small water reservoirs	Planned up to 31.12.1990 (million m^3)	Situation in 1997 (million m^3)	Situation in 1985 with siltation of 30% (million m^3)
West	57	29.4	101	24.2		20.58
Middle	49.7	6.7	44	36.2		4.69
East	51.3	5.5	47	45.4		3.85
Total	158	41.6	192	105.8	56.4	29.12

areas covered by wetland plants have greater water losses by evapotranspiration than by surface evaporation. The evaporation of the surface water depends mainly on the stage of vegetation development. In the literature [6], approximate values were given for the evapotranspiration of plants on the banks and in the littoral zones of selected SWRs in Slovakia (Table 3).

4 Current Situation in Slovakia

Construction of SWRs for agricultural purposes in Slovakia stopped around 1990. However, although SWR construction still continues, the purpose is different, e.g., ponds for fish breeding, water areas for recreation, and water for winter sports resorts. New reservoirs are generally private and therefore they are not included in the official statistics for Slovakia (Table 4).

To assess the development of water supply in Slovakia and to forecast future needs, we obtained available data on already built reservoirs. In the analysis of the water reservoirs built to date, we obtained information about 198 SWRs in the country. Some data are incomplete, and we will try to complete these in the near future. Figure 3 shows a planned view of a dam and its reservoir.

The dam must have the potential to be filled with runoff or to store a sufficient volume of water that will fill the reservoir between runoff events. It is essential that the dam and reservoir have sufficient depth and volume to supply water through extended periods of drought.

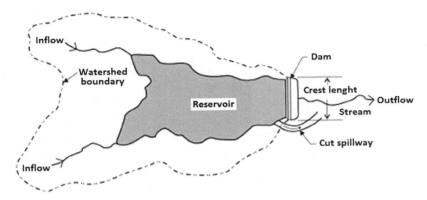

Fig. 3 Planned view of dam and reservoir (http://www1.agric.gov.ab.ca/$department/deptdocs. nsf/all/agdex4613)

The catchment area is the source of water that flows into the reservoir and allows the holding of enough water in the storage zone. The sizes of the catchment areas for the SWRs varied widely. The smallest catchment area of an SWR was found to be 1.2 km^2 and the largest (Vodná nádrž Jatov) was 117.12 km^2. The surface of the catchment is also the basis for creating the maximum design flow Q_{100} for determining the dimensions of the safety spillway and the flood wave flow through the reservoir. The maximum flow varied between 3.3 m^3 s^{-1} and 114 m^3 s^{-1} (Tulčík–Záhradné). The decisive factor for the use of irrigation water is the storage volume of the SWR, and this volume ranged from 2,100 m^3 in the smallest SWR to 3.352 million m^3 in the biggest. Differences in the water surfaces of the SWRs are not so crucial. The smallest water surface area was 1.08 ha and the largest was 74.2 ha. The total water surface area of all SWRs in Slovakia is 2,102 ha, and the average surface area is 11.24 ha per reservoir.

The width of the dam crest is also significant. In Slovakia, 18 SWRs have a dam crest width of less than 3 m. Most of the reservoirs have a dam crest width of 3 to 4 m. The width of the crest directly determines the necessary volume of earth required for the construction of the dam. Extending the width of the dam crest from 3 to 4 m increases the volume of earth in the dam by more than 25%. Therefore, only a few dams have a crest width of 5 m or even 6 m. The overall heights of dams in the SWRs have very different values, as the design of the dam is based on the shape of the valley and the required volume of the reservoir. Only 19 dams in SWRs in Slovakia have a height of less than 3.5 m. In 36 SWRs, the height of the dam varies from 3.5 to 4.5 m and a similar number of SWRs (32) have dam heights of 7.5 to 12.5 m. The biggest group (43 SWRs) has dam heights between 5 and 7 m. Several dams are even higher than 12.5 m. For a dam height of up to 6 m, the shape of the dam is a simple trapezium and for the upper dam is used a double trapezium. Extension of the ditch in the middle of the height means that there is e.g. 1 m wide pavement in the middle of the dam.

Detailed data on the construction of SWR dams in Slovakia has not been available. However, we found that most of the SWRs were built with homogeneous earthen dams and only a few have heterogeneous dams. Several dams consist of mixtures of the different soils that were available at the construction site.

Many ideas have been advanced to solve the problems of mechanical stability, imperviousness, and internal erosion in dam walls. Two main solutions are employed today; their usage is determined according to the height of the dam. For relatively low dams, homogenous cross-sections of impervious materials are complemented by a drainage mechanism that consists of sandy materials, which may collect water in the case of a possible leakage, but which avoids the erosion of the impervious materials. Alternatively, the upstream part of the dam is made of impervious material and the downstream part is made of more pervious material (Fig. 4). The following materials should be avoided: organic material, including topsoil and decomposing material; material with a high mica content; calcitic clays; fine silts; schists and shales; cracking clays; and sodic soils. Material containing roots or stones should also be avoided.

Two types of constructions are used almost exclusively for water outflow. In SWRs with a water depth of up to 3 m, a outflow is almost exclusively used for the outflow from reservoir to stream; the outflow is located above the point at which the stream enters the culvert. For bigger reservoirs, outflow chambers with sluice gates are used.

The spillway is a critical part of dam construction. Emergency spillways should be provided for all dams to carry out large flood flows safely through the embankment, unless the principal spillway is large enough to pass the design discharge without overtopping the dam. The maximum design flow (Q_{100}) for determining the dimension of the safety spillway flood wave flow through an SWR in already built SWRs, e.g., the one at Tulčík–Záhradné (Ternianka stream), is 3.3–114 $m^3 s^{-1}$ [7].

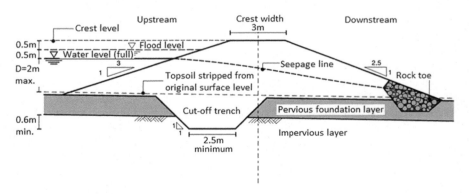

Fig. 4 Section view of dam (https://www.aboutcivil.org/imajes/Cross-section-earth-dam.PNG)

4.1   Sediments and Eutrophication of Small Water Reservoirs

From the technical and operational points of view, the main problem with the operation of SWRs is their siltation by sediments from the surrounding slopes – the products of erosion processes in the catchments. The permanent process of transport and sedimentation of the particles brought in by the inflows and the discharges from the surroundings affects the qualitative and quantitative character-istics of SWRs. The deposition of sediments has a number of negative impacts on SWRs, the most important of which are reductions of useful water volumes in the SWR, limitations on the functionality of the handling equipment, deterioration of the quality of the accumulated water, and deterioration of the ecological conditions in the surrounding landscape.

The main causes of reservoir siltation are bank erosion, internal fouling, and siltation by inflow. When estimating the rate of sedimentation in water reservoirs, it is necessary to analyze the supply of sediment from the catchment, using calculation methods that take into account the storage efficiency of the reservoir [8].

Together with sediments, various chemicals, especially nutrients (nitrogen and phosphorus), are transferred to SWRs. The result is eutrophication of the reservoir. Therefore, measures to prevent eutrophication need to be addressed in tainted reservoirs.

The most effective measures against eutrophication that are generally applicable to all types of reservoirs are measures that can reduce the concentrations of nutrients in the reservoir. Phosphorus should be limited to a level that does not support the growth of primary producers. Eutrophication is reliably restricted when the concen-tration of biologically available phosphorus entering the reservoir does not exceed a value of about 0.03 mg l^{-1}, which is the natural concentration in water from the catchment and precipitation.

The last substantial removal of sediments in SWRs in Slovakia was done before 1990. Therefore, today, the amount of sediment at the bottoms of the SWRs is significant, and generally reaches 30–40% of the storage capacity of the SWRs.

In general, therefore, Slovakia has about 35% less water storage than the amount that was intended in the SWRs. To analyze the state of reservoir siltation, we chose to examine the Bodrog and Hornád river basins according to the SWME documen-tation. For example, the Sigord Reservoir was completely revitalized in recent years, owing to failure. Table 5 shows the sediment contents of selected reservoirs within the scope of the Bodrog River Basin Management [9].

The total original water volume of the SWRs within the Bodrog River Basin Administration was 3,731,700 m³. Based on our estimation of the siltation rate, the sediment volume was calculated to be 1,328,895 m³, and therefore the approximate actual volume of water was 2,402,805 m³. Thus, about 30% of these SWRs contain sediment deposits and the water is temporarily lost when significant droughts occur. A similar situation is probably seen in the other river basins in the country. Of note, Kubinský et al. [10] mention that comparisons of historical data and research out-comes show that all the SWRs they investigated experienced gradual accumulation

Table 5 State of sediments in small water reservoirs in the Bodrog river basin (according to information from SWME)

Reservoir	Stream	Storage capacity (thousand m³)	Siltation (%)	Lost volume (thousand m³)	Residual volume (thousand m³)
Bidovce	Trstianka	30	35	10.5	19.5
Košické Olšany	Olšiansky p.	18	35	6.3	11.7
Seňa	Belžiansky p.	50	35	17.5	32.5
Trstené p/Hornád	Trstenský p.	28	35	9.8	18.2
Vyšná Kamenica	Svinický p.	24	35	8.4	15.6
Sigord–Kokošovce	Delňa	95.3	35	33.355	61.945
Tulčík–Záhradné	Ternianka	12	35	4.2	7.8
Šemša	Šemšianský p.	15	35	5.25	9.75
Oreské	Turský p.	52	35	18.2	33.8
Vyšná Rybnica	Okna	346	35	121.1	224.9
Pozdišovce	Pozdišovský p.	254	35	88.9	165.1
Klčov	Klčovský p.	54	35	18.9	35.1
Žakovce	Vrbovský p.	530	35	185.5	344.5
Štrba	Mlynica	30	35	10.5	19.5
Vrbov III.	Vrbovský p.	57.4	35	20.09	37.31
Rakovec	Batovec	28	35	9.8	18.2
Tovarné	Tovarniansky p.	231	35	80.85	150.15
Nový Ruskov	Drienovec	85	35	29.75	55.25
Veľké Ozorovce	Číža	1,105	40	442	663
Zemplínska Teplica	Číža	162	40	64.8	97.2
Parchovany	Manov kanál	28	40	11.2	16.8
Nižný Žipov	Žipovský p.	178	30	53.4	124.6
Byšta	Byšta	152	25	38	114
Stropkov	Chotčianka	95	20	19	76
Hervart–Klušov	Tisovec	72	30	21.6	50.4
Total volume (thousand m³)		3,731.7		1,328.895	2,402.805

SWME Slovak Water Management Enterprise, *p.* creek

of sediment load, to a greater or smaller extent. To obtain a state strategy for drought protection, it is necessary to examine not only the construction of new sources but also the reconstruction of already built reservoirs.

5 Current Problems in the Management of Small Water Reservoirs

The state of SWRs built in Slovakia and surrounding countries is not always optimal. Several of these SWRs are now no longer operative because of their poor technical condition, and some have been damaged during floods; also, their current purpose is different from the purpose for which they were designed.

The difficulties with the current state and use of SWRs can be summarized as:

- problems in water management,
- technical problems,
- ecological problems,
- economic problems,
- property ownership problems,
- legislative problems.

The water management problems are caused by changes in the design parameters of reservoir construction, owing to insufficient initial surveys or owing to non-compliance with the care specified for operating the reservoir. The water management problems are mainly related to changes in the real values of the maximum and minimum flow rates and also to changes in the quality of the inflowing water.

An important element in the filling of SWRs is sediment, which:

- is related to soil erosion,
- contains many nutrients and toxic substances,
- reduces the usable volume of the SWR,
- changes the functionality of the SWR by its deposits.

Other problems related to sediments in SWRs are changes in volume, which lead to changes in hydraulic function; areas (in the case of decreased water levels) that rapidly become silted; and shallow areas that support the growth of wetland vegetation [11].

The most dangerous sediments are deposits near a safety spillway. Wetland plants develop rapidly in this space, and this can significantly affect its safety and thus the safety of the whole dam and reservoir. Removal of sediments costs much more than regular maintenance expenses.

There are technical problems related to the management of the SWRs in Slovakia. As mentioned above, their ownership has changed in the past 20 years. But substantial numbers of owners have not received enough money from the state for the necessary routine repairs, so, in general, the situation of the SWRs has worsened. However, because of recurrent floods in recent years, rapid repair of SWRs is required.

The need for repair has arisen because most of the SWRs were built in the 1960s and 1970s and their present state corresponds to their age. The most common technical problems are:

- safety spillways that are clogged and enclosed
- poor state of discharge equipment
- non-sustainable coastal vegetation
- wet places below the dam
- damage to the dam caused by the growth of bushes and trees
- uneven dam crests.

6   Conclusions

The value of water in nature and in society is becoming higher and higher, and its absence and short-term excesses during flood are regarded as natural disasters. Humans always try to modify the landscape to avoid both such disasters, because of the high costs of the damage caused. Floods have occurred in urban and rural areas many times in the past, but the damage was not as severe as it is nowadays.

One possibility of avoiding flooding is to decrease the peak flow upstream of SWRs. It is possible to do this by flooding part of the landscape or increasing water retention in the reservoirs. The water stored during a flood can be used during the dry season. Therefore, the SWR is now an important element of water management plans in catchment management, and small towns and villages also have an interest in this topic. We can assume that the numbers and importance of SWRs will increase during this century.

The design and management of SWRs in Slovakia has not been a matter of interest. Most of the experts who designed and built the reservoirs have now retired or are employed in different fields of the building industry. A whole generation of such experts is now missing. It is important to motivate the older experts to teach and prepare a new generation of engineers and experts.

7   Recommendations

New research has provided information about the importance of SWRs in augmenting the availability of water to meet increased demands in the landscape and for agriculture that will arise owing to climate change in the near and far future. It is necessary to confirm the function, utility, and calculated volume of water in SWRs. Therefore:

- engineers (designers) of new SWRs will need appropriate and secure rules for their design, e.g., technical standards that take account of the latest knowledge,
- farmers should include irrigation in their production plans to produce stable yields each year.

If the above-mentioned recommendations are applied, SWRs will maintain their place in the landscape, which is now slowly being lost. It is important to change the quantitative categorization of SWRs, as well as to renew technical standards for their design and provide funding for monitoring and reconstruction.

Acknowledgments This chapter was supported by the following grants and projects:

- APVV - Slovak Science and Research Agency-16-0278; Use of hydromelioration structures for mitigation of the negative extreme hydrological phenomena effects and their impacts on the quality of water bodies in agricultural landscapes
- KEGA - Cultural and Educational Grant Agency 028SPU-4/2017; Monitoring of Elements of Environment – practical course.

References

1. Jurík Ľ, Húska D, Halászová K, Bandlerová A (2015) Small water reservoirs - sources of water or problems? J Ecol Eng 16(4):22–28
2. STN (Slovak technical standard) 73 6824: Malé vodné nádrže (Small water reservoirs) (in Slovak)
3. Marton D, Paseka S (2017) Uncertainty impact on water management analysis of open water reservoir. Environments 4(1):10. http://www.mdpi.com/2076-3298/4/1/10/htm
4. STN (Slovak technical standard) 73 6815: Vodohospodárske riešenie vodných nádrží (Water utilization planning of reservoirs) (in Slovak)
5. ON (Branch technical standard) 83 0635: Potřeba závlahové vody při doplňkové závlaze (Need of irrigation water in supplemental irrigation) (in Czech)
6. Jurík Ľ (2013) Vodné stavby (Hydraulic structures). Slovenská poľnohospodárska univerzita. Nitra (in Slovak)
7. Klementová E (2001) Malé vodné nádrže Slovenska (Small water reservoirs of Slovakia). In: Kamenský J (ed) Malé vodné diela a alternatívne zdroje energie. Proceedings of international conference LITERA Košice (in Slovak)
8. Vrána K, Beran J (2005) Rybníky a účelové nádrže (Ponds and water reservoirs) ČVUT. Praha (in Czech)
9. Prevádzková dokumentácia (Operation documentation) SVP, š.p., OZ Košice, Správa povodia Bodrogu, Trebišov (in Slovak)
10. Kubinský D, Weis K, Fuska J, Lehotský M, Petrovič F (2015) Changes in retention characteristics of 9 historical artificial water reservoirs near Banská Štiavnica, Slovakia. Open Geosci 7(1):880–887
11. Verebová (Šoltísová) A (2010) Assessment of the influence of the erosion - transport and sedimentation processes in the catchment basin for the small water reservoirs. Doctoral thesis. Košice (in Slovak)

Part III
Soil and Water

Interaction Between Groundwater and Surface Water of Channel Network at Žitný Ostrov Area

P. Dušek and Y. Velísková

Contents

Abstract Surface water-groundwater interaction is a dynamic process which can be influenced by many factors most associated with the hydrological cycle. Besides the fluctuation of surface water and groundwater levels and their gradient, this interaction is also influenced by the parameters of the aquifer (regional and local geology and its physical properties). The next significant factors are precipitation, the water level regime of rivers or reservoirs in the area of interest, and last but not the least the properties of the riverbed itself. The investigation of the interaction between the surface water and groundwater was applied utilizing modern numerical simulations on the Gabčíkovo-Topoľníky channel, one of the main channels of irrigation and drainage channel network at Žitný Ostrov. Žitný Ostrov area is situated in the southwestern part of Slovakia, and it is known as the biggest source

P. Dušek and Y. Velísková (✉)
Institute of Hydrology, Slovak Academy of Sciences, Bratislava, Slovakia
e-mail: veliskova@uh.savba.sk

A. M. Negm and M. Zeleňáková (eds.), *Water Resources in Slovakia:*
Part I - Assessment and Development, Hdb Env Chem (2019) 69: 135–166,
DOI 10.1007/698_2017_177, © Springer International Publishing AG 2017,
Published online: 14 December 2017

of groundwater in this country. For this reason, experts give it heightened attention from different points of view. The channel network was built up in this region for drainage and safeguarding of irrigation water. The water level in the whole channel network system affects the groundwater level and vice versa. With regard to the mutual interaction between channel network and groundwater, it has been necessary to judge the impact of channel network silting up by alluvials and the rate of their permeability to this interaction. The aim of this contribution was to collect the available data from the area of interest for simulation of real and theoretical scenarios of interaction between groundwater and surface water along the Gabčíkovo-Topoľníky channel. The obtained results give valuable information about how the clogging of the riverbed in the channel network influences the groundwater level regime in the area.

Keywords Channel network, Groundwater, Interaction, Numerical simulation, Surface water, Žitný Ostrov (Rye Island)

1   Introduction

Management of surface water and groundwater and their sources requires a quantitative understanding of the interaction between river and aquifer [1]. Fluctuation of the surface water can significantly affect the water table regime of groundwater in the surrounding aquifer. This effect is primarily seen in lowland areas where there is less variation in the subsurface geological heterogeneity. From this point of view, the area of Žitný Ostrov is very interesting. In this area, there is an existing network of irrigation and drainage channels which was built because of the use of land for agricultural purposes and at the same time for the protection of this area from floods, due to the very low terrain slope of the whole Žitný Ostrov area.

Numerical models are an all-round, often used, and (assuming their correct application) accurate tools for studying the interaction between surface water and groundwater. The investigation of this interaction in the selected area of interest was carried out using a three-dimensional simulation of groundwater flow applying available measured data and data acquired from other institutions.

2   State of the Art

Interaction of surface water and groundwater happens in multiple zones. In general, we recognize three zones corresponding to the motion of water, specified as local flow, intermediate flow, and regional flow. Local flow is impacted by short-term climatic changes, while regional flow is separated from the short-term changes [2, 3]. Interaction of surface water and groundwater is strongly affected by the morphology and geology of the investigated area [1]. Interaction can have different forms, related to the exchange of water between aquifer and rivers, lakes, wetlands, seas, and oceans. Every form can have different types of water flow, and the processes occurring in them affect the chemical and biological cycle of nutrients [4].

The interaction between channel and aquifer happens on the boundary of the water body and the aquifer or the aquifer's unsaturated zone. This is defined by the difference in the water table elevations of both systems and by the physical properties of the subsoil in the area of interaction [5]. In 1856 Henry Darcy proved how can the head difference between two points and a clogging layer (defined by saturated hydraulic conductivity) affect the groundwater flow. This equation is defined as Darcy's law as follows:

$$q = -k \frac{\Delta h}{\Delta l} \tag{1}$$

where k is the coefficient of saturated hydraulic conductivity, q is the flow rate, h is the height of the level, and l is the distance between the measured points. Darcy's law applies to a natural porous environment where local accelerations in the fluid are much smaller than the viscous forces that are often observed for a Reynolds number greater than 10 [6].

Interaction is influenced by the following parameters: river sediments, river geometry (cross-section profiles, flow direction), channel length, water level, and groundwater level. Interaction is also influenced by spatial changes in saturated hydraulic conductivity values and fluctuation of levels in the channel and the aquifers. Surface and groundwater interactions take place in three basic cases. The water from the aquifer flows into the channel (gaining channel), water flows from the channel into the watercourse (losing channel), or a combination of both. The losing channel can either be fully connected (saturated zone) or disconnected (with an unsaturated zone between the channel and the groundwater level) [7].

The interaction between aquifer and surface flow is continuous, and the flow from/to the channel and from/to the aquifer can vary depending on the difference in the level of the water. Strong level fluctuations may occur due to torrential rainfall, snow melting, or drainage of a weir or a reservoir. In the case of high or flood flows with a relatively short duration, it may temporarily change the direction of flow of water between the channel and the aquifer, forming a bank storage [8], the volume of which returns into the system after the surface water level falls.

Interaction of the surface channel with the aquifer is quantified by the increase in lateral seepage. This is a value spatially distributed along the length of the channel. For specifying the lateral seepage, we use the continuity equation in this form [9]:

$$\frac{\partial S}{\partial t} + \frac{\partial Q}{\partial x} = q \tag{2}$$

The lateral increase of inflow q on the right side of the equation is based on two parts: the surface flow from the land q_{sw} and the inflow from or to the streambed q_{gw}.

$$q = q_{sw} + q_{gw} \qquad (3)$$

The amount of this seepage defines the scale of effect of the surface channel on the water table regime of the groundwater and vice versa. The ratio between the surface and subsurface part of q is variable and depends on the local climatic, geological, and geographical conditions. When and what parts are needed in the computations depends on the target of the computations in the channel or river network and on the conditions in the area of interest [9].

2.1   Streamflow

The fluid flow is divided into the following:

- Unsteady (nonstationary), where the discharge Q, mean cross-section velocity v, and depth y depend on the length coordinate x and time t.
- Steady (stationary) flow is characterized by hydraulic properties which are not time-sensitive. The cross-section velocity is only a function of the length coordinate, as it changes along the channel or remains constant. Steady flow can be uneven, with velocity changes along the length, or even, with a constant velocity of flow [10].

Open channels are categorized based on their flow profile as:

- Prismatic channels with constant geometrical properties along the length
- Non-prismatic channels with changing flow profile along the length, while the changes in shape can be mathematically defined
- Natural channels with non-regular shapes of the flow profile, with changes along the length [11]

The flow in natural channels which have irregular flow profile shape or slope along their length is uneven. The most efficient way to assess the surface water table is to use the section method. The calculation is based on the Bernoulli equation.

2.2   Groundwater Flow

In the case of groundwater flow, this section will focus on the steady flow of groundwater with free surface water, as this is the most common case for solving the course of groundwater. Steady flow is characterized by independence of the filtration velocity vector from time, i.e., it is only a function of location in the flow area. Unconfined flow is filtration without pressure, as the aquifer is directly in contact with the unsaturated zone. The calculation is then based on the principle of volume conservation and Darcy's law.

2.3   Boundary Conditions for Groundwater Flow

Basic differential equations of steady filtration flow are not sufficient to solve specific groundwater flow problems. The problem has to be characterized by boundary conditions. In solving the problems described by differential equations, the boundary conditions are assigned by the value of the function (or its derivation) whose solution for the territory outside the boundary we are looking for. The main task of groundwater hydraulics is to investigate the pressure (commonly expressed by groundwater level, piezometric height). The boundary condition, which is set at piezometric altitude level, is called the boundary condition of the first order or the Dirichlet's boundary condition ($h = h(x,y)$ or $h =$ const). The boundary condition determined by piezometric height or gradient of groundwater is the boundary condition of the second order or Neumann's boundary condition, as it expresses specific seepage across the boundary of the filtration area. The linear combination of conditions of the first and second orders is the condition of the third order or mixed boundary condition $q = f(H)$. In some cases, other dependencies of boundary conditions can be set, e.g., depending on time or flow type. The basic type of delimitation of the aquifer consists of river banks, water reservoirs, canals, and impervious boundaries at the contact of the environment with poorly permeable or impermeable rocks schematized as impermeable. In unconfined aquifers, the upper boundary is an unconfined phreatic (groundwater) water table. Also, there may be an internal boundary, e.g., the boundaries of zones with different permeability, or the existence of pumping or infiltration devices.

2.4   Filtration Properties of Soils

An environment which is continuously filled with cohesive or incoherent soils is called a porous environment. The properties of this environment associated with groundwater flow (the filtration properties) depend on the mechanical properties of the soil and on the flow of liquid in this environment. The soil mechanical properties are summarized in a grain line which graphically represents the percentage of grains of a certain size and indicates to what extent the soil is homogeneous. This line is the result of granulometric soil analysis. It indicates as a percentage of the weight ratio of particles of a given diameter to the weight of the entire soil sample.

The basic soil mechanical characteristics are:

- The curve of granularity and the amount of grain nonuniformity, defined by the above ratio
- Volume porosity defined as the ratio of pore volume V_p to total soil volume V
- The permeability factor, which expresses the ability of the porous environment to allow liquid or gas to pass through it
- The coefficient of saturated hydraulic conductivity, which expresses the ability of the soil to drain water

The coefficient of saturated hydraulic conductivity is determined primarily using direct (laboratory and field) but also indirect (computational) methods. Indirect empirical methods of specifying the coefficient of saturated hydraulic conductivity are based mainly on the results of the granulometric analysis. A significant number of empirical equations exist, e.g., Hazen I and II; Orechová; Seelheim; Zieschang; Beyer; Zauerbrej; Kozeny I and II; Zamarin I, II, III, and IV; Schlichter I, II, and III; Krűger; Palagin; Carman-Kozeny; Špaček; Beyer-Schweiger; etc. Their use is limited however by the conditions of validity [12].

The groundwater flow equation includes a dependent variable, i.e., the height of groundwater calculated using the model, and independent variables including spatial coordinates x, y, and z, time t, and other parameters. These include parameters of material properties describing the hydraulic characteristics of the porous environment and the hydrological parameters representing the load on the model environment [13]. The parameters of material properties needed to define the continuous numerical simulation of the acquired environment include:

- Specific storativity S_s describing the volume of water which outflows from a unit volume of aquifer while the head drops by a unit height
- Specific yield S_y the volume of water which outflows from a unit area of aquifer while the head drops by a unit height

Because it was not possible to acquire the values of specified storativity and specified yield of the geological materials on site, general values of these parameters for appropriate geological materials were used, taken from the literature [14, 15].

3 Mathematical Methods of Solving the Groundwater Flow Problem

3.1 Analytical Methods

Partial differential equations describing the three-dimensional nonstationary flow of groundwater are hard to solve in their basic form using analytical methods. Some assumptions are therefore made to simplify the problem (e.g., neglecting the vertical component of the flow), enabling at least an approximate solution for the specific problem. Most of the time, the analytical solution is only possible for cases where the area of interest has a simple shape, the environment is homogeneous and isotropic, and the starting boundary conditions are set as constants. One of the basic problems which can be solved using the analytical method is inflow and outflow from a well. Theis [16] solution of nonsteady flow from the river to aquifer impacted by pumping from a vertical well is based on an array of assumptions, e.g., that the river streambed is based on the impermeable layer and that between the river and the aquifer and there is no divide with different hydraulic properties. In 1965 Hantush included a clogging layer placed in the streambed [17].

The basis of analytical methods is the theory of potential laminar flow. The velocity potential is a product of saturated hydraulic conductivity and the piezometric head with a negative sign, whereby partial derivations in the direction of coordinates define the components of the filtration velocity vector in this direction; it is a function of the location of the point in the filtration area.

3.2 Numerical Models

With the development of numerical mathematics and computing, numerical methods have been introduced in solving the problems described by partial differential equations. The most commonly used numerical methods are the finite-difference method (FDM) and finite element method (FEM). The advantage of simulation models is that they do not require a regular shape of the boundary of the area to be solved and the environment may not be homogeneous or isotropic. The next one is that different boundary conditions may apply to different parts of the boundary or there may be sources and sinks with a time variable value of inflow or outflow. Modeling of surface water and groundwater should not be applied separately, because they are interconnected components of the hydrological cycle and ecosystem, especially in the case of river basins with the occurrence of large deposits of river sediments [18]. It is important to understand the river-aquifer interaction for integrated water resources management [19]. Possible problems with defining input parameters of the model can be solved through so-called stochastic modeling, where model parameters are randomly changed in series for a large number of numerical simulation implementations. Subsequently, it is possible to select the most suitable parameter set that most closely corresponds to the hydrological and hydrogeological ratio in the area of interest [20, 21]. It is important to distinguish between numerical simulations on the local or regional scale. In practice, in terms of interaction modeling, the scale is divided into four main subgroups [22]. This produces a point scale where it is possible to precisely quantify the interaction between flow and water based on physical environmental parameters. The modelled environment is usually only a part of an aquifer and a river. Most frequently, one-dimensional or two-dimensional models, predominantly in the cross section of the flow and the aquifer, are used for this problem. The most frequently used models are HYDRUS [23]. For the numerical simulation MODFLOW, an application of HYDRUS code exists for this environment as one of the optional packages (HYDRUS package) [24]. The local scale includes the entire cross section of the flow and its length, along with the surrounding drainage and local geology. The subbasin scale from a hydrological point of view describes a closed system, but groundwater flow in most cases is not affected by the boundary of the subbasin. On a regional scale, problems with inconsistency in underlying data may arise, mainly from different sources and from different time periods. The parameters of anthropogenic activity (water structures, surface manipulation, industry, agriculture) also enter into regional models. Increasingly, integrated modeling systems are being created for regional modeling tasks including all components of the hydrological cycle for a given region [25].

TRIWACO is a computational system for the quasi-three-dimensional simulation of groundwater flow [26], based on the finite element method (FEM). This program was built to solve the groundwater flow in the horizontal plane. The model is capable of simulating groundwater flow in several permeable layers (aquifers) separated by semi-pervious layers. The main advantage is the modular structure and flexibility. The environment contains several separate programs which use their own specific input files and which produce output files that can be analyzed by different applications for data management. The program is capable of simulating both steady and unsteady groundwater flow. At the same time, it is possible to simulate the zone of unsaturated flow [27].

AEM (*analytical element method*) was created at the end of the 1970s by Otto Strack at the University of Minnesota [28]. This method skips the discretization of the area of interest in the network of elements. Elements of the surface water network are inputted directly using hydrologic boundary conditions. Traditionally, the superposition of analytical functions has been considered to be limited to a homogeneous groundwater collector of constant permeability. With appropriate application, the method of analytical elements is applicable also to heterogeneous environments and for confined and unconfined flow [28]. AEM is applied in several simulation models, namely, MODAEM [29], WhAEM [30], and GFLOW [31].

The *MODFLOW* model [32, 33] is capable of solving the simulation of both confined and unconfined aquifers. Horizontal flow in individual aquifers is solved separately. The interaction of the layers is expressed by the vertical drop from one layer to the next, which is either directly entered or is quantified by the vertical hydraulic conductivity of the adjacent layers. Other physical and hydraulic environmental parameters entering the model are horizontal hydraulic conductivity, storativity, and porosity. To solve the basic differential equation describing the flow of groundwater, the finite differential method is used with the nodes located in the centers of the rectangular grid. The flow area is clearly defined by the position of the lower and upper edges of each aquifer and the boundary of the area of interest. At each node of the network, it is possible to specify boundary conditions of the first to third order. The following modules are available for the user of the program: well, drainage, evapotranspiration, infiltration, channel without flow control, channel with specified flow, and general pressure boundary condition. The model simulates steady as well as the unsteady flow of groundwater. The basic outputs are maps of isolines of hydraulic height and maps of isolines of the reduction in hydraulic heights for the individual aquifer layers, in the case of the transient flow for individual pressure and time levels. A water balance can be evaluated for a given pressure condition, time step, and defined location. This means that when groundwater directly communicates with surface water, it is possible to find out how much water is drained by the recipient in the defined section or how much water is infiltrated from it into groundwater. These data are of great importance for the calibration of the model.

Another part of MODFLOW is the PEST (parameter estimation) program [34]. This is an automated estimation of numerical simulation parameters. During the PEST parameter estimation process, it looks for optimal parameter values for which the sum of squared deviations between observed and calculated values is minimal. Parameter estimations are governed by the Gauss-Marquardt-Levenberg algorithm. The modular MT3D program, dealing with the transport of contaminants, was developed in 1990 and was further developed until the MT3DMS version [35]. It simulates changes in contaminant concentrations in groundwater, due to advection, dispersion, diffusion, and chemical reactions. It also deals with the transport of contaminants from external sources, wells, drains, watercourses, and surface pollution. MODFLOW results can be used as input data for the MODPATH program [36] to calculate particle trajectories in a given area.

4 Description of the Area of Interest

The Žitný Ostrov area is located in the southwestern part of Slovakia, on the border with Hungary. Its boundaries are formed in the south by the banks of the Danube, in the north by the branches of the Little Danube, and on a short stretch in the east, it is bounded by the river Váh. The territory belongs geographically to the Low Danube Plain. The situation of Žitný Ostrov within Slovakia is schematically shown in Fig. 1. The island has an elliptical shape, its length is 84 km, the width ranges between 15 and 30 km, and the total area is 1,885 km^2 [37]. With its dimensions, this island is the largest river island in Europe. The territory of the island is of a flat character. The longitudinal slope of the area reaches only 0.25‰ [38], with a decreasing tendency in the southeast direction. This small slope was created by the gradual deposition of gravel, sand, and flood sludge. The highest point on the Žitný Ostrov area is located near Šamorín (134 m above sea level), and the lowest is the area at Komárno (105 m above sea level). The altitude of the terrain in the locality is 108.4 m a.s.l. up to 121.5 m a.s.l. The terrain is lower from the Danube watercourse to the Little Danube and at the same time from the west or northwest boundary of the territory to the east or southeastern boundary.

The area of interest is geologically included in the area of the Holocene floodplain of the Žitný Ostrov. The geological structure is characterized by the emergence of fluvial sediments. In their overburden, they are strata of fluvial

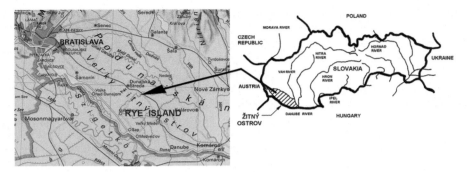

Fig. 1 Location of Žitný Ostrov within Slovakia

sediments of the Quaternary, whose deepest positions reach a thickness of 200–500 m [39, 40]. Hydrogeological ratios in the studied area are determined by the geological structure and the Danube. The poorly permeable Danube floodplains are filled with water and form a massive phreatic horizon. The groundwater level is affected by the Danube water fluctuation, and the difference between the lowest and highest observed groundwater levels is 250–600 cm. The groundwater level is dependent on the water level in each channel and varies according to the overall water level in the drainage system linked to the Little Danube. In the core of the island, there are sandy sediments reaching a thickness of up to about 300 m in the central, tectonically falling part of the island. Gravel sediments range from 50 or 70 cm below the surface of the terrain (in the central and upper parts of the island) up to 6 or 8 m (mostly in the lower part of the island) [41]. Due to its predominantly gravel foundation, Žitný Ostrov is an important collector of groundwater which is extensively used as drinking water.

The geological structure of the Žitný Ostrov interface is characterized by great heterogeneity. Gravels or sandy gravels are covered by younger alluvial loamy to loamy sand sludge sediments, less sandy clay, and clay sediments. There are predominantly clays or sand in the subsoil of 8–20 m from the Quaternary period. The hydrogeological conditions here are influenced by the great thickness of the sandy gravel sediments of the Quaternary. Depending on the grain composition and the sand fraction, the values of saturated hydraulic conductivity range from 10^{-2} to 10^{-6} m s^{-1} [42]. The flow rate of the drained collectors is very high. The River Danube is the source of constantly replenishing groundwater supplies; water infiltrates the rock environment all year round.

The Danube on the territory of Žitný Ostrov creates an extensive branch system. The natural character of the river is altered by embankments and equalizing parts of the watercourse. This has also changed the natural hydrological conditions: the Danube's branches and meanders are separated from the main stream by the embankments. The current hydrological conditions are strongly influenced by the building of

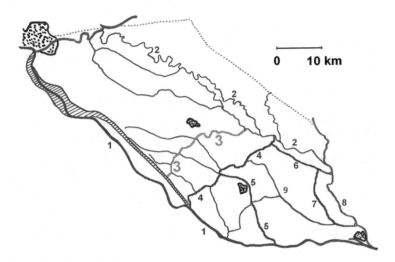

Fig. 2 Scheme of the Žitný Ostrov channel network (1, Danube; 2, Little Danube; 3, Gabčíkovo-Topolníky channel; 4, Chotárny channel; 5, Čalovo-Holiare-Kosihy channel; 6, Aszód-Čergov channel; 7, Čergov-Komárno channel; 8, Váh; 9, Komárňanský channel) [38]

the Gabčíkovo water management project (VD Gabčíkovo). The channel network of Žitný Ostrov (see Fig. 2) consists of six main partially interconnected channels: Gabčíkovo-Topolníky channel, Chotárny channel, Čalovo-Holiare-Kosihy channel, Aszód-Čergov channel, Čergov-Komárno channel, and Komárňanský channel [43]. The total area covered by the current drainage system is 1,469 km². The area of drainage with a built-up channel network is 1,252 km². The total length of the channel network is almost 1,000 km. Its density is about 1 km/1.25 km². The most important channels in the drainage system are the Chotárny and Gabčíkovo-Topolníky channels, which are connected to the Little Danube. The Gabčíkovo-Topolníky channel is connected with the Danube by an inflow structure and leads to the Klátovský branch of the Little Danube. The Chotárny channel is supplied with water from the VD Gabčíkovo surplus water channel. Its tributaries are the Gabčíkovo-Nárad channel, Čilížsky channel, Jurová-Veľký Meder channel, Kračany-Bohelov channel, Belský channel, and Býč channel. From the Chotárny channel, water is used for irrigation through another network of channels, connected to the network by floodgates.

The area of interest for further measurements of hydrodynamic parameters (velocity, flow rate, conductivity, thickness, and composition of bottom sediments) and for modeling of surface and groundwater interactions was selected. It is bounded by the Gabčíkovo-Topolníky channel, the Klátovský branch of the Little Danube, and the left-side seepage channel of the VD Gabčíkovo supply channel. The choice of the territory was influenced by the appropriate soil structure with regard to the assessment of the interaction of surface and groundwater and by the availability of basic background data and good accessibility to the actual watercourses in the case of in situ measurements.

The *Gabčíkovo-Topolníky* (*S VII*) *channel* is part of the drainage network of Žitný Ostrov. It takes off the surface and seepage water from the area of Gabčíkovo to the Klátovský branch, from there to the Little Danube or via the Bele-Kurti (Belský) and Palkovičovo-Aszód channels to the Aszód pumping station (Little Danube). This channel was built in the 1960s as part of the "Drainage of Central Žitný Ostrov and Medzičiližie" investment scheme, the purpose of which was to improve the water management of the area by drainage and to bring water for the irrigation of agricultural areas. At present, it provides for the removal of excess seepage water from the left-side seepage channel of VD Gabčíkovo, with the possibility of overgrading from the upper section of the VD Gabčíkovo navigation chamber via the SHPP VII-Malé Gabčíkovo. It is supplied with water from the left-side seepage channel of VD Gabčíkovo between the villages of Baka and Gabčíkovo. The inflow facility for the Gabčíkovo-Topolníky canal is situated in the left-hand barrier of the derivation channel in front of the navigation chambers. It consists of a takeoff with a drop to the elevation of 126.00 m and supply ducts $2 \times$ DN 1,400 mm. The water collected by the inflow facility is used in the Malé Gabčíkovo small hydroelectric power plant – S VII. The canal enters the Klátovský branch of the Little Danube between the villages of Topolníky and Trhová Hradská. Its length is 28.7 km. There are two water meter stations: the Gabčíkovo station at river km point 25.7 with a basin of 10.7 km^2 and the Topolníky at river km point 0.30 with a basin of 349.27 km^2, both operated by the Slovak Hydrometeorological Institute. The Gabčíkovo-Topolníky channel has side channels called AVII, BVII, and CVII.

The left-side seepage channel of the supply channel (ĽPKPK) with a length of 16.65 km is supplied with water from the left-side seepage channel of the Hrušov reservoir and directs the seepage from the left-side dike to the pass-through at river km point 4.0. There are outlets supplying water for irrigation at river km points 2.914 (A VII), 8.800 (B VII), and 14.588 (C VII).

The *Vojka-Kračany channel* (*A VII*) is supplied with water through an outlet at river km point 2.912. The cross-sectional profile of the channel is trapezoidal; the channel banks are reinforced with natural vegetation cover. The channel bottom width is 2.0 m, and the slope is 1:2. The channel drains surface water and seepage from the area of the left-side seepage channel of VD Gabčíkovo. The water is not polluted by wastewater and is therefore suitable for improving the supply of the Hroboňov ponds.

The *Šulany-Jurová channel* (*B VII*) is supplied through a collection facility at river km point 8.800 from LPKPK. The transverse profile of the channel is trapezoidal in shape, and the channel banks are fortified with vegetation cover. The bottom of the channel has a width of about 3.3 m, and the slope is 1:2.2.

The *Trstená-Baka channel* (*C VII*) is supplied from the LPKPK through a collection facility at river km point 14.588. The transverse channel profile is trapezoidal in shape.

The floodgate at the S VII and A VII channel junction serves to regulate channel levels and to provide water supply to the Boheľovský channel. In summer mode, the level is automatically maintained at a maximum operating level of 112.50 m. The

structure is not technically ready for winter operation, so all the restraining structures are open during the winter and water passes freely through the structure.

The hydrological regime of Žitný Ostrov is monitored in a network of water
stations in the Danube, Váh, and Little Danube basins. The minimum daily flow in
the Danube in the period 2008–2015 occurred on 30.11.2011 at the bridge station
Medveďov with a value of 743.5 m^3 s^{-1}. The maximum daily flow occurred on
7.6.2013 with a value of 10,020 m^3 s^{-1} at the same station. The minimum monthly
flow rate was 917.3 m^3 s^{-1} in November 2011 also at Medveďov. The maximum
monthly flow occurred in June 2013 at the Iža station with a rate of 5,527 m^3 s^{-1}.
Minimum flows occur mainly in the winter months, with annual lows mostly at the
end of November. Maximum flow rates occur mainly in the summer months, with
annual maxims mostly in June. Channel network flows are dependent on the
manipulation of the water structures on the channel network.

The Slovak Hydrometeorological Institute (SHMI), which has a built-up network of observation points (probes and sources), operates and monitors long-term
mode observation of the quantity and quality of the groundwater to detect the
occurrence and assess the quantitative and chemical status of the groundwater. In
2012, in the area of the Danube (right and left side up to Komárno), 251 points of
the state hydrological groundwater network [42] were monitored. There are more
than 80 monitoring probes in the area of the Žitný Ostrov channel network. All
monitoring points are located in Quaternary sediments. The values of weekly
measurements in the groundwater observation network in the area of interest
were provided by the SHMI.

The groundwater level time series for SHMI probes in 2014 are shown in Figs. 3
and 4. These probes form the northwest and southeast boundary condition of a
constant level in the assembled model. At the same time, daily sums of precipitation

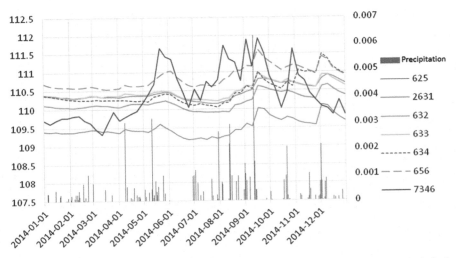

Fig. 3 Groundwater levels (left y-axis [m amsl]) in SHMI probes and rainfall (right y-axis [m])
(southeastern boundary condition of the model)

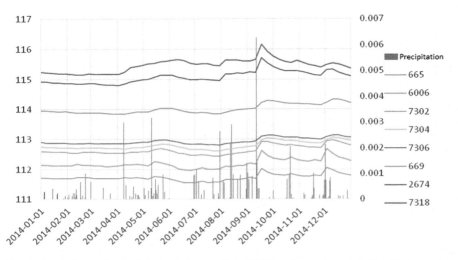

Fig. 4 Groundwater levels (left *y*-axis [m amsl]) in SHMI probes and precipitation (right *y*-axis [m]) (northwestern boundary condition of the model)

are displayed on the secondary axis. In the case of the southeast boundary condition of the model (Fig. 3), the fluctuation of the groundwater level is noticeable especially for probe 7,346 located near the Danube main stream. The SHMI probes are de facto deployed along a cross section of Žitný Ostrov perpendicular to the Danube's flow itself. The fluctuation of the levels for the southeast boundary condition probes ranges approximately from 109 m amsl up to 112 m amsl, while in the case of the southeastern boundary condition, this scatter is from 111.5 m amsl up to 116 m amsl.

The area of interest is predominantly in the climatic area characterized as warm, dry, with mild winters, and longer sunshine [44]. The territory is one of the warmer regions of Slovakia and is classified as having a lowland climate. Average January temperatures range from −4 to −1°C, and average July temperatures from 19.5 to 20.5°C. The territory is characterized by an upper interval of annual precipitation sum of 650 mm and a lower interval of annual rainfall sum of 530 mm [45]. The most important climatic element affecting Žitný Ostrov's water regime is precipitation.

Like all streams and stream segments with small slopes, the channel network of Žitný Ostrov is prone to deposition of sediment in channel waterbeds because of slowly flowing water. Due to the increasing tendency of sediment volumes, this parameter should be included in the input parameters of the numerical simulation. The sediment can form a substantial portion of the cross-sectional area of the bed, thereby greatly reducing the flow capacity of the stream. At the same time, these river sediments have a predominantly fine grain composition and high values of the coefficient of saturated hydraulic conductivity, thus adversely affecting the interaction between the flow and the aquifer. The complete sediment measurement in the Gabčíkovo-Topoľníky channel was carried out in May and June 2014 and represents the present state of clogging in the channel. Measurements were made at each river km point. Levels and thickness of sediments were measured at each channel width meter in the cross section. The measured sediment thicknesses ranged from 0.09 to 1.5 m. The grain composition of the samples was determined, and the values of the coefficient of saturated hydraulic conductivity were calculated using the empirical formulas of Bayer-Schweiger and Špaček [46]. These values, along with the sediment thicknesses and cross-sectional channel widths/cross sections, were used as the MODFLOW River (RIV) input parameter to calculate the conductivity [29].

5 Modeling of Surface Water and Groundwater Interaction

5.1 Parameters of the Area of Interest

The groundwater flow model was calibrated according to the selected measured time series of groundwater levels. The following parameters were part of the model calibration: specific yield, conductivity parameter, flow through the bottom of the streambed, and the coefficient of saturated hydraulic conductivity of the model layers and geological materials.

In defining the geological structure of the subfields in the interest area of a numerical model, there are two basic options for definition: a homogeneous or heterogeneous environment. A homogeneous environment is an aquifer with the same parameters or physical properties throughout the model space. Since the geological structure of the subsoil is generally heterogeneous, it is a significant simplification of the problem which can be used only for specific problems in small areas or where one geological material significantly exceeds the proportions of other materials. Among sites with similar characteristics in the territory of the Slovak Republic, it is possible to include Žitný Ostrov, where the geological structure is predominantly represented by the Danube gravel sediments of great thickness [41]. In contrast, heterogeneous environments are represented by several materials of different powers, with various, often very distinct physical characteristics. The heterogeneous environment is predominantly defined in the MODFLOW model using the available geological wells in the interest area. In Aquaveo GMS [47, 48], the heterogeneity of the borehole environment can be defined using transverse profiles and horizons or using the geo-statistical simulation of T-PROGS [49]. Most of these methods can be highly automated, which significantly reduces the time needed to define the geological structure of the interest area.

The value of the coefficient of saturated hydraulic conductivity for the geological aquifer materials in the area was determined based on the data contained in the reports of engineering geology and hydrogeological wells in the SGIDS (State Geological Institute of Dionýz Štúr) database (Fig. 5). By processing a larger amount of data including the spatial stratification of the wells in the area of interest

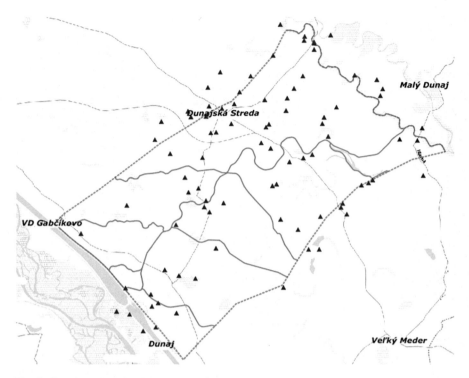

Fig. 5 Situation of the SGIDS boreholes in the interest area

and based on the grain curves of the borehole materials, it was possible to calculate the coefficient of saturated hydraulic conductivity of the material.

The GEOFIL software was used for this task. In the GEOFIL program, it is possible to calculate the coefficient of saturated hydraulic conductivity of the material on the basis of the grain curves given. Because of the presence of many materials, often with a negligible share, the material composition of the aquifers was reduced to four basic materials occurring in the main interest area: sand, loam, gravel, and clay (Table 1). Using available data from SGUDŠ geological reports, grain curves of materials from multiple available sites were read. The value used for the coefficient of saturated hydraulic conductivity is the result of averaging the available values for a particular material. The minimum and maximum values from the available data serve as the upper and lower limit values for the coefficient of saturated hydraulic conductivity when estimating input parameters for numerical simulation.

Regional geology was applied to a three-dimensional network of finite differential elements by the application of hydrogeological units (the so-called solids name in the English literature). Figure 6 compares the result with the model of the geological composition of the interest area.

Table 1 Coefficient of saturated hydraulic conductivity of materials in the area of interest

	Sand	Loam	Gravel	Clay
Average (m s^{-1})	9.48×10^{-5}	5.47×10^{-7}	5.43×10^{-2}	2.62×10^{-9}
Min (m s^{-1})	2.26×10^{-5}	5.07×10^{-9}	3.10×10^{-4}	1.87×10^{-9}
Max (m s^{-1})	2.69×10^{-4}	1.10×10^{-6}	1.89×10^{-1}	3.89×10^{-9}

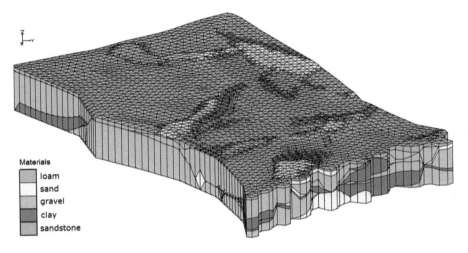

Materials
- loam
- sand
- gravel
- clay
- sandstone

Fig. 6 Three-dimensional regional geology

5.2 Boundary Conditions of Numerical Simulation

The boundary conditions of numerical simulation of the area of interest are defined as the level regime at the boundaries of the model territory. Boundary conditions have two parts.

The first part consists of the river boundary condition (solid lines), which is a mixed boundary condition as defined by the RIV (river) module; all significant rivers and channels in the modelled territory are defined by this boundary condition. On the northeastern border of the model is the Klátov arm of the Little Danube, and at the southwestern border of the model, it is LKKPK. The river boundary conditions at the center of the model copy the direction of the Gabčíkovo-Topolníky, Gabčíkovo-Ňárad, Jurová-Veľký Meder, Kračany-Boheľov, Vojka-Kračany (A VII), Šulany-Jurová (B VII), and Baka-Gabčíkovo (C VII) channels. The second part is the boundary condition of the constant level, defined by the time-variant specified-head boundary (CHD) module. This is the first-order Dirichlet boundary condition applying in the northwestern and southeastern parts of the model. The level mode is defined by groundwater level monitoring probes (dotted lines) (see Fig. 7).

Rivers can provide water to the aquifer or draw water from the aquifer into the river, depending on the gradient between the surface water and groundwater level.

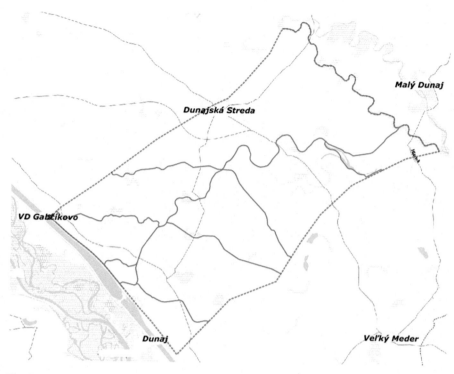

Fig. 7 Boundary conditions of numerical simulation: dotted line, CHD; full line, RIV

The purpose of the river (RIV) module [50] is to simulate the influence of surface flow in the river on the interaction between river and aquifer. Seepage between river and aquifer is simulated between each river segment and the model cell in which the segment is located. The RIV module does not simulate the flow of water itself. It only simulates the seepage between the river and the aquifer. For MODFLOW, other modules simulate flow as well as seepage, such as stream (STR1) [51, 52] and streamflow-routing (SFR1, SFR2) modules [51, 53, 54].

It is assumed that significant water level losses occur only through the more permeable bottom sediment of the river bed. At the same time, it is assumed that the model cell under the less permeable sediment remains fully saturated, that is, the groundwater level does not fall below the bottom of the bed. Considering these assumptions, the flow between river and aquifer is calculated as follows:

$$QRIV_n = CRIV_n(HRIV_n - h_{i,j,k}) \qquad (4)$$

where $QRIV_n$ is the seepage between the river and the aquifer, in positive value toward the aquifer, $CRIV_n$ is the coefficient of conductance, $HRIV_n$ is the surface water level, and $h_{i,j,k}$ is the groundwater level in the model cell below the river segment.

Fig. 8 Conceptualization
of bottom sediment and
seepage coefficient in
model cell

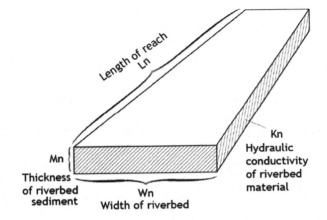

Figure 8 depicts the isolated bottom sediment and the parameters entering into the calculation of the seepage coefficient in the individual cell of the model. The length (L_n) of the bottom sediment is the length of the river flowing through the model cell, W_n is the flow width, M_n is the bottom sediment thickness, and K_n is the coefficient of saturated hydraulic conductivity of the bottom sediment. The seepage coefficient is then calculated as:

$$\mathrm{CRIV}_n = \frac{K_n L_n W_n}{M_n} \tag{5}$$

Equation (4) provides an appropriate approximation of the interaction of the river and the aquifer with a certain spread of groundwater levels. In most cases, if the groundwater level in the aquifer falls below a certain value, the seepage from the river will cease to depend on the level of the water in the aquifer. If the groundwater level is higher than the bottom of the river bed, the seepage through the sediment layer is directly proportional to the difference in levels between the river and the aquifer. If the groundwater level drops below the bottom of the bed, an unsaturated layer in the aquifer beneath the bottom of the bed is created [55]. Since MODFLOW considers the saturated environment in the model cell, the groundwater level will be equal to the elevation of the river bed. If this height is denoted as RBOT_n, the seepage through the bottom sediment will be

$$\mathrm{QRIV}_n = \mathrm{CRIV}_n(\mathrm{HRIV}_n - \mathrm{RBOT}_n) \tag{6}$$

where QRIV_n, CRIV_n, and HRIV_n are defined as in Eq. (4). If the groundwater level drops below the bottom of the bed RBOT_n, then there is no increase in flow through the bottom sediment to the aquifer, and the seepage will remain at a constant value until the groundwater level again rises above the $RBOT_n$. This approach to the calculation is expressed using equations such as:

$$QRIV_n = CRIV_n(HRIV_n - h_{i,j,k}), \qquad h_{i,j,k} > RBOT_n \qquad (7)$$
$$QRIV_n = CRIV_n(HRIV_n - RBOT_n), \qquad h_{i,j,k} < RBOT_n \qquad (8)$$

Figure 9 shows the flow through the bottom sediment of the river. The seepage is zero if the groundwater level is equal to the level of the surface water ($HRIV_n$). For a groundwater level higher than $HRIV_n$, the flow passes from the aquifer into the river, indicated by a negative value of the inflow into the aquifer. At values of h lower than $HRIV_n$, the seepage is positive, i.e., toward the aquifer. Positive seepage increases linearly with the drop in h until h reaches $RBOT_n$. Once $RBOT_n$ is reached, the seepage value remains constant. The concept of surface water and groundwater interaction in the MODFLOW model assumes that the interaction is independent of the position of the flow segment in the model cell and that the level of the surface water level is constant throughout the river section and the time step. It is assumed therefore that the surface water flow in the river does not change significantly within one-time step, i.e., that the stream does not suddenly dry up or overflow its banks. It is assumed that these events will be so short that they do not affect the interaction between the flow and the aquifer.

The constant head boundary condition (CHD) is used as a fixed boundary condition, i.e., this condition maintains the groundwater level in the cell at a constant height regardless of any fluctuation of the groundwater level in the surrounding cells without this boundary condition. Consequently, the constant level boundary condition functions as an endless source of water which can flow into the system or as an endless overflow through which the water flows out of the

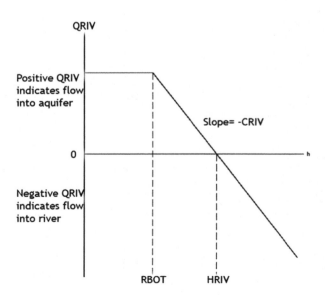

Fig. 9 The scheme of seepage from the river into the aquifer through the bottom sediment

model. For this reason, this boundary condition has a significant effect on the course of the groundwater level in the model. If not used appropriately, it can lead to unrealistic simulations, for example, when this boundary condition is used near a place of interest or site where we try to model a realistic groundwater level. Unlike other MODLFOW boundary conditions, in the case of constant level, the groundwater levels can be linearly interpolated between the time steps, so the specified levels can vary between the different stress periods of the simulation.

5.3   Regional Groundwater Level Regime in the Area of Interest

Groundwater in the area of interest, its movement and the level regime, is primarily affected by the Danube and the Little Danube, precipitation, subsurface geology, slope ratios, and last but not the least the channel network of Žitný Ostrov and its manipulation. Numerical simulation of the interest area was created for two cases:

1. Only with the influence of the main channel in the interest area, i.e., the Gabčíkovo-Topoľníky channel, which has the most significant influence on the surface regime of groundwater
2. With the influence of the secondary channels in the territory which are connected to the Gabčíkovo-Topoľníky channel

The direction of groundwater flow can generally be defined as a flow from west to east, with the groundwater level decreasing with the fall in the terrain. Figures 10, 11, and 12 show groundwater hydroisolines for March, August, and December of 2014.

In a further set of figures (see Figs. 13, 14, and 15), differences in groundwater level height are shown for the same time horizons, considering the lateral channels of the Žitný Ostrov channel network. The results of the simulation show that the presence of the lateral channel network is manifested mainly by a decrease in the groundwater level in the vicinity of the channels, except the spring period, when it is possible to monitor a partial rise in GW levels in the upper channel network.

The GW level regime was then simulated for the various stages of clogging using the conductivity values. The conductance parameter, as already mentioned, is one of the parameters directly affecting the rate of interaction between river and aquifer. For one value, the width of the flow, the bottom sediment thickness in the river, and the value of the saturated hydraulic conductivity of the sediment are included. To assess the impact of the channel network on the extent of surface and groundwater interactions at the site, three conductivity values were used:

– $8.11 \text{ m}^2 \text{ d}^{-1} \text{ m}^{-1}$, which corresponds to the actual degree of clogging
– $0 \text{ m}^2 \text{ d}^{-1} \text{ m}^{-1}$, which means completely impermeable sediment
– $500 \text{ m}^2 \text{ d}^{-1} \text{ m}^{-1}$, which represents the channel bed without sedimentation

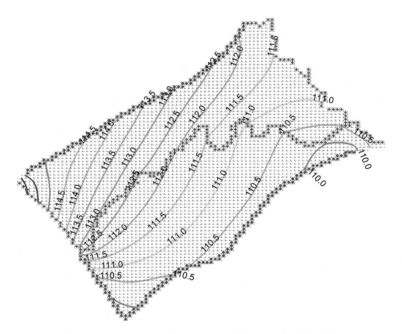

Fig. 10 Course of the groundwater level in the area of interest as of 5.3.2014 – homogeneous environment

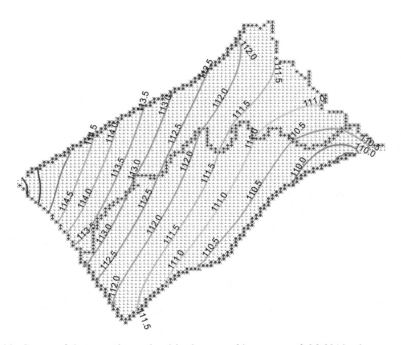

Fig. 11 Course of the groundwater level in the area of interest as of 6.8.2014 – homogeneous environment

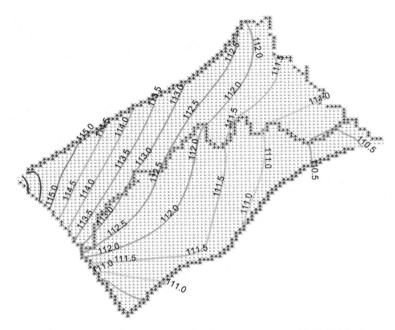

Fig. 12 Course of the groundwater level in the area of interest as of 3.12.2014 – homogeneous environment

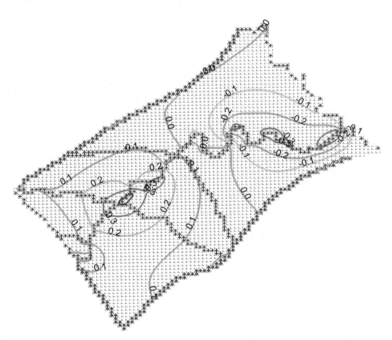

Fig. 13 Difference in simulated groundwater level considering the side channel network impact as of 5.3.2014

Fig. 14 Difference in simulated groundwater level considering the side channel network impact as of 6.8.2014

Fig. 15 Difference in simulated groundwater level considering the side channel network impact as of 3.12.2014

Figure 16 shows the course of the groundwater level for March 2014 and the conductance value of 8.11 m^2 d^{-1} m^{-1}. It is possible to observe a relatively low river connection to the aquifer, as the low permeability of the sediment prevents natural interaction between the river and the aquifer. For channel conductivity value at the level of 0 m^2 d^{-1} m^{-1}, the groundwater level course in the area is very similar (Fig. 17). In the case of removing of the sediment (conductivity parameter of 500 m^2 d^{-1} m^{-1}), the channel effect is evident (see Fig. 18).

To compare the effect of this parameter on the interaction rate and thus the seepage, it was necessary to compare the simulated outputs with the measured groundwater levels. For this purpose, three SHMU observation probes with a weekly measuring step were selected, namely, the 657-Mád, 662-Vrakúň, and 663-Kútniky Povoda probes, which are situated in close proximity to the Gabčíkovo-Topolníky channel. Groundwater level values were compared with simulated outputs for the different conductivity values for the relevant channel segment; besides the conductivity value which is close to the real environment, situations with its higher and lower values were also simulated, i.e., 8.11, 0, and 500 m^2 d^{-1} m^{-1}, the latter corresponding to a saturated hydraulic conductivity coefficient of approximately 5 × 10^{-4} m s^{-1}. The last simulated case was the value of the constant head boundary condition (CHD). Since the conductivity parameter does not enter this boundary condition, it is possible to consider this condition as the numerical equivalent of a river without the presence of less permeable sediment. The results of the simulation and the

Fig. 16 Course of the groundwater level in the area of interest as of 5.3.2014 (8.11 m^2 d^{-1} m^{-1})

Fig. 17 Course of the groundwater level in the area of interest as of 5.3.2014 (0 m^2 d^{-1} m^{-1})

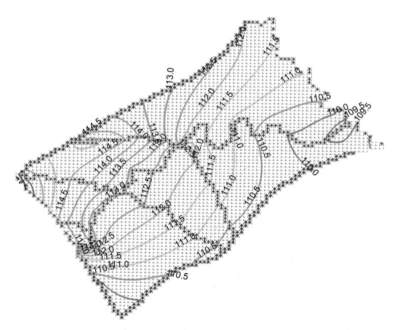

Fig. 18 Course of the groundwater level in the area of interest as of 5.3.2014 (500 m^2 d^{-1} m^{-1})

comparison with the measured values are presented in the form of graphs (Figs. 19, 20, and 21). The results show that the simulated groundwater level is closest to the real course of the groundwater level for conductance, which was calculated from the coefficient of saturated hydraulic conductivity of the bottom sediment of the Gabčíkovo-Topolníky channel.

The simulation of impenetrable sediment resulted in a rise in the groundwater level by approximately 40 cm. A drop in the groundwater level and thus greater connection of the river to the aquifer occurred in the simulation of the higher conductance value or when simulating the constant head boundary condition, which is almost identical.

The next step was to create a series of observation probes in the model serving as a tool for determining the groundwater level in a given model cell. GW level

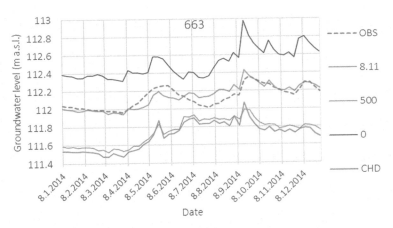

Fig. 19 Measured and simulated groundwater levels for observation probe 663

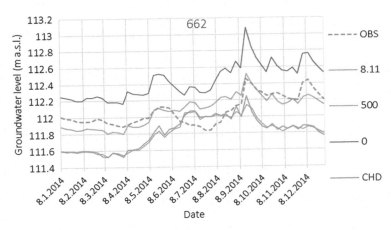

Fig. 20 Measured and simulated groundwater levels for observation probe 662

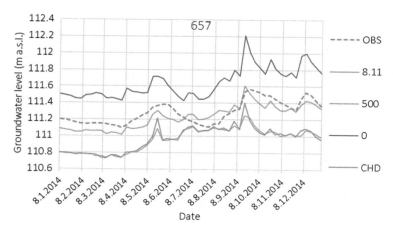

Fig. 21 Measured and simulated groundwater levels for observation probe 657

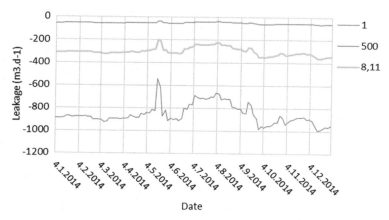

Fig. 22 Seepage for different conductance values ($m^2 d^{-1} m^{-1}$)

elevation values were then compared to the surface water level in the river and also to the seepage values for the different conductances. In addition to the calculated conductivity value of 8.11 $m^2 d^{-1} m^{-1}$, the values of 500 and 1 $m^2 d^{-1} m^{-1}$ (impermeable sediment) were used. For the zero conductivity value, the seepage would be zero as well.

Subsequently, the time series of seepage for the equivalent of the current clogged state were compared with lower (500 $m^2 d^{-1} m^{-1}$) and higher (1 $m^2 d^{-1} m^{-1}$) clogging of the channel network (Fig. 22). At a theoretically higher degree of clogging of the channel network, the seepage volume is in the range of 16.8–17.6% of the seepage volume for the current clogging state. At a lower clogging of the channel network, the seepage amounts to 2.73–3.02 times the flow seepage volume for the current channel clogging state (273–302% of volume).

6   Conclusion and Recommendations

The aim of this chapter was to quantify changes in the interaction between surface and groundwater in the area of Žitný Ostrov with changes in flow conditions in the channels running through this area. Žitný Ostrov is part of the Danube Plain, and in this region, a channel network has been built in the past which serves for irrigation purposes in agricultural cultivated areas and also for draining the lowland area of Žitný Ostrov in case of floods or high-level conditions in the River Danube. Of the possible partial sites, the area around the Gabčíkovo-Topoľníky channel was selected, which has water meters installed at its north and south ends for flow and surface water level monitoring with a daily measurement step.

The MODFLOW numerical model is considered as one of the appropriate tools for the implementation of numerical simulation in a given locality. It is globally the most used three-dimensional numerical simulation model. The availability of literature dealing with modeling in the finite differential network is very high (SCOPUS contains more than 1,500 articles with the keyword MODFLOW as of May 2017). Last but not the least, there is the availability of commercial and noncommercial graphical user interfaces for this simulation code.

The disadvantages in simulating the interaction between groundwater and surface water include the simplification of the calculation and parameterization of the flow between river and aquifer, where seepage is a primary function of the conductance parameter, which is expressed by the same value for infiltration and drainage from or into the river. This partial simplification can be replaced with a combination of two boundary conditions for one river [56]. In addition to this limitation, the bottom sediment parameters do not take into account the complex sediment heterogeneity of bottom sediment [57, 58] in most modeling environments (not only MODFLOW). Instead, the bottom of the channel is idealized as a homogeneous geological structure, the parameters of which are achieved by the numerical model calibration itself. This simplification is mainly applied due to the demanding exact quantification of heterogeneity of river sediment in the field. Despite these shortcomings, numerical simulations are, in the long run, a comprehensive and appropriate tool for quantifying the interaction between river and aquifer.

The influence of environmental heterogeneity on the groundwater/surface water regime was examined for an equivalent heterogeneous environment created in a numerical simulation environment based on regional geology. However, the graphical outputs of groundwater isolines in the area of interest suggest that regional heterogeneity does not have a significant impact on the general flow of groundwater. Inserting the heterogeneity parameter into the model causes the groundwater level to rise between 0.05 and 0.35 m. In the case of the regional geology of Žitný Ostrov, however, where high permeability of the gravel base prevails, the presence of less permeable clay or sand layers causes predominantly local depressions of groundwater, which are not significant in regional simulations. Simulation results show that the interaction between groundwater and surface water in the locality is significantly affected by the deposition of sediment layers in the channel network. The theoretical outputs of a more favorable status in the channel network can serve as an illustration of the potential for influencing groundwater level through manipulation of the fluctuation of levels in the channels across the interest area.

Acknowledgment The chapter was created with support from VEGA project no. 2/0058/15 and APVV-14-0735. This publication is also the result of the implementation of the ITMS 26240120004 project entitled Centre of Excellence for Integrated Flood Protection of Land supported by the Research and Development Operational Programme funded by the ERDF.

References

1. Woessner WW (2000) Stream and fluvial plain ground water interactions: rescaling hydro-geologic thought. Ground Water 38(3):423–429
2. Toth J (1970) A conceptual model of the groundwater regime and the hydrogeologic environment. J Hydrol 10:164–176
3. Schaller MF, Fan Y (2009) River basins as groundwater exporters and importers: implications for water cycle and climate modeling. J Geophys Res 114:1–21
4. Schwarzenbach RP, Westall J (1981) Transport of nonpolar organic compounds from surface water to groundwater. Laboratory sorption studies. Environ Sci Technol 15(11):1360–1367
5. Rushton KR, Tomlinson LM (1979) Possible mechanisms for leakage between aquifers and rivers. J Hydrol 40:49–65
6. Selker JS, Keller CK, McCord JT (1999) Vadose zone processes. CRC Press, Boca Raton
7. Irvine DJ, Brunner P, Franssen HH, Simmons CT (2012) Heterogeneous or homogeneous? Implications of simplifying heterogeneous streambeds in models of losing streams. J Hydrol 424–425(2012):16–23
8. Winter TC (1999) Relation of streams, lakes, wetlands to groundwater flow systems. Hydrogeol J 7:28–45
9. Dulovičová R, Kosorin K (2007) Determination of lateral additions of discharges by interaction between open channels and groundwater. Acta Hydrologica Slovaca 8(2):245–253. (in Slovak)
10. Mäsiar E, Kamenský J (1989) Hydraulics for civil engineers II. Alfa, Bratislava. (in Slovak)
11. Květon R (2012) Mathematical modeling of flow in open channels. STU Publisher. (in Slovak)
12. Šebová E (2011) Interaction between surface water and groundwater at Žitný ostrov – current results and experience. Acta Hydrologica Slovaca 12(2):151–157. (in Slovak)
13. Andersen MP, Woessner WW, Hunt RJ (2015) Applied groundwater modeling – simulation of flow and advective transport, 2nd edn. Elsevier, Amsterdam
14. Domenico PA, Mifflin MD (1965) Water from low-permeability sediments and land subsidence. Water Resour Res 1(4):563–576
15. Morris DA, Johnson AI (1967) Summary of hydrologic and physical properties of rock and soil materials, as analyzed by the hydrologic laboratory of the U.S. Geological Survey, 1948–60, Water Supply Paper 1839-D, US Geological Survey
16. Theis CV (1941) The effect of a well on the flow of a nearby stream. EOS Trans Am Geophys Union 22(3):734–738
17. Krčmář D (2012) Modelling of surface and ground water interaction. Podzemná voda XVIII (1):1–13. (in Slovak)

18. Saleh F, Flipo N, Habets F, Ducharne A, Oudin L, Viennot P, Ledoux E (2011) Modeling the impact of in-stream water level fluctuations on stream-aquifer interactions at the regional scale. J Hydrol 400(2011):490–500

19. Baalousha HM (2011) Modelling surface–groundwater interaction in the Ruataniwha basin, Hawke's Bay, New Zealand. Environ Earth Sci 66:285–294

20. Bosompemaa P, Yidana SM, Chegbeleh LP (2016) Analysis of transient groundwater flow through a stochastic modelling approach. Arab J Geosci 9:694

21. Yidana SM, Addai MO, Asiedu DK, Banoeng-Yakubo B (2016) Stochastic groundwater modeling of a sedimentary aquifer: evaluation of the impacts of abstraction scenarios under conditions of reduced recharge. Arab J Geosci 9:694

22. Barthel R, Banzhaf S (2016) Groundwater and surface water interaction at the regional-scale – a review with focus on regional integrated models. Water Resour Manag 30:1–32

23. Šimůnek J, van Genuchten MT, Šejna M (2016) Recent developments and applications of the HYDRUS computer software packages. Vadose Zone J 15(7):25

24. Twarakavi NKC, Šimůnek J, Seo S (2008) Evaluating interactions between groundwater and vadose zone using the HYDRUS-based flow package for MODFLOW. Vadose Zone J 7(2): 757–768

25. Welsh WD, Vaze J, Dutta D, Rassam D, Rahman JM, Jolly ID, Wallbrink P, Podger GM, Bethune M, Hardy MJ, Teng J, Lerat J (2013) An integrated modelling framework for regulated river systems. Environ Model Softw 39(2013):81e102

26. Royal Haskoning (2004) Triwaco user's manual. Royal Haskoning, Amersfoort. https://www.royalhaskoningdhv.com/

27. Baroková D (2006) Determination of structures impact on groundwater level regime and possibilities of regulations. STU Publisher. (in Slovak)

28. Strack ODM, Haitjema HM (1981) Modeling double aquifer flow using a comprehensive potential and distributed singularities. Water Resour Res 17:1535–1549

29. Aquaveo (2013) WMS user manual (v9.1) – the watershed modeling system. Aquaveo, Provo. http://www.aquaveo.com/

30. Haitjema HM, Wittman J, Kelson V, Bauch N (1994) WhAEM: program documentation for the wellhead analytic element model. US Enviromental Protection Agency

31. Haitjema HM (1995) Analytic element modeling of groundwater flow. Academic Press, Cambridge

32. Harbaugh AW, Banta ER, Hill MC, McDonald MG (2000) MODFLOW-2000, the U.S. Geological Survey modular ground-water model – user guide to modularization concepts and the ground-water flow process. US Geological Survey, Reston

33. Harbaugh AW (2005) MODFLOW-2005, the U.S. Geological Survey modular ground-water model – the ground-water flow process. US Geological Survey, Reston

34. Doherty J (1994) PEST – model independent parameter estimation. Watermark Numerical Computing, Corinda

35. Zheng C (2010) MT3DMS v5.3 – supplemental user's guide. The University of Alabama, Tuscaloosa

36. Pollock DW (2012) User guide for MODPATH version 6 – a particle-tracking model for MODFLOW. US Geological Survey, Reston

37. Pásztorová M, Vitková J, Jarabicová M, Nagy V (2013) Impact of Gabčíkovo waterworks on soil water regime. Acta Hydrol Slovaca 14(2):429–436. (in Slovak)

38. Velísková Y, Dulovičová R (2008) Variability of bed sediments in channel network of Rye Island. In: IOP conference series: earth and environmental science

39. Káčer Š et al (2005) Digital geological map of Slovak Republic in scale M 1:50,000 and 1:500,000. SGIDŠ, Bratislava

40. Maglay J et al (2009) Geological maps of Slovakia – map of quaternary layer thickness, M 1: 500,000. SGIDŠ, Bratislava

41. Malík P, Bačová N, Hronček S, Ivanič B, Káčer Š, Kočický D, Maglay J, Marsina K, Ondrášik M, Šefčík P, Černák R, Švasta J, Lexa J (2007) Set up of geological maps at a scale of 1:50,000 for the needs of integrated landscape management. Manuscript – Archive of Geofond Union ŠGIDŠ, Bratislava, p 552

42. Gavurník, J, Bodácz B, Čaučík P, Paľušová Z (2011) Danube – refilling source of grounwater. SHMI report (in Slovak)
43. Dulovičová R, Velísková Y, Koczka Bara M, Schűgerl R (2013) Impact of silts distribution along the Chotárny channel on seepage water amounts. Acta Hydrologica Slovaca 14(1):126–134. (in Slovak)
44. Faško P, Štastný P (2002) Atlas krajiny Slovenskej republiky. Ministerstvo životného prostredia SR, Bratislava
45. Malík P et al (2011) Comprehensive geological information base for the needs of nature conservation and landscape management (GIB-GES). SGIDŠ report (in Slovak)
46. Dulovičová R, Velísková Y (2005) Saturated hydraulic conductivity of silts in the main channels of the Žitný ostrov channel network. Acta Hydrologica Slovaca 6(2):274–282. (in Slovak)
47. Aquaveo (2011) MODFLOW – conceptual model approach I. Aquaveo, Provo. http://www.aquaveo.com/
48. Aquaveo (2011) GMS user manual (v9.1), the groundwater modeling system. Aquaveo, Provo. http://www.aquaveo.com/
49. Carle SF (1999) T-PROGS: transition probability geostatistical software version 2.1. Hydrologic Sciences Graduate Group University of California, Davis
50. McDonald MG, Harbaugh AW (1988) A modular three – dimensional finite – difference ground – water flow model. US Geological Survey, Reston
51. Prudic DE, Konikow LF, Banta ER (2004) A new Streamflow-Routing (SFR1) package to simulate stream-aquifer interaction with modflow-2000. US Geological Survey, Reston
52. Aquaveo (2011) MODFLOW – STR package. Aquaveo, Provo. http://www.aquaveo.com/
53. Niswonger RG, Prudic DE (2010) Documentation of the Streamflow-Routing (SFR2) Package to include unsaturated flow beneath streams – a modification to SFR1. US Geological Survey, Reston
54. Aquaveo (2011) MODFLOW – SFR2 package. Aquaveo, Provo. http://www.aquaveo.com/
55. Marino MA, Lithin JN (1982) Seepage and groundwater. Elsevier, Amsterdam, p 489. ISBN 0-444-41975-6
56. Zaadnoordijk JW (2009) Simulating piecewise-linear surface water and ground water interactions with MODFLOW. Ground Water 47(5):723–726
57. Fleckenstein JH, Niswonger RG, Fogg GE (2006) River–aquifer interactions, geologic heterogeneity, and low-flow management. Ground Water 44(6):837–852
58. Doppler T, Franssen HJH, Kaiser HP, Kuhlman U, Stauffer F (2007) Field evidence of a dynamic leakage coefficient for modelling river–aquifer interactions. J Hydrol 347(1–2): 177–187

Impact of Soil Texture and Position of Groundwater Level on Evaporation from the Soil Root Zone

M. Gomboš, D. Pavelková, B. Kandra, and A. Tall

Contents

Abstract The lower boundary of unsaturated soil zone is formed by groundwater level. At this level, water from unsaturated soil zone flows to groundwater and vice versa. Groundwater penetrates the unsaturated zone. By capillary rise, groundwater can supply water storage in the root zone and thus influence on actual evaporation in this soil layer. The degree to which this occurs depends on the given soil texture and the groundwater level position with regard to the position of lower root zone boundary.

The paper quantifies the impact of soil texture on the involvement of groundwater in the evaporation process. The results were obtained by numerical experiment on GLOBAL model. The measurements used for model verification and numerical simulation were gained in ESL (East-Slovakian Lowland).

Keywords Actual evapotranspiration, Groundwater level, Particle size distribution

1 Introduction

Water evapotranspiration is a thermodynamic process during which mass converts from solid or fluid phase to gaseous phase. It is the most decisive regulator of energy flow in the hydrologic cycle. The process of evaporation from plants and

---

M. Gomboš (✉), D. Pavelková, B. Kandra, and A. Tall
Institute of Hydrology, Slovak Academy of Sciences, Bratislava, Slovak Republic
e-mail: gombos@uh.savba.sk

A. M. Negm and M. Zeleňáková (eds.), *Water Resources in Slovakia:*
Part I - Assessment and Development, Hdb Env Chem (2019) 69: 167–182,
DOI 10.1007/698_2017_181, © Springer International Publishing AG 2018,
Published online: 25 February 2018

water or soil surface is called evapotranspiration. Maximal possible evaporation from land covered by vegetation under particular meteorological conditions is called potential evapotranspiration (ET_0). Real evaporation from land covered by vegetation is called actual evapotranspiration (ET_a). Evapotranspiration is one of the key elements in water balance in nature [1, 2]. It crucially affects the biomass creation and water storage in the unsaturated zone of the soil profile. The unsaturated zone (UZ) is the water source for the biosphere. If there is enough water in a soil profile, then $ET_0 = ET_a$. If $ET_0 > ET_a$, it indicates the water deficit in the root zone of a soil profile and the beginning of soil profile drying [3–6]. Unsaturated zone is determined by surface runoff and, on the lower boundary, by a position of groundwater level (GWL). Apart from evaporation and rainfall, the amount of water in UZ in lowland conditions, and also root zone of a soil profile, is influenced by groundwater level [7–9]. For some time during rainless periods, groundwater can supply water storage in the root zone of a soil profile by capillary rise [10]. Thereby water availability for plants improves and actual evapotranspiration intensity rises [11]. In consequence of the water transfer, GWL lowers, and therefore unsaturated zone is enlarged. When the groundwater level drops under a certain critical level, the water transfer from GWL to the root zone is negligible. During the long-lasting rainless period, the intensity of the actual evapotranspiration slowly decreases due to the lack of water in the root zone [12–15]. Surface soil horizon and consequently the whole root zone is getting into the state of soil drought. It is a state in which the creation of biomass decreases and physiological activity of the plants is focused merely on survival due to the water deficit [16, 17]. It is defined as a threshold point (TP) for potential pF = 3.3 on the retention curve.

The aim of the chapter is to quantify the impact of GWL on the course and intensity of actual evapotranspiration in vegetation period (VP – April to September). Apart from this, the aim is to determine the threshold values of GWL in the examined areas. The threshold values are the values indicating that the impact of groundwater on actual evapotranspiration is negligible. The impact of GWL was examined in lowland areas with plant cover, where the main part of their root zone is located less than 1 m under the surface. Research works are based on field measurements and numerical simulation on mathematical model global.

2 Experiment Description

Research works were carried out on East-Slovakian Lowland, in the Kamenec area ($\phi = 48°21'02.9''$; $\lambda = 21°48'52.6''$; 95 m) and Horeš area ($\phi = 48°22'32.4''$; $\lambda = 21°53'54.4''$; 94 m). Figure 1 shows the situation of the selected localities.

In both areas, winter wheat was grown during the examined year. The major part of the root zone of winter wheat is less than 1 m under the ground. In terms of soil types, the first one is predominantly silty-clayey loam (Fig. 2) and in the second one is clayey (see Fig. 3).

Fig. 1 Situation of the selected localities

To examine the issues in question, the vegetation period of the year 2007 was chosen. In terms of soil water storage, this period was one of the driest periods in the last 30 years. The investigation was carried out by way of numerical experiment on the mathematical model GLOBAL.

The GLOBAL is a simulation mathematical model of soil water transfer which enables the calculation of moisture potential distribution or soil moisture in real time [18]. The principle of the model is a numerical calculation of the following nonlinear partial differential equation of water movement in the aeration zone:

$$\frac{\partial h_w}{\partial t} = \frac{1}{c(h_w)} \frac{\partial}{\partial z} \left[k(h) \left(\frac{\partial h_w}{\partial z} + 1 \right) \right] - \frac{S(z,t)}{c(h_w)} \qquad (1)$$

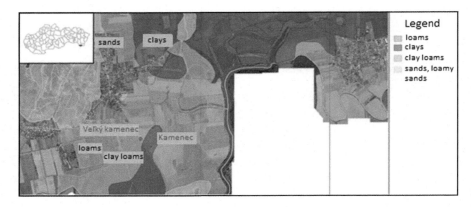

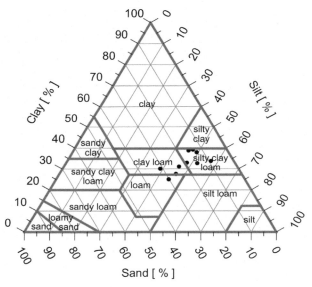

Fig. 2 Specification of soil types using triangular classification diagram by a vertical line of the soil profiles Kamenec into the depth of 1 m by 0.1 m layers

h_w soil moisture potential; z vertical coordinate; $k(h_w)$ unsaturated hydraulic conductivity of the soil; $S(z, t)$ intensity of the water takeoff by the plant's roots from unit soil volume per time unit $(cm^3/cm^3)\ d^{-1}$; θ bulk soil moisture (cm^3/cm^3).

The model GLOBAL enables the simulation to be executed with one-day time-step. Daily values are used as the basic inputs for setting up the boundary conditions. One-day time-step inputs for the meteorological and vegetation parameters are used. Hydrophysical characteristics of the soil (retention curves; saturated and unsaturated hydraulic conductivity of the soil; hydrolimits and some physical properties of the soil as porosity, density, and bulk density; the moisture of saturated soil) also enter the model GLOBAL. Moisture retention curve is described by the formula of van Genuchten [19].

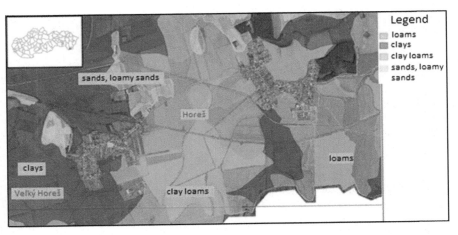

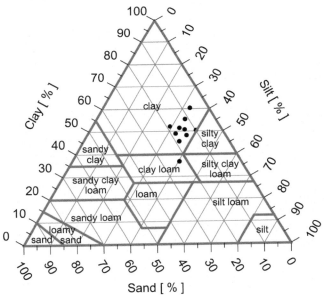

Fig. 3 Specification of soil types using a triangular classification diagram by a vertical line of the soil profiles Horeš into the depth of 1 m by 0.1 m layers

Potential evapotranspiration ET_0 is calculated according to FAO, by Penman's method of Monteith [20]. For determining the actual transpiration or evaporation intensities, the method developed on IH SAS was used. According to this method, the evapotranspiration structure depends on the value of leaf area index (LAI). The intensity of potential evaporation E_{eo} is calculated from the value of potential evapotranspiration ET_0 using the formula:

$$E_{\mathrm{eo}} = \mathrm{ET}_0 \cdot \exp(-m_1 \cdot \mathrm{LAI}) \tag{2}$$

The value of empirical coefficient ($m_1 = 0.463$) was gained by field measurements in the wheat plant cover. Calculation of the actual evapotranspiration intensities and its structure is based on the knowledge of potential evapotranspiration ET_0 and the relationship between $E_{\mathrm{eo}}/\mathrm{ET}_0$ and moisture of the soil profile, i.e.:

$$\mathrm{ET}_r = E_{\mathrm{eo}}/\mathrm{ET}_0 = f(\theta) \tag{3}$$

The used calculation method is based on the assumption that means value of soil moisture in the root zone depends on the intensity of evaporation. The higher the evaporation intensity is, the higher is the value of θ_k, in which evaporation starts to decrease. This method was verified using the model GLOBAL. Modelling shows soil moisture distribution and soil moisture potential, a daily interception and evapotranspiration rates and their elements, infiltration, existing water deficiency in the soil, and more information. Results of monitoring of water storage into the depth of 0.8 m are available for both localities. Model GLOBAL was verified in 2007 by these results in two localities.

Experimental research on the impact of GWL on actual evapotranspiration rate is based on actual evapotranspiration rate quantification for different simulated positions of average GWL during vegetation period of the year 2007. Variability of GWL, as well as hydrometeorological and other input data, remains same. Calculation process during the numerical experiment was the same in both cases, following these steps:

1. Average GWL in vegetation period of the year 2007 was calculated.
2. Course of GWL in vegetation period 2007 (lower boundary condition) was shifted by vertical so that average values of GWL_i^k in vegetation period had in every k^{th} shift different characteristic positions of $\mathrm{GWL}_{\mathrm{VP}}^k$.

where:

GWL_i^k is the position of GWL in i^{th} day of the vegetation period for k^{th} shift that is k^{th} average GWL.

$\mathrm{GWL}_{\mathrm{VP}}^k$ is the average GWL during the vegetation period in k^{th} shift.

Values $\mathrm{ET}_{a,i}^k$ and $\mathrm{ET}_{a,\mathrm{VP}}^k$ were calculated for every GWL_i^k and $\mathrm{GWL}_{\mathrm{VP}}^k$.

where $\mathrm{ET}_{a,i}^k$ is the value of actual evapotranspiration ET_a in the ith day of the vegetation period in the kth average GWL.

$$\mathrm{ET}_{a,\mathrm{VP}}^k = \sum_{i=1.4.2007}^{30.9.2007} \mathrm{ET}_{a,i}^k \tag{4}$$

3. Three basic levels as characteristic position of $\overline{\text{GWL}_{\text{VP}}^{k}}$ were considered:

The lower edge of the root zone of a soil profile, 1 m deep under the $\overline{\text{GWL}_{\text{VP}}^{1}}$, average $\overline{\text{GWL}_{\text{VP}}^{2}}$ in the vegetation period of the year 2007, and average value $\overline{\text{GWL}_{\text{VP}}^{3}}$ at which is the influence of GWL on ET_a negligible (threshold level of GWL position). In addition, two other positions of GWL were chosen in the interval $\left\langle \overline{\text{GWL}_{\text{VP}}^{k}}; \overline{\text{GWL}_{\text{VP}}^{2}} \right\rangle$ or representation of the course of the dependency. Threshold level of groundwater level $\overline{\text{GWL}_{\text{VP}}^{3}}$ was identified by progresive selection of subsequent groundwater level GWL so that $\overline{\text{GWL}_{\text{VP}}^{k}} < \overline{\text{GWL}_{\text{VP}}^{2}}$. Threshold value of groundwater level $\overline{\text{GWL}_{\text{VP}}^{3}}$ is identified after fulfilling the condition:

$$\left(\text{ET}_{a,\text{VP}}^{k} - \text{ET}_{a,\text{VP}}^{k-1} \leq 0.01 \times \text{ET}_0\right) \wedge \left(\overline{\text{GWL}_{\text{VP}}^{k}} - \overline{\text{GWL}_{\text{VP}}^{k-1}} \leq 0.5\right)$$
$$\Rightarrow \overline{\text{GWL}_{\text{VP}}^{k}} = \overline{\text{GWL}_{\text{VP}}^{3}} \tag{5}$$

4. The following dependencies were gained in the examined areas:

$$\text{ET}_{a,\text{VP}} = f\left(\overline{\text{GWL}_{\text{VP}}}\right) \tag{6}$$

3 Results and Significances

Figure 2 shows that heavy soils occur in Kamenec locality. Two calculation material layers were considered for improvement of the calculation precision during the simulation. There are very heavy soils in the locality of Horeš (Table 1 and Fig. 3). Three calculation profiles were considered here.

Table 2 shows the basic characteristics of GWL position in the examined areas during the vegetation period in 2007, as well as the total volume of precipitation (P) during the vegetation period in question.

During the period under consideration, water regime and its components were simulated to the depth of 4 m. They were analyzed in detail to the lower boundary of the root zone and to the depth of 1 m under the surface. Calculation of time-step was 1 day. Hydrometeorological inputs and lower boundary condition (GWL) were entered into the calculations accordingly.

Figures 4 and 5 show the results of the verification of the model GLOBAL. In both areas, the verification was executed using field measurement water storage, to the depth of 0.8 m under the surface.

From the course of measured and calculated values, it is obvious that, in terms of soil water storage, it is a dry year. Soil water storage into the depth of 0.8 m was

Table 1 Basic hydrophysical characteristics of soils

Locality	Veľký Kamenec		Veľký Horeš		
Type of soils	Heavy soils		Very heavy soils		
Layers	0–40	40–100	0–50	50–60	60–100
Alpha	0.0103	0.0093	0.0154	0.0158	0.0131
n	1.4143	1.4655	1.3130	1.3149	1.3888
Theta s	0.396	0.413	0.442	0.483	0.4895
Theta r	0.0788	0.0806	0.0919	0.0988	0.0966
Ks	3.37	5.09	7.48	10.7	15.3
Available WC	224		200		

Table 2 Average position of GWL and the total volume of precipitation in the examined areas during the vegetation period in 2007

Locality	$\overline{GWL}_{VP}^{2007}$ (m a.s.l.)	$\overline{GWL}_{VP}^{2007}$ (m under the surface)	Standard deviation (m)	GWL_{max}- GWL_{min} (m)	ΣP_{VP}^{2007} (mm)
Veľký Kamenec	94.87	2.34	0.62	0.65	308.80
Veľký Horeš	95.49	2.15	0.48	0.86	308.80

during the whole vegetation period in between the wilting point and threshold point. This confirms evapotranspiration deficiency (ED):

$$ED = ET_0 - ET_a \tag{7}$$

During the vegetation period 2007, ED was 289 mm in Kamenec area and 293 mm in Horeš area. The deficit represents 62 and 63% of total potential evapotranspiration during the whole vegetation period. Figures 6 and 7 show the course of the total daily values of ET_a, ET_0 and precipitations in VP of 2007.

Figure 8 is a graphic representation of the results of the calculation of dependency characterized by the formula (6). It shows that the GWL position affects the evapotranspiration rates in both areas where winter crop is grown. Three phases can be identified in the curve of dependency characterized by the formula (6) – linear phase, nonlinear phase, and residual phase.

Linear phase starts at the GWL position at $\overline{GWL}_{VP}^{1}$. When GWL drops under a certain point, the dependency becomes nonlinear. At this phase, the impact that groundwater has on the actual evapotranspiration rates decreases. Hydraulic connection between the root zone and groundwater is negligible. The dependency (6) then becomes residual. At the residual stage, the impact line asymptotically approaches the line parallel to the vertical axis of GWL depths. Considering the condition described in (5), the residual phase starts at the interval between two lowest points. In this case, the lowest points of the interval are located in the depth of 3.73 m. Under the conditions of the profiles in question, these are the threshold levels of $\overline{GWL}_{VP}^{3}$. Water transfer from GWL to the root zone of a soil profile at the

Fig. 4 Monitored and calculated (by model GLOBAL) soil water storage into the depth of 0.8 m in the localities of Kamenec and hydrolimities FWC (pF = 2.52), TP (pF = 3.3), and WP (pF = 4.18)

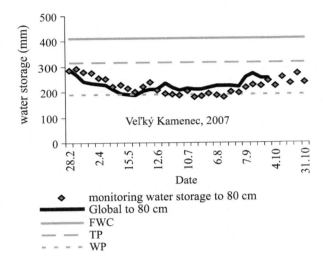

Fig. 5 Monitored and calculated (by model GLOBAL) soil water storage into the depth of 0.8 m in the localities of Horeš and hydrolimities FWC (pF = 2.52), TP (pF = 3.3), and WP (pF = 4.18)

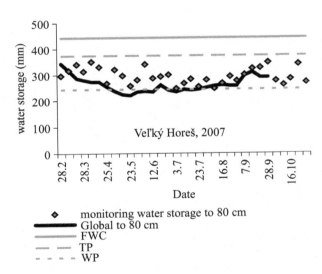

depth of 3.73 m is negligible. The impact on the actual evapotranspiration and biomass production is minimal as well. The course of the dependency and the limits of the individual phases are influenced mainly by hydrological characteristics of the environment and variations of GWL position.

Figure 9 shows the monitored dependence of GWL on the water supply in the root zone for different soil types. In 2007, GWL ranged from 3.5 to 2 m in monitored profiles. The figure shows that the heavier the soil is, the lower is the position of beginning of the influence of GWL on water supply in the root zone of the soil profile. In terms of retention, the influence of GWL is stronger in heavy soils

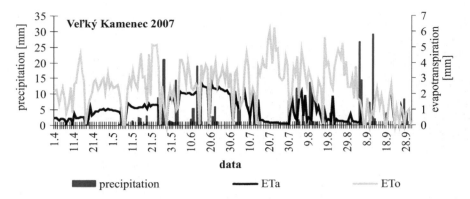

Fig. 6 The course of total daily values of ET_O, ET_a, Z in the vegetation period of 2007 in Kamenec area

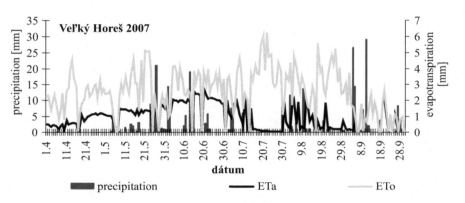

Fig. 7 The course of total daily values of ET_0, ET_a, Z in the vegetation period of 2007 in Horeš area

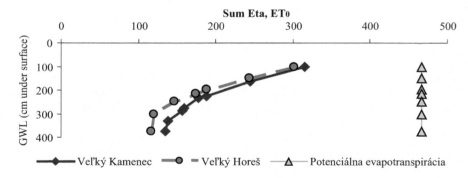

Fig. 8 The course of the dependency (6) and ET_0 in Kamenec and Horeš areas

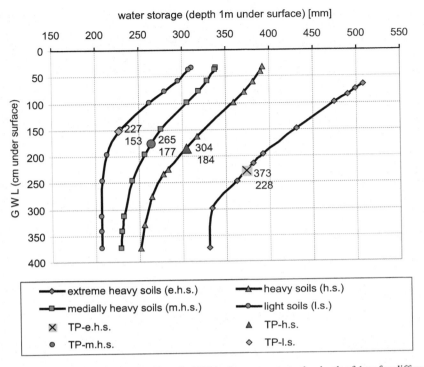

Fig. 9 Influence of position of GWL to the WS in the root zone to the depth of 1 m for different soil types

with high content of clay elements. The shape of dependence curves was also analyzed. The detailed analysis of the waveform shows that the beginning of a nonlinear course appears when the average of GWL at a distance of 3σ (σ-standard deviations of the vertical movement of GWL) is below the bottom border of the evaluated soil layer. At the distance of 2σ, the course of lines of dependences significantly changes, and this indicates the intense influence of GWL on the water supply in the monitored layer. It follows that the level of GWL increasingly encroaches into the soil profile. If the average position of the GWL under the balance layer of soil is 2σ, then in case of the Gaussian random variable (position of GWL), 4.5% of positions of GWL above its average value directly affects the balance layer of soil. As an average value of GWL approaches to the bottom border of evaluated soil layer, the influence of GWL to the water storage is increasing.

From the course of given dependencies, threshold limits on the levels of GWL were identified. That is defined as the average position of GWL at which the water supply of balanced layer of soil oscillates around threshold point (TP) under the given hydrophysical conditions.

From the above graph, it results that in case of light soil, the water supply to the depth of 1 m at the TP is 227 mm. To this moisture, condition corresponds the

threshold limit level of GWL 1.53 m below the surface. This level of GWL is below
the bottom border of evaluated soil layer at a distance of size 1σ. For the medially
heavy soils (TP = 265 mm), the threshold value of GWL is located at a depth of
1.77 m below the surface. GWL is then at a level below the σ or 2σ bottom border of
the evaluated soil layer lowered for. For the heavy soils (TP = 304 mm), GWL is
located at a depth of 1.84 m below the surface. As well in this case, the location of
GWL is below the σ or 2σ bottom border of evaluated soil layer lowered for. For the
very heavy soils (TP = 373 mm), it is located at a depth of 2.28 m below the
surface. GWL is then at a level between 2σ and 3σ.

Figures 10 and 11 illustrate the courses of daily values of actual and potential
evapotranspiration during the vegetation period in question in the positions
$\overline{GWL_{VP}^1}$, $\overline{GWL_{VP}^2}$, and $\overline{GWL_{VP}^3}$. It is obvious that the GWL position has a great
influence on water evaporation from soil. At the threshold position $\overline{GWL_{VP}^3}$, the
actual evapotranspiration in both areas was 135 and 117 mm. Actual $ET_{a, VP}$ at
$\overline{GWL_{VP}^2}$ was 178 and 74 mm.

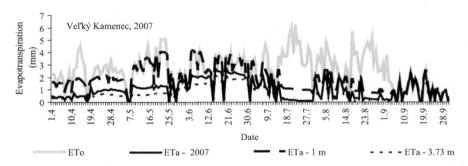

Fig. 10 The course of daily values of ET_0 and total daily volumes of ET_a at the positions of
$\overline{GWL_{VP}^1}$, $\overline{GWL_{VP}^2}$, and $\overline{GWL_{VP}^3}$

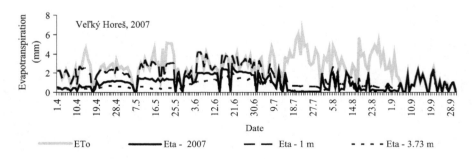

Fig. 11 The course of daily values of ET_0 and total daily values of ET_a at the positions of $\overline{GWL_{VP}^1}$,
$\overline{GWL_{VP}^2}$, and $\overline{GWL_{VP}^3}$

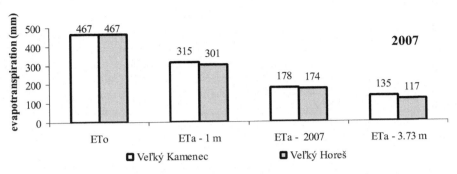

Fig. 12 The sum of potential evapotranspiration rates during the whole vegetation period and the sums of actual evapotranspiration rates at the positions $\overline{GWL_{VP}^1}$, $\overline{GWL_{VP}^2}$, and $\overline{GWL_{VP}^3}$

The real impact of groundwater on $ET_{a,\ VP}$ in the examined areas was 24 and 33% from the total amount of evaporation during the vegetation period (Fig. 12).

4   Conclusions

The experiment confirmed the significant role of groundwater in the hydrologic cycle. The impact of GWL position on the course of evapotranspiration rate, as well as the values of the actual evapotranspiration, was quantified. The impact of groundwater on water evaporation from soil was divided into three phases – linear, nonlinear, and residual phase. The boundary between linear and nonlinear phase was determined as GWL threshold position. It is a position, where water transfer from GWL to the root zone is negligible, and its impact on evaporation is minimal. In the particular conditions of the experiment, GWL threshold position was located 3.7 m under the ground. For the purposes of the experiment, extremely dry vegetation period of 2007 was chosen. The evaporation deficit during this period was 62 and 63% of the potential evaporation. It was shown that 24–33% of the water evaporated from the surface of the ground came from groundwater supply.

The results show that the impact of groundwater on the evaporation rates is directly proportionate to the hydrophysical characteristics of soil and GWL variations. GWL regulation could be an effective measure of soil water regulation during the periods of soil drought.

5   Recommendations

Hydrologic cycle is understood as multiple entities composed of many parts from the system point of view. Bonds exist between these parts. The cycle is divided into atmosphere, crop cover, unsaturated zone, and groundwater. Interaction processes run between each subsystem. Resultant of this processes influences soil water storage and its availability for plant cover. It manifests mainly during periods of meteorological drought. Groundwater influences water storage in the unsaturated zone of the soil profile and vapor through interaction processes. Intensity of this influence is dependent on the height of the groundwater level above the critical level and on the difference between ET_0 and Et_a. It is necessary to develop knowledge of interaction processes between elements of hydrologic cycle from the inscribed reasons. Gaining of new knowledge to mentioned issue was also the subject of this chapter.

Following tasks arise from submitted chapter to continue the research works:

1. Study the impact of texture on critical depth of groundwater level, threshold water storage in the root zone of soil profile, dependency course $\overline{WS}_{VO} = f(\overline{HPV}_{VO})$, and influence of groundwater on actual evapotranspiration in given hydropedological conditions. Quantification of the impact of each fraction or their combinations on said parameters is designed within the range of this research.
2. To verify identified dependencies $\overline{WS}_{VO} = f(\overline{HPV}_{VO})$ on the basis of monitoring the root zone water storage of soil profile (1 m) and position of groundwater level. To use for verification apart from own measurements and also partial results of actual monitoring of groundwater level and water storage.
3. To identify localities on East-Slovakian Lowland in which the groundwater level has small, average and higer impact on actual evapotranspiration and water storage of the soil profile into the depth of 1 m.

Weaknesses of the proposed approach:

The proposed approach is difficult in gaining of necessary data and subsequent data processing, numerical simulation, and interpretation. Data from field measurements are necessary.

The strength of this approach is gained benefits. Their importance is that it will be able to gain information about important parameters that are hard to quantify on the basis of easily measurable parameters. In other words, it will be able to gain information about their characteristic parameters and decisive balance components on the basis of easy and cheap measurements.

Acknowledgments The authors would like to thank for the kind support of the project VEGA 2/0062/16.

This contribution is the result of the project implementation: Centrum excelentnosti pre integrovaný manažment povodí v meniacich sa podmienkach prostredia/Centre of excellence for the integrated river basin management in the changing environmental conditions, ITMS code 26220120062; supported by the Research & Development Operational Programme funded by the ERDF.

References

1. Novák V (1995) Vyparovanie vody v prírode a metódy jeho určovania (Evaporation of water in nature and methods of determining). Veda, Bratislava, p 260
2. Wang X, Huo Z, Feng S, Guo P, Guan H (2016) Estimating groundwater evapotranspiration from irrigated cropland incorporating root zone soil texture and moisture dynamics. J Hydrol 543:501–509
3. Behrman KD, Norfleet ML, Williams J (2016) Methods to estimate plant available water for simulation models. Agric Water Manag 175:72–77
4. Kandra B, Tall A (2011) Determining the intensity and duration of soil drought by the method of effective precipitation. Crop Prod 60:373–376
5. Mathias SA, Sorensen JPR, Butler AP (2017) Soil moisture data as a constraint for groundwater recharge estimation. J Hydrol 552:258–266
6. Rodný M, Šurda P (2010) Stanovenie indexov meteorologického sucha a ich spojitosť s vodným režimom pôdy lokality Báč na Žitnom ostrove. Hydrologické dny 2010: Hradec Králové, sborník příspěvků. Nakladatelství Český hydrometeorologický ústav, Praha, pp 109–116
7. Csafordi P, Szabo A, Balog K, Gribovszki Z, Bidlo A, Toth T (2017) Factors controlling the daily change in groundwater level during the growing season on the Great Hungarian Plain: a statistical approach. Environ Earth Sci 76:675–675
8. Gomboš M, Pavelková D (2011) The impact of groundwater level position on the actual evapotranspiration in heavy soils in Eastern-Slovakian Lowland, vol 13. Ovidius University Annals Constantza – Civil Engineering XIII, pp 65–71
9. Stojkovová D, Orfánus T (2015) Occurence of drought in the regime of ground water accumulated in quarter sediments of western and Eastern Slovakia. Crop Prod 64:225–228
10. Kotorová D, Mati R (2008) The trend analyse of water storage and physical properties in profile of heavy soils. Agriculture 54:4155–4164
11. Štekauerová V, Skalová J, Nováková K (2010) Assignment of hydrolimits for estimation of soil ability to supply plants by water. Crop Prod 59:195–198
12. Kandra B (2010) The creation of physiological stress of plants in the meteorological conditions of soil drough. Crop Production 59:307–310
13. Mati R, Kotorová D, Gomboš M, Kandra B (2011) Development of evapotranspiration and water supply of clay-loamy soil on the East Slovak Lowland. Agric Water Manag 7:1133–1140
14. Orfánus T, Stojkovová D, Nagy V, Nemeth T (2016) Variability of soil water content controlled by evapotranspiration and groundwater-root zone interaction. Arch Agron Soil Sci 62:1602–1613
15. Šútor J, Vitková J, Rehák Š, Stradiot P (2014) Vplyv evapotranspiračného deficitu na dynamiku zásob vody v pôde v podmienkach Záhorskej nížiny. Acta Hydrol Slovaca 15:15–23
16. Sun S, Chen H, Wang G et al (2016) Shift in potential evapotranspiration and its implications for dryness/wetness over Southwest China. J Geophys Res Atmos 121:9342–9355
17. Šútor J, Štekauerová V, Nagy V (2010) Comparison of the monitored and modeled soil water storage of the upper soil layer: the influence of soil properties and groundwater table level. J Hydrol Hydromech 4:279–283
18. Majerčák J, Novák V (1994) GLOBAL, one-dimensional variable saturated flow model, including root water uptake, evapotranspiration structure, corn yield, interception of precipitations and winter regime calculation: research report. Institute of Hydrology S.A.S, Bratislava, p 75
19. Van Genuchten MT (1980) A closed equation for predicting the hydraulic conductivity of unsaturated soil. Soil Sci Soc Am J 44:892–898
20. FAO (1990) Annex V: FAO Penman-Monteith formula. Report from the expert consultation on revision of FAO methodologies for crop water requirements, Rome, 28–31 Mar 1990

Part IV
Water Quality

Assessment of Water Pollutant Sources and Hydrodynamics of Pollution Spreading in Rivers

Y. Velísková, M. Sokáč, and C. Siman

Contents

Abstract Water is a necessary component of the human environment, as well as all vegetal and animal ecosystems. Unfortunately, water quality not just in Slovakia but also in other countries of the world, worsened in the course of the twentieth century, and this trend has not been stopped even at present. Current legislation evaluating the quality of water bodies in Slovakia is based on the implementation of the Water Framework Directive (2000/60/ES). The Directive requires eco-morphological monitoring of water bodies, which is based on an evaluation of the rate of anthropogenic impact. This does not refer only to river beds but also the state of the environs of each

Y. Velísková (✉) and C. Siman
Institute of Hydrology, Slovak Academy of Sciences, Bratislava, Slovakia
e-mail: veliskova@uh.savba.sk

M. Sokáč
Faculty of Civil Engineering, Slovak University of Technology in Bratislava, Bratislava, Slovakia

A. M. Negm and M. Zeleňáková (eds.), *Water Resources in Slovakia:*
Part I - Assessment and Development, Hdb Env Chem (2019) 69: 185–212,
DOI 10.1007/698_2017_199, © Springer International Publishing AG 2018,
Published online: 24 May 2018

stream. While in the past point sources of pollution were considered as the most significant source of pollution in surface streams, after the installation of treatment plants for urban and industrial wastewater, non-point sources of pollution emerged as the critical sources of pollution in river basins. This contribution deals with the distribution and quantity assessment of pollutant sources in Slovakia during the period 2006–2015. The primary point sources evaluated are the ones representing higher values than the 90 percentile of the empirical distribution of total mass and also the mass of applied manures and fertilisers as non-point pollutant sources.

The development of computer technologies enables us to solve ecological problems in water management practice very efficiently. Mathematical and numerical modelling allows us to evaluate various situations of spreading of contaminants in rivers without immediate destructive impact on the environment. However, the reliability of models is closely connected with the availability and validity of input data. Hydrodynamic models simulating pollutant transport in open channels require large amounts of input data and computational time, but on the other hand, these kinds of models simulate dispersion in surface water in more detail. As input data, they require digitisation of the hydro-morphology of a stream, velocity profiles along the simulated part of the stream, calculation of the dispersion coefficients and also the locations of pollutant sources and their quantity. The highest extent of uncertainty is linked with the determination of dispersion coefficient values. These coefficients can be accurately obtained by way of field measurements, directly reflecting conditions in the existing part of an open channel. It is not always possible to obtain these coefficients in the field, however, because of financial or time constraints. The other aim of this contribution is to describe the methodology of this coefficient calculation and to present the value range obtained. The results and obtained knowledge about values of longitudinal dispersion coefficients and dispersion processes can be applied in numerical simulations of pollutant spreading in a natural stream.

Keywords Dispersion, Mixing, Numerical modelling, Pollution, Stream, Water quality

1   Introduction

Water is an essential component of the human environment, as well as all vegetal and animal ecosystems. The development of industry, transportation and agriculture, increase in living standards, an extension of urban areas and subsequent increase in storm water volumes transported by sewer systems all significantly influence the environment. Pollutants from point and area sources worsen the quality not only of water but also of soil and the atmosphere.

In the classification process of water sources, it is insufficient to classify the capacity (quantity) alone because the water quality is a determining factor for many applications (water supply, food industry, the pharmaceutical industry, irrigation). Water quality is defined as a representative dataset which defines the physical, chemical and biological water attributes from the possibilities of water use for different purposes (drinking water, recreational, industrial, agriculture, power generation, transport).

In the field of surface and groundwater protection, the situation in the Slovak Republic has significantly changed due to the accession of the country to the EU. Since that accession process, practically all the Slovakian legislation concerning water has been changed due to the acceptance of the principles of the EU Water Framework Directive (WFD 2000/60/EC), including the basic water management activity: protection of the quality of water sources. The Water Framework Directive requires as an obligatory goal to achieve and maintain "good water quality" status within the defined period (for Slovakia this period was set up to the year 2015). For surface water, the main criterion is the level of ecological and chemical quality.

In connection with the adoption of measures for improving the surface water quality, numerical simulation models are very useful tools which can simulate the consequences of the adopted measures, i.e. their suitability and efficiency, or otherwise to show the inefficiency or unsuitability of proposed measures.

2 Water Quality

Water quality is construed as affording the possibility of using water for the required purpose. However, the purpose itself is neither precisely defined nor essential. In practice, this means that it is not the chemical purity but its desirable properties that determine the quality of water. For instance, distilled water can be used for filling accumulator batteries, but it is not suitable for drinking purposes. Conversely, drinking water is not suitable as accumulator filler. Hence, what is an inappropriate component of water (e.g. minerals) in one case is precisely the desirable component in another.

It is necessary to bear in mind that achieving or maintaining good water status is the purpose of water use (see the definition in WFD 2000/60/EC, for instance). It is also necessary to take into account the degree of toxicity for waterborne organisms, or organisms bound to aquatic ecosystems, as well as the degree of toxicity for the environment in general.

It is also necessary to realise that the notion of water quality is a relative one, i.e. it changes in time and space.

Water protection is considered to be a basic water management activity towards which most of the activities performed in water management are directed. The integrated protection of water resources, which currently constitutes one of the limits to the development of human society, is the goal of this activity.

Politics and economic interests may also play a negative role in the area of water protection. Economic forecasters predict that, just as there are currently wars for oil, in the future there will be wars for water, which is becoming a restricting factor on the development of society in some locations, due to the depletion or deterioration of water resources.

The statement that life is not possible without water sounds like a platitude, but it nevertheless remains valid. However, it should be added that life is not possible without good-quality water, i.e. water not meeting the requirements for its use in terms of its quality: either as drinking, utility, irrigation or other water uses. Thus, water quantity and quality become the basic parameters of the utility value of water; this can be expressed as the resultant product of these two parameters. Hence, if one of these two parameters (water quantity or quality) is zero, the total utility value of the water is also zero (there is more water in a small pure spring than in a dirty river).

The role of water managers is not to maintain water in nature in an absolutely pure condition; after all, that is probably not even possible (except areas with strict nature and landscape protection). Their role in terms of sustainable development is rather to maintain water quality at an adequate level. It means to maintain water quality at such a level as to ensure the exploitation of water resources for the required purpose, or to ensure universal water protection, including aquatic ecosystems and ecosystems dependent on water. Generally, it means the improvement of water status and the effective and economical utilisation of waters. It is necessary to recognise that the requirement of "returning to the original state" is no longer feasible today, not to mention that it is not possible to define the "original state" of waters.

The development of human society down the centuries has also led to pressure on water quality protection, not only to ensure basic human requirements (drinking water) but also to utilise water in other spheres of human activity (e.g. industry, recreation, urban sanitation). At the turn of the nineteenth and twentieth centuries, water managers virtually became some of the earliest protectors of nature and also users of biotechnologies (wastewater treatment processes). The traditional philosophy was based on the principle of protecting people from nature. The increased sensitivity of the population to essential nature protection and the popularisation of environmentally friendly perspectives have also been reflected in the ambit of water protection, hence in the introduction of a new concept of nature protection. This means that new, opposing opinions on environmental protection prevail today, as well as the related requirements of protection of nature against people.

3 EU Legislation in the Area of Water Protection and Its Basic Principles

One of the first documents adopted jointly by the European Community was the European Water Charter [1]. It was prepared by the European Committee for the Conservation of Nature and Natural Resources of the Council of Europe and adopted on 6th May 1968 in Strasbourg. The Water Charter defines the basic principles of water protection and management which were later reflected in the overall EU policy.

1. There is no life without water. It is a treasure indispensable to all human activity.
2. Freshwater resources are not inexhaustible. It is essential to conserve, control and, wherever possible, increase them.
3. To pollute water is to harm humans and other living creatures which are dependent on water.
4. The quality of water must be maintained at levels suitable for the use to be made of it and, in particular, must meet appropriate public health standards.

5. When used water is returned to a common source, it must not impair the further uses, both public and private, to which the common source will be put.
6. The maintenance of adequate vegetation cover, preferably forest land, is imperative for the conservation of water resources.
7. Water resources must be assessed.
8. The wise husbandry of water resources must be planned by the appropriate authorities.
9. Conservation of water calls for intensified scientific research, training of specialists and public information services.
10. Water is our common heritage, the value of which must be recognised by all. Everyone must use water carefully and economically.
11. The management of water resources should be based on their natural basins rather than on political and administrative boundaries.
12. Water knows no frontiers: as a common resource, it demands international cooperation.

In the following part, the basic principles of EU legislation related to water protection are explained.

With regard to the scope of the individual legal documents, we may divide EU legislation into two basic groups. The first is called horizontal legislation which covers the entire environmental area (e.g. EIA, nature and landscape protection regulations), while the second is called vertical (specific) legislation which is focused more on the individual components of the environment (e.g. water, soil, air quality protection).

In the area of water protection, the EU legal system uses the following three forms of legislative documents:

- Directive
- Regulation
- Decision

An EU Directive expresses an endeavour to introduce universal legal norms whereby, however, it is possible to maintain traditional practice and adapt to the degree of development in the given countries. A Directive is a legal document which does not take precedence over the legislation of a Member State. However, Member States are required to "indirectly" apply a Directive, i.e. to apply the principles of the Directive which have to be absorbed into the legislation of the given Member State. However, the principle is that a Member State may adopt measures going "beyond the framework" of the respective Directive, but it must not adopt less strict criteria than those stipulated by the given Directive. Currently, within the EU, only Directives are being applied in water management. Absorbing the principles of EU Directives into the national legislation is referred to as transposition of the law; implementing the adopted measures is referred to as an implementation of the law.

An EU Regulation is a generally valid legislative document directly applicable within the territory of all Member States. From the legal perspective, a Regulation takes precedence over national law.

A Decision is a highly specific legal document, directly binding only for those for which it is intended. It is usually issued only when one of the Member States violates the provisions of EU legislation; it can be compared to a court decision.

It is evident that harmonisation of the legislative requirements of EU Member States is quite a demanding task, due not solely to differences of opinion on environmental protection but mainly due to the varying levels of protection in the individual Member States, which are related to their levels of economic and social development. For this reason, the EU bodies have adopted the principle of the so-called lowest common denominator, i.e. the primary determination of the minimal environmental protection requirements which are common and acceptable to all Member States. Following approximation and implementation, these minimal requirements will be increased incrementally until they achieve the target status (protection level) standard for all Member States.

EU environmental legislation recognises the following universal principles:

- Environmental protection must not encroach upon the protection of the EU internal market, nor constrain competition within the EU.
- Prevention is emphasised.
- Greening of the economy and social policy (ultimately of all activities).
- "Polluter pays" principle (PPP).
- Harmonisation and unification of Member States' legislation.
- Right of citizens to information on the status of the environment.

In addition to the above principles, specific principles also apply to water management and water protection. We list at least some of these principles here:

- Payment of all costs incurred by activities in the area of water management (WM) must be self-fundable.
- WM activities to be pursued based on natural river basins.
- Achievement (or maintenance) of the so-called good status of water bodies within the EU.

4 Legislation in Slovakia

4.1 Water Framework Directive (WFD, 2000/60/EC)

The Water Framework Directive (WFD, 2000/60/EC) is the primary legislative document for water quality (but also for the entire EU water management policy).

The WFD introduces a new approach to water management based on river basins, or natural geographical and hydrological units; it imposes specific deadlines on the EU Member States to develop river basin management plans including programmes of measures. The new approach to water protection makes it possible to create a unified system for water evaluation within the EU Member States, affording reliable and comparable results of the condition of water bodies in any European region. The next asset is the application of same procedure for the determination of objectives and implementation of all necessary measures for the protection and improvement of water status, as well. The WFD deals with surface waters (rivers, lakes); transitional, coastal waters; groundwaters; and, under certain specific conditions, also terrestrial ecosystems dependent on water and wetlands. The WFD introduces several innovative approaches to water management, such as public participation in planning and integration of economic approaches to the planning and integration of water management with other economic sectors.

The main objective of the WFD is to achieve the so-called good status of waters in the Member States, which will ensure the protection and improvement of the state of aquatic ecosystems and sustainable, balanced and equitable water use. This status should have been achieved by 2015 or must be achieved by 2027.

The European Commission developed a basic document for the EU Member States: the WFD Common Implementation Strategy adopted by the Member States in May 2001. This Strategy is regularly updated at 2-year intervals for the subsequent period.

4.2 Council Directive 91/271/EEC Concerning Urban Wastewater Treatment

The main objective of this Directive concerning urban wastewater treatment is the protection of aquatic ecosystems from the adverse effects of discharges of untreated or insufficiently treated urban wastewater.

The requirements of this Directive can be characterised as follows:

• The requirement to build a public sewage system and two-stage wastewater treatment plant in agglomerations of over 2,000 population equivalents (p.e.).
• Each discharge of wastewater must be permitted by the relevant authority.
• More stringent criteria in agglomerations of over 10,000 p.e., in the food industry and in sensitive areas (elimination of nutrients, nitrogen (N) and phosphorus (P)).
• Permits for wastewater discharges are subject to review.
• Emphasis on the reduction or disposal of pollution at the point of origin and reuse of treated water.
• Sludge must not be disposed of in surface waters, and it should be recycled.

The emission requirements of Directive 91/271/EEC on urban wastewater treatment are complemented by qualitative immission water protection requirements which are formulated in the related directives, mainly [2]:

• Directive 76/160/EEC concerning the quality of bathing water
• Directive 75/440/EEC concerning the quality required of surface water intended for the abstraction of drinking water
• Directive 78/659/EEC on the quality of freshwaters requiring protection or improvement in order to support fish life

Based on the requirements of this Directive, it is quite evident that the implementation of these requirements demands major measures and costs. There are mainly investments in the construction of new sewage systems and wastewater treatment plants (WWTP) and in the renovation of the existing systems and reconstruction of existing WWTPs (alteration of technologies to extended disposal of bionutrients).

5   Water Pollution Sources

Sources of pollution are considered to be any activity or phenomenon resulting in a deterioration of water quality. Based on the geographical form, each water pollution source can be categorised as follows:

- Point sources (e.g. sewage system outflow, oil spillage)
- Line sources (e.g. transport structures or pipelines)
- Diffuse sources (e.g. numerous leaking cesspits in a village)
- Areal sources (e.g. agricultural pollution such as fertilisers, pesticides, herbicides; exhaust gases, precipitation)

Significant water pollution sources are usually included in tabular or map form in the basic water management land-use planning documents.

A point source of water pollution is a pollution source with a concentrated input of pollution into waters which is limited to a relatively small area or almost confined to a single geographical point. These pollution sources are usually precisely quantifiable, so it is usually easy to monitor them, and the impact of every individual source can be accurately determined. As a result of diffusion and transport of the pollutant, linear or areal contamination of groundwater or surface waters can occur.

Line sources of pollution usually consist of leaks of pollutants along transport and traffic structures such as highways or railways, or along with other transport facilities such as oil pipelines or large sewage collectors.

In the literature, diffuse sources of pollution are usually understood as several point sources of pollution together, whereby it is not possible to determine the impact or effect of the individual (point) source. A village with leaking cesspits or septic tanks which, in combination, act almost as an areal source of pollution but where, in this case, there are several point sources of pollution is a typical example.

Areal sources of pollution are those where the pollutant is input over a large area. It is usually not possible to quantify the pollutant nor to accurately demarcate the point of penetration of the pollutant. In these cases, this is primarily groundwater pollution. Agricultural activities are a typical example, e.g. areal application of fertilisers and pesticides.

5.1   Water Quality and Pollution Source Deployment in Slovakia

Surface water quality at all monitored sites complied in each year with the limits for selected general indicators and the radioactivity indicators. Exceeded limit values were recorded mainly for synthetic and non-synthetic substances, hydro-biological and microbiological indicators and nitrite nitrogen. Until 2007, surface water quality was assessed according to STN 75 221 in five quality categories and eight indicator groups. In the years 1995–2007, 40–60% of abstraction sites showed the fourth and fifth quality categories for the groups of F (micro-pollutants) and E (biological and microbiological indicators) [3].

In line with the requirements of WFD 2000/60/EC, water quality is expressed in terms of the ecological and chemical balance of surface water bodies. Adverse and critically adverse ecological situations are recorded in approx. 4–8% of water bodies, and approx. 3–10% do not reach good chemical balance.

Monitoring for groundwater chemical balance is carried out as part of basic monitoring (171 stations) and operational monitoring (295 stations). Both types of monitoring show exceeded values for set contamination limits. In 1995–2006, groundwater quality was assessed according to STN 75 7111 in 26 water management significant areas.

Major sources of contamination of water bodies include residential agglomerations, industry and agriculture. The main point sources of surface water pollution comprise industrial plants and wastewater treatment plant outlets. Applications of fertilisers in agriculture represent an areal source of pollution [4, 5].

The contamination of surface water bodies is characterised in general by chemical oxygen demand by dichromate (COD_{Cr}), biochemical oxygen demand (BOD), total nitrogen amount and insoluble substances (IS). Distribution of the main producers by these parameters in Slovakia is shown in Figs. 1, 2, 3 and 4.

Fertilisers applied in agriculture are divided into two groups: industrial fertilisers based on chemicals (N-P-K) (nitrogen, phosphorus, potassium) and organic fertilisers. Their consumption is summarised by the district in each year. The total amounts of applied N-P-K and organic fertilisers in each district during the 10-year period (2006–2015) are shown in Figs. 5 and 6. The consumption of these fertilisers can also be monitored in kilogrammes per hectare of agricultural soil, as shown in Figs. 7 and 8. As can be seen from these figures, the distribution of applied amounts of fertiliser is different in partial districts in this case. So, from the point of view of water contamination evaluation, it is important to pick out suitable and comparable parameters and units.

The total annual consumption of specific kinds of fertiliser was different in each year. It turns out that the minimum amount of industrial fertilisers was applied in 2009 and 2010 (Fig. 9). From these years onwards, the N-P-K fertiliser consumption has increased in each year. In contrast, organic fertiliser application has slightly decreased.

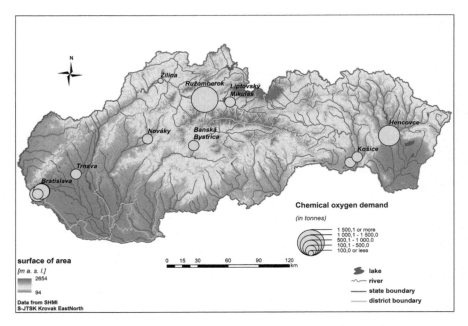

Fig. 1 Deployment of the main producers of surface water contamination based on the average discharge amount of COD_{Cr} (in the period 2006–2015)

6   Hydrodynamic Numerical Models of Pollutant Transport

The development of computer technologies enables us to solve ecological problems in water management practice very efficiently. Mathematical and numerical modelling allows us to evaluate various situations of contaminant spreading in rivers without immediate destructive impact to the environment.

A lot of mathematical and numerical models have been developed to simulate water quality (e.g. WQMCAL, AGNSP, CORMIX, QUAL2E, SWMM, P-ROUTE, MIKE1, ZNEC, MODI, HSPF, SIRENIE). These models are based on various approaches, including hydrodynamic, statistical and balanced (reviewing past similar events). These models can simulate the real situation in streams. However, the range of reliability and accuracy of the results is vast [4, 6–12].

Problems of dependability and correctness of dispersion numerical models are wide-ranging, and it is not possible to cover them only in one chapter or study. For this reason, part of this chapter is focused on simulation models based on the hydrodynamic approach, i.e. models based on numerical solution of the advection-dispersion equation and determination of one of the leading characteristics of mixing processes in streams: the longitudinal dispersion coefficient.

Simulation models can describe the transport of contaminants even three-dimensionally. On the other hand, the input is labour intensive, and the development

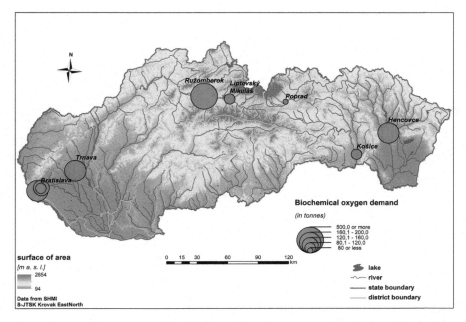

Fig. 2 Deployment of the main producers of surface water contamination based on the average discharge amount of BOD (in the period 2006–2015)

of the model structure is time-consuming as well. However, this lengthy approach is important for deep reservoirs with thermal stratification. Two-dimensional modelling of the transport processes is reasonably accurate if applied to shallow reservoirs without strong thermal stratification. Otherwise, it can be used in detailed studies of the movement of pollutants into the surface water before complete mixing of transported substances across the section of flow occurs (the so-called mixing length). After this moment it is sufficient to apply a one-dimensional model of water quality.

The reliability of models is influenced by the fact that numerical simulations always mean some simplification of the complicated natural conditions. Finally, the rate of reliability is closely connected with the level of input availability and validity [13].

Hydrodynamic models simulating pollution transport in open channels require a good deal of input data and computation time, but on the other hand, these kinds of models simulate dispersion in surface water in more detail. As input data, they require digitisation of the hydro-morphology of the stream bed, velocity profiles along the simulated part of a stream, calculation of the dispersion coefficients and also the positions of pollutant sources and their massiveness. Calculation of dispersion coefficient values has the highest extent of uncertainty. These coefficients can be exactly obtained by way of field measurements, directly reflecting conditions in an existing part of an open channel. It is not always possible to obtain these coefficients in the field though, because of financial or time reasons. Several authors [8, 11, 12, 14–22] have

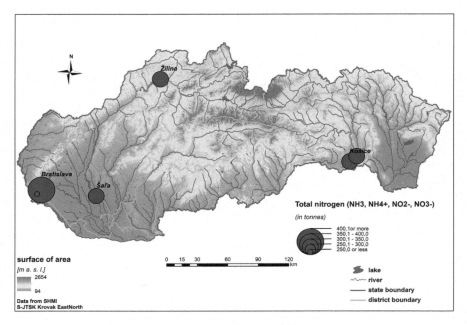

Fig. 3 Deployment of the main producers of surface water contamination based on the average discharge amount of total nitrogen (in the period 2006–2015)

tried to get the empirical relations, especially for the longitudinal dispersion coefficient. Their studies could be used to estimate an approximate value of the dispersion coefficient. The results of studies [11, 23–26] on the conditions of Slovakian rivers show that the formulae derived in this way are often not applicable for those rivers. The reasons behind this are the longitudinal slopes used in the formulae are very flat, or the precondition of stream channel roughness is not suitable or the shape of the cross-section profile is inappropriate. So, for this reason, we need to try to obtain the applicable range of this dispersion coefficient values.

7   Longitudinal Dispersion

7.1   Basic Theoretical Terms

Dispersion, from the hydrodynamic point of view, is the spreading of mass from highly concentrated areas to less concentrated areas in flowing fluid. Mass in flowing water is not transported only in the reach of the stream line, but it also gradually spreads outside that line as a consequence of velocity pulsations and mass concentration differences. Mass dispersion with advection is the basic motion mechanics of particles transported in water. Reductions in maximum concentrations are the result

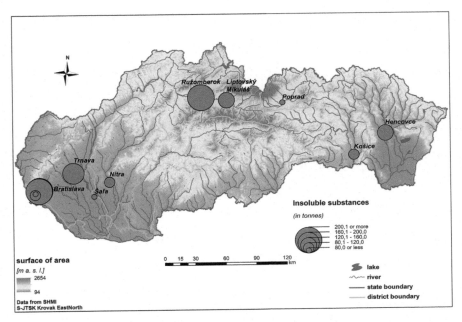

Fig. 4 Deployment of the main producers of surface water contamination based on the average discharge amount of IS (in the period 2006–2015)

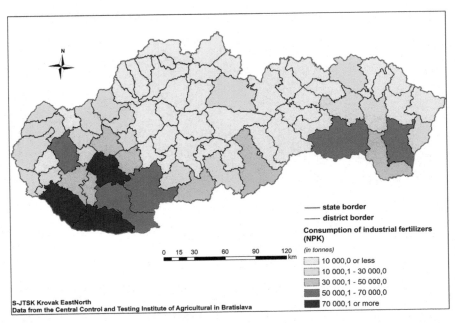

Fig. 5 Total amount of applied N-P-K fertilisers in districts during the 10-year period (2006–2015)

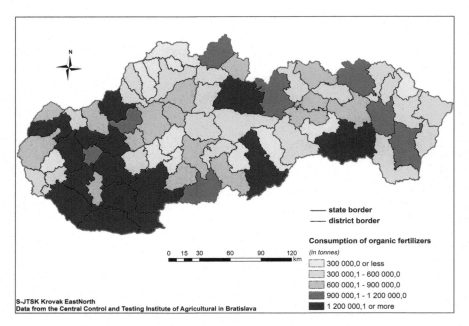

Fig. 6 Total amount of applied organic fertilisers in districts during the 10-year period (2006–2015)

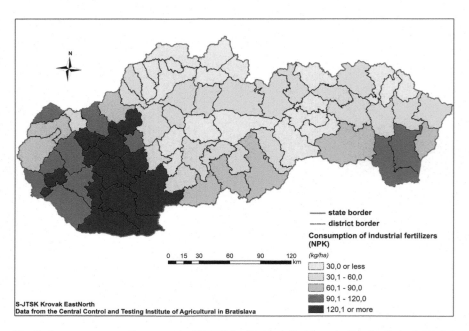

Fig. 7 Average consumption amount of N-P-K fertilisers applied to agricultural soils in the period 2006–2015

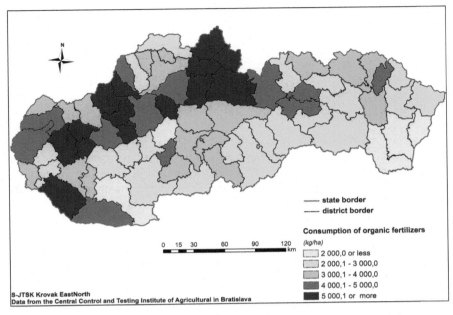

Fig. 8 Average consumption amount of organic fertilisers applied to agricultural soils in the period 2006–2015

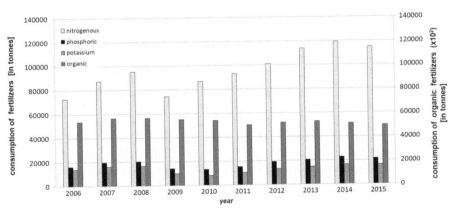

Fig. 9 Trend in fertiliser application during the period 2006–2015

of the effects of those mechanics. The main characteristics of dispersion are dispersion coefficients in relevant directions. Identification of these dispersion characteristics is the key task for solving the problem of pollutant transport in streams and for modelling of water quality.

The most straightforward description of mass spreading in water is the one-dimensional advection-dispersion equation, which takes the phenomenon in longitudinal direction x (well-proportioned distribution of mass concentration is required along the depth and across the width of a stream). The form of this equation is [25, 27]

$$\underset{\text{I}}{\frac{\partial Ac}{\partial t}} + \underset{\text{II}}{\frac{\partial Qc}{\partial x}} - \underset{\text{III}}{\frac{\partial}{\partial x}\left(AD_L\frac{\partial c}{\partial x}\right)} = \underset{\text{IV}}{-AKc} + \underset{\text{V}}{c_s q} \tag{1}$$

where c is the mass concentration (kg m^{-3}), D_L is the longitudinal dispersion coefficient (m^2 s^{-1}), A is the discharge area in the stream cross-section (m^2), Q is the discharge in the stream (m^3 s^{-1}), K represents the rate of growth or decay of contaminant (s^{-1}), c_s is the concentration of the contaminant source, q is the discharge of the source, x is the distance (m) and t is time (s).

Part I in Eq. (1) expresses pollutant concentration change in time, part II represents pollutant transport through the velocity field, part III describes pollutant transport by diffusion and dispersion, part IV means chemical or biological nonconservation of pollutant, and part V represents pollutant sources in the stream.

Equation (1) covers two basic transport mechanisms:

- Advection (or convection) transport by fluid flow
- Dispersion transport by the concentration gradient

This one-dimensional approach is applicable for rivers or streams with comparatively non-wide channels, or for sewers, for example. In this case, the pollutant spreading has markedly one-dimensional character. However, this assumption is not acceptable for reservoirs, where the spreading phenomenon has three-dimensional character, meaning that the hydraulic characteristics and their values vary with the width as well as the depth of the discharge cross-section.

As the dispersion coefficient value is affected by the turbulence intensity in the given stream section, its magnitude depends upon its main hydraulic characteristics: the shape and magnitude of its cross-section profile, its flow velocity and its longitudinal slope. For this reason, the relationships derived by several authors for calculating the coefficient use the same characteristics (see Table 1).

Most of the published relationships used for calculating D_L are based on experimental results from laboratory physical models, or directly from field measurements on the rivers themselves. Such relationships are often expressed in the following form [9, 12, 25, 27]:

$$D_L = ph u_* \tag{2}$$

where p is the empirical dimensionless coefficient, h is the mean river section depth (m) and u_* is the friction velocity (m s^{-1}).

The empirical dimensionless coefficient p acquires values, according to the authors concerned, in a fairly wide range, depending on the particular local conditions. This can be documented based on the results of Elder [16], Krenkel and Orlob [21] (laboratory conditions), as well as those of Říha et al. [9, 12, 15], Pekárová and

Table 1 Relationships for assignment of D_L

Author	Relationship	D_L (m^2 s^{-1})
Parker [28]	$D_L = 14.28 R^{3/2} \sqrt{2 \cdot g \cdot i}$	0.345
Elder [16]	$D_L = 5.93\, u_* h$	0.101
Yotsukura and Fiering [29]	$\frac{D_L}{h \cdot u_*} = f\left(\frac{u}{u_*}\right)$	0.222
Krenkel and Orlob [21]	$D_L = 9.1\, u_* h$	0.155
Thackston and Krenkel [30]	$D_L = 7.25\, u_* h \left(\frac{u}{u_*}\right)^{1/4}$	0.113
Fischer [17]	$D_L = 0.011\, \frac{u^2 W^2}{u_* h}$	11.37
McQuivey-Keefer [31]	$D_L = 0.058\, \frac{Q}{i \cdot W}$	1.323
Kosorin [20]	$D_L = \dfrac{0.0696\, W^2 u_{\max}^2}{H \sqrt{g H i}}$	97.78
	$D_L = \frac{0.278\, W^2 u^2}{H \sqrt{g H i}}$	183.96
	$D_L = \frac{0.688\, Q^2}{H^3 \sqrt{g H i}}$	268.78
	$D_L = \frac{0.172\, W^2 C^2 \sqrt{i}}{\sqrt{g H}}$	309.70
Říha et al. [12, 15]	$D_L = \frac{0.001617}{p} \cdot \frac{W^2 C^2 \sqrt{i_e}}{\sqrt{g H}}$	
Kashefipour-Falconer [32]	$D_L = 10.612\, \frac{h u}{u_*}$	
Sahay–Dutta [33]	$D_L = 2 h u \left(\frac{W}{h}\right)^{0.96} \left(\frac{u}{u_*}\right)^{1.25}$	

R hydraulic radius, g acceleration of gravity, i longitudinal slope of stream, u flow velocity, W width, Q discharge, H maximum depth in cross-section profile, C coefficient by Chézy

Velísková [11], Brady and Johnson [34], and Glover [35]. The latter results were derived from field experiments on natural streams. The conditions of field measurements are briefly given in Table 2.

The reliability of models is influenced by the fact that the numerical simulations always involve simplification of the complicated natural conditions. Ultimately the rate of reliability is intimately connected with the level of input availability and validity.

The problem of contaminant spreading is current not only in the case of modelling of water quality in natural streams but also in urbanistic structures, i.e. in sewer networks. For that reason, it is necessary to pay attention to this fact. One aspect of our interest is, therefore, the calculation of the longitudinal dispersion coefficient for prismatic channels, in this case for sewers, from field experiments.

7.2 Field Measurements

Our field measurements were done at the experimental hydrological base of the Institute of Hydrology in Liptovský Mikuláš. Part of a sewer network built in

Table 2 Values of dimensionless dispersion coefficient p and longitudinal dispersion coefficient D_L from experimental measurements

Author	Conditions	D_L (m^2 s^{-1})	p
Říha et al. [9, 12]	Svitava river $B = 11.5$ m; $H = 1.1$ m; $Q = (3.4–3.7)$ m^3 s^{-1}; $u = (0.4–0.5)$ ms^{-1}	7.2–8.1	15–23
Říha et al. [9]	Svratka river $B = (20–24)$ m; $H = (0.85–1.9)$ m; $Q = 12$ m^3 s^{-1}; $u = (0.25–0.45)$ ms^{-1}	7.2–9.5	21–32
Brady and Johnson [34]	River Wear $B = (20–27.6)$ m; $h = (0.45–1.85)$ m; $Q = (1.62–3.93)$ m^3 s^{-1}; $u = (0.07–0.15)$ ms^{-1}; $i = (0.004–0.17)\%$	4.4–87.22	94.9–2,200
Velísková and Pekárová [11]	Ondava river (upper part) $B = 12$ m; $H = 0.28$ m; $Q = 0.904$ m^3 s^{-1}; $u = (0.35–0.51)$ ms^{-1}; $i = 0.0033$	0.84–1.36	49.3–80
Glover [35]	South Platte River Meandering river $R = 0.46$ m; $Q = 15$ m^3 s^{-1}; $u = 1.33$ ms^{-1}; $n = 0.028$	15.7	500
Glover [35]	Mohawk River Complicated flow conditions, power station, reservoir, inflows $h = 6$ m; $Q = 30$ m^3 s^{-1}	6.0	800
Fischer [17]	Rectangular laboratory flume	0.0072–0.063	8.7–30
Fischer [17]	Triangular laboratory flume	0.123–0.415	190–640

2004–2005 as part of the ISPA project entitled "Development of the Environment in the Liptov Region" (more specifically the connecting collector between Liptovský Hrádok and Liptovský Mikuláš) was selected for field measurements.

The collector has a profile of DN 500 or 600 mm, and it lies in an area of low slopes (from 2 to 9.5‰), which are near to minimal slopes. After more detailed reconnaissance, two collection parts were selected for measurements: the first is above Podtureň village and the second part is just above the Borová Sihoť campsite, both near Liptovský Hrádok. In the first part (Fig. 10), distributions of tracer concentration in time were measured at various distances in a straight line.

In the second part, there is some curving of the sewer track (30°, 45° and 90° trajectory diversion) in which the distributions of tracer concentration in time were measured in various parts of the sewer (Fig. 11). The aim was to determine the influence of these trajectory diversions on the magnitude of the dispersion coefficient.

Common salt (NaCl) was used as a tracer, and this influenced the variation in wastewater flow conductivity. Fluorescein dye was added to the tracer to monitor the course of the tracer substance along the mensural profile. The dosage of tracer was 5 L, which was discharged into the sewer in a single injection (Fig. 12).

The measurement of conductivity was performed with a portable conductivity metre in a mensural manhole. The conductivity metre probe was situated in the

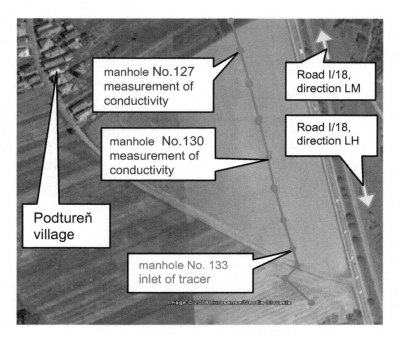

Fig. 10 Sewer collector – straight part (above Podtureň village)

centre of the wastewater flow. The conductivity values were shown on the conductivity metre display in the digital form. A stopwatch was located next to the conductivity metre (see Fig. 13). Fluctuation in the values was recorded by the camcorder, and the record of measurement was manually digitalised subsequently.

The output of measurements is the record of tracer concentration distribution in time in terms of the distribution of wastewater conductivity in the sewer. Examples of graphical expression of this record are given in Figs. 14, 15 and 16.

Evaluation of the experimental results consists in the simulation of the tracer experiment (concentration distribution) for various values of the longitudinal dispersion coefficient.

The basis for this numerical simulation is the analytical solution of Eq. (1) for instantaneous injection of tracer [36]:

$$c\left(x,t\right) = \frac{G}{2A\sqrt{\pi D_L t}} \cdot \exp\left[-\frac{\left(x - \bar{u}t\right)^2}{4D_L t}\right] \tag{3}$$

where $c(x,t)$ is the mass concentration at one place and time, D_L is the longitudinal dispersion coefficient, A is the discharge area in the stream cross-section, G is the mass of the tracer, u is the mean velocity and x is the distance and t is the time.

The difference between the measured and simulated values was evaluated. The minimum difference determined the value of the longitudinal dispersion coefficient

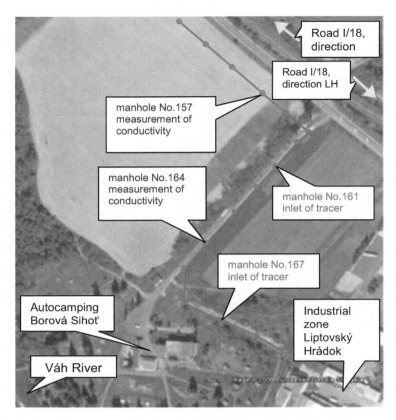

Fig. 11 Sewer collector – incurving of sewer track (near Borová Sihoť campsite)

Fig. 12 Tracer preparation, conductivity measurement

Fig. 13 Measuring device in a manhole

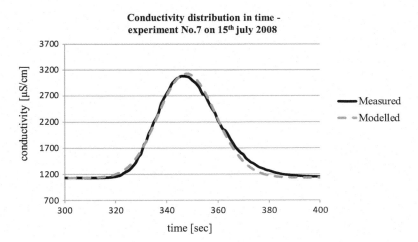

Fig. 14 Behaviour of conductivity in mensural profile (manhole no. 127), experiment no. 7, 15th July 2008 (injection of tracer at manhole no. 133)

for each of the experiments. Although the probe was located in the streamline, it ignored the irregular distribution of concentration (conductivity) across the width of the mensural profile. This was the reason for adding a correction coefficient to the model calculations, which was derived from the ratio of inflow tracer volume and outflow tracer volume.

For the experiment series with lower discharge (and thus also lower water level), the measurement processing revealed that in the given measurement section, there was a stream flow obstacle, creating a kind of "dead zone" in it. Most significantly

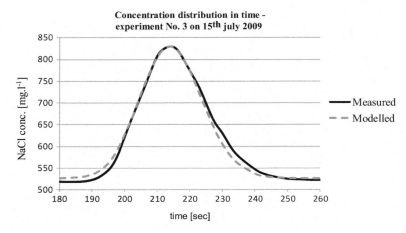

Fig. 15 Behaviour of concentration distribution in mensural profile (manhole no. 164), experiment no. 3, 15th July 2009 (injection of tracer at manhole no. 167)

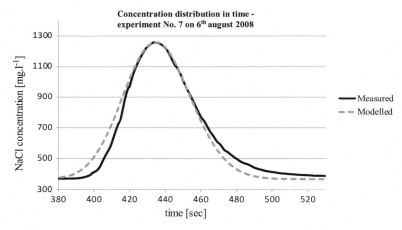

Fig. 16 Behaviour of concentration distribution in mensural profile (manhole no. 127), experiment no. 7, 6th August 2008 (injection of tracer at manhole no. 133)

this phenomenon showed up in the section with the 90° bend (there were also the lowest water depths). Figure 15, therefore, incorporates one of the conductivity examples exactly from this section. The tracer accumulated in this dead zone and was released gradually later. That distorted the conductivity distribution curve, making it asymmetrical and giving it a "tail" because of the later tracer release. This can be seen in the left part of Fig. 16. For this reason, for evaluation, we considered as decisive the concentration rising wave part.

The results of field measurements in the sewer network show the values of longitudinal dispersion coefficient ranging between 0.09 and $0.12 \text{ m}^2 \text{ s}^{-1}$ in the straight part and between 0.03 and 0.07 $m^2 s^{-1}$ in part with modifications to the sewer track direction.

Field measurements were also performed in various surface water bodies in Slovakia. Two typical cases of stream types in Slovakia were selected. The first one (further marked as "case A") is a typical lowland stream (Malá Nitra stream), where the water velocity and turbulence are very low. For contrast, in the second case ("case B") we picked out a typical mountain stream (upper reach of the river Hron) with a high degree of turbulence.

7.2.1   Case Study A, Malá Nitra Stream

Field measurements were performed along an approximately 400 m section of the stream Mala Nitra, close to the village of Veľký Kýr (Fig. 17). This stream is situated in the south-western part of Slovakia; measurements were taken in Veľký Kýr settlement region (N +48° 10′ 50.02″, E +18° 9′ 19.60″). The regime of the stream discharge is affected by flow regulation in the form of a weir located 15 km upstream at the bifurcation point with the Nitra River. Cross-sections were initially been trapezoidal, but the discharge area along the stream has been slightly changed by natural morphological processes over the years. The longitudinal bed slope was 1.5‰. Measured discharge values during field experiments were within the interval $(0.138–0.553)$ $m^3 s^{-1}$.

7.2.2   Case Study B, River Hron

The mainstream flowing through the town Brezno is the river Hron. This river has a partially alpine runoff regime with maximum flows in April and minimum flows in January. The average annual discharge of the Hron in the Brezno profile, at river km point 243,200, is about 8 $m^3 s^{-1}$.

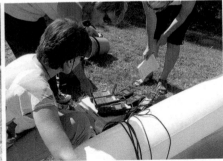

Fig. 17 Measurements along the Malá Nitra stream (measurements of conductivity in cross-section profiles on the left; comparison and calibration of used conductometers on the right)

Table 3 Values of longitudinal (D_L) and dimensionless (p) dispersion coefficient

		Case study			
		A (Malá Nitra stream)	B (Hron river)	Sewer pipe (prismatic channel bed)	
				Straight part	Part with modifications of track direct
Analytical solution (Eq. 3)	D_L (m² s⁻¹)	0.12–0.18	1.05–1.45	0.09–0.12	0.03–0.07
	p (−)	5–11	21–53	11–25	5–39

The adjusted river bed has a trapezoidal cross-section shape, and the river bed is partially stabilised with stone backfill. Over the years, the shape of the cross-section profile has been partially modified by natural morphological processes (as well as by the stone backfill). The average cross-sectional velocity was 0.64 m s⁻¹, but locally the velocity reached values up to 1 m s⁻¹. Measured discharge values during the fieldwork ranged from 4.2 to 5.3 m³ s⁻¹. The longitudinal bed slope was more than 3‰.

To determine the longitudinal dispersion coefficient, a tracer with a known quantity and concentration was discharged into the geometric centre of the stream width at the beginning of the measured section in a single injection. A solution of common salt (NaCl) was used again as the tracer, causing a change in the flowing water conductivity.

The velocity distribution and discharge were measured at each cross-section for all tracer experiments. Subsequently, the time courses of tracer concentration were monitored at each measured cross-section of the stream. Measured cross-section profiles were distributed evenly along the length of the examined section. Conductivity measurements were completed with portable conductivity metres, located in the centre or evenly across the cross-section width. Measurements were always carried out from the beginning of the increase in conductivity values (front of the tracer wave) until the original (background) conductivity values were restored in each cross-section profile. Each run of the tracer experiment was repeated at least two times.

The same methodology was used here for calculation of D_L values as in the case of measurements in the sewer pipe. Values of the longitudinal and dimensionless dispersion coefficient from all field experiments are summarised in Table 3.

It can be seen that the higher values of longitudinal dispersion coefficient are typical for a natural stream with a higher degree of turbulence. Despite this, D_L values obtained from the lowland stream (case A) are only slightly higher than those from the prismatic channel/sewer pipe measurements. Moreover, the results show that it is necessary to consider the influence of transverse flow in the assignment of longitudinal dispersion values in curved parts of the stream.

8 Conclusions and Recommendations

The issue of water quality is the key to sustainable human development. This chapter gives basic information about terms linked with questions and problems of water quality and contaminant spreading in natural streams in Slovakia. Brief information is also given about legislation in the area of water resource protection valid in

Slovakia nowadays. Despite all the activities concerning water quality protection, several challenges still face us.

By the EU Water Framework Directive, the protection of water resources in use or resources prepared for use to meet a water management need should be construed as the integrated protection of quality and quantity of groundwaters and surface waters. The issue of the sources of water pollution, with either direct or indirect impact on water resources, is the decisive factor in the protection of water resources quality. Water quality protection is based on maintaining the possibility of utilising the water (water resources) for the required purpose. Accordingly, the objective is not to prevent the transport of pollutants into the waters but to maintain their quantity and concentration at such a level that the long-term utilisation of water is rendered possible.

If humans are to manage water resources, it is necessary to know the demands for water from various aspects, and the possibilities, dangers and risks involved, but also the processes of water flow and pollutant transport and spreading. Knowledge of these processes helps us to predict the future status of water resources and design suitable hedges against damage to water resources. These processes in natural conditions are so complex that without using numerical methods and computers, this would be impossible. On the other hand, the outputs of simulation models are only as reliable as their inputs.

Nevertheless, numerical models are useful tools for resolving water quality issues. They need a particular volume of input data, but they also make it possible to evaluate various alternatives of precautions and remedies. One of the crucial parameters or inputs of models simulating contaminant spreading in natural streams is the dispersion coefficient. Its value strongly influences the simulation and calculation results. It is necessary therefore to determine its value as correctly as possible.

One of the ways of determining the dispersion coefficient values is through tracer field measurements. This method covers all typical peculiarities in evaluated conditions or backgrounds in the field. For comparison of different conditions, we performed tracer experiment in a sewer pipe and in two natural streams. The sewer pipe was selected as a model of a prismatic channel with/without modifications to track direction, and in the case of natural streams, we tried to choose different types of streams. The Malá Nitra is a typical lowland stream with low velocities and turbulence, whereas the river Hron is a mountain stream with high velocities and turbulence. Both streams are typical for specific regions in Slovakia.

Results from our sewer pipe experiments show that in comparison with the values gained from the straight line section, it is evident that the values of D_L in the curved part are lower than in the straight, direct line. In contrast, the values of the dimensionless coefficient p show more significant diffusion, and the values specifically in the curved part are higher. These results show and confirm that it is necessary to consider the influence of transverse flow too in the assignment of longitudinal dispersion values in curved sections of a stream. In this case of curved line route, the transverse flow influences the dispersion mechanism in the watercourse. The values of D_L determined using empirical formulas with the inclusion of this assumption also correspond to measured values better than relations which are derived only from longitudinal diffusion process assumptions. Since the shapes of measured curves of distribution of conductivity show the occurrence of "dead zones" in several parts, the empirical relations which were also used included this fact. The range of values of D_L calculated using these empirical relationships for the hydraulic parameters of each channel and given discharge was nearly identical with measured values.

There were three angles of curvature in the measured route part: 90°, 135° and 105°. The influence of the angle of curvature of longitudinal dispersion coefficient value has so far not been traced down from implemented measurements.

Comparison of measured and calculated longitudinal dispersion coefficient values in curved stream sections confirms that the measured values are near to the values found in laboratory conditions. This fact results from similarity of flow conditions in the selected sewer section to those in a laboratory channel.

Results from tracer experiments in natural stream conditions show that despite careful selection of the investigated stream section, in both cases the hydraulic conditions did not meet ideal flow conditions and that the investigated channel parts were not so prismatic as we supposed. In the case of study A, the reasons were sediments, zones with relatively thick silts or other objects deforming the velocity field and retention in so-called dead zones, which caused deformation of the tracer cloud. In the case of study B, the reason was the irregular distribution of large rocks (boulders) in the river bed, which formed areas with significantly different flow velocities. Such areas with different flow velocities generate a "meandering" stream line with significant deformations. All results show that the spread of tracer was not optimal, preferential flows were established and thus distortions of tracer cloud occurred.

However, as expected, the values of dispersion coefficients were higher in case of study B on the river Hron.

The obtained results and experience can be used for numerical simulation and prediction of water quality and contaminant transport in the investigated streams or similar types by using the values of the dimensionless dispersion coefficient. Using numerical models, it is also possible to design alternative solutions for treated wastewater release back into recipient water bodies without any risks to the water quality and biota.

Acknowledgement This chapter was created with support from VEGA project no. 1/0805/16. This contribution/publication is the result of the project implementation ITMS 26220120062 Centre of Excellence for the Integrated River Basin Management in the Changing Environmental Conditions, supported by the Research and Development Operational Programme funded by the ERDF.

References

1. European Water Charter. International Environmental Agreements Database Project. https://iea.uoregon.edu/treaty-text/1968-europeanwatercharterentxt. N.p., n.d. Web. 6 Nov 2017
2. Hansen W, Kranz N (2003) EU water policy and challenges for regional and local authorities. In: Background paper for the seminar on water management, Ecologic Institute for International and European Environmental Policy, Berlin and Brussels, Apr 2003. https://www.ecologic.eu/sites/files/download/projekte/1900-1949/1921-1922/1921-1922_background_paper_water_en.pdf. N.p., n.d. Web. 6 Nov 2017

3. State of the environment reports of the Slovak Republic from 2005 to 2015. http://enviroportal. sk/spravy/kat21
4. Rankinen K, Lepisto A, Granlund K (2002) Hydrological application of the INCA model with varying spatial resolution and nitrogen dynamics in a northern river basin. Hydrol Earth Syst Sci 6(3):339–350
5. Rhodes AL, Newton RM, Pufall A (2001) Influences of land use on water quality of a diverse New England watershed. Environ Sci Technol 35(18):3640–3645
6. Abbott MB (1978) Commercial and scientific aspects of mathematical modelling. Applied numerical MODELLING. In: Proceedings of 2nd international conference, Madrid, Sept 1978, pp 659–666
7. Gandolfi C, Facchi A, Whelan MJ (2001) On the relative role of hydrodynamic dispersion for river water quality. Water Resour Res 37(9):2365–2375
8. Jolánkai G (1992) Hydrological, chemical and biological processes of contaminant transformation and transport in river and lake systems. A state of the art report. UNESCO, Paris, 147 pp
9. Julínek T, Říha J (2017) Longitudinal dispersion in an open channel determined from a tracer study. Environ Earth Sci 76:592. https://doi.org/10.1007/s12665-017-6913-1
10. McInstyre N, Jackson B, Wade AJ, Butterfield D, Wheater HS (2005) Sensitivity analysis of a catchment-scale nitro-gen model. J Hydrol 315(1–4):71–92
11. Pekárová P, Velísková Y (1998) Water quality modelling in Ondava catchment. VEDA Publishing, Bratislava. ISBN 80-224-0535-3. (in Slovak)
12. Říha J, Doležal P, Jandora J, Ošlejšková J, Ryl T (2000) Water quality in surface streams and its mathematical modelling. NOEL, Brno, p 269. ISBN 80-86020-31-2. (in Czech)
13. Sanders BF, Chrysikopoulos CV (2004) Longitudinal interpolation of parameters characterizing channel geometry by piece-wise polynomial and universal kriging methods: effect on flow modeling. Adv Water Resour 27(11):1061–1073
14. Bansal MK (1971) Dispersion in natural streams. J Hydrol Div 97(HY11):1867–1886
15. Daněček J, Ryl T, Říha J (2002) Determination of longitudinal hydrodynamic dispersion in water courses with solution of Fischer's integral. J Hydrol Hydromech 50(2):104–113
16. Elder JW (1959) Dispersion of marked fluid in turbulent shear flow. J Fluid Mech 5(Part 4):544–560
17. Fischer HB, List J, Koh C, Imberger J, Brooks NH (1979) Mixing in inland and coastal waters. Academic Press, New York, p 483
18. Chen JS, Liu CW, Liang CP (2006) Evaluation of longitudinal and transverse dispersivities/ distance ratios for tracer test in a radially convergent flow field with scale-dependent dispersion. Adv Water Resour 29(6):887–898
19. Karcher MJ, Gerland S, Harms IH, Iosjpe M, Heldal HE, Kershaw PJ, Sickel M (2004) The dispersion of 99Tc in the Nordic seas and the Arctic Ocean: a comparison of model results and observations. J Environ Radioact 74(1–3):185–198
20. Kosorin K (1995) Dispersion coefficient for open channels profiles of natural shape. J Hydrol Hydromech 43(1–2):93–101. (in Slovak)
21. Krenkel PA, Orlob G (1962) Turbulent diffusion and reaeration coefficient. J Sanit Eng Div, ASCE 88(SA2):53–83
22. Swamee PK, Pathak SK, Sohrab M (2000) Empirical relations for longitudinal dispersion in streams. J Environ Eng 126(11):1056–1062
23. Velísková Y (2001) Characteristics of transverse mixing in surface streams. 1. Preview of experimental values. Acta Hydrol Slovaca 2(2):294–301. ISSN 1335-6291. (in Slovak)
24. Velísková Y (2002) Impact of geometrical parameters of cross-section profile on transverse mixing. In: Transport vody, chemikálií a energie v systéme pôda-rastlina-atmosféra. X. International Poster Day, Bratislava 28. 11. 2002, Bratislava, pp 489–495. ISBN 80-968480-9-7
25. Veliskova Y (2004) Statement of transverse dispersion coefficients at upper part of Hron River. Hydrol Hydromech 52(4):342–354
26. Velísková Y, Sokáč M, Dulovičová R (2009) Determination of longitudinal dispersion coefficient in sewer networks. In: Popovska C, Jovanovski M (eds) Eleventh international symposium on water management and hydraulic engineering: proceedings. University of Ss. Cyril and Methodius, Faculty of Civil Engineering, Skopje, pp 493–498. ISBN 978-9989-2469-6-8

27. Velísková Y, Sokáč M, Halaj P, Koczka Bara M, Dulovičová R, Schügerl R (2014) Pollutant spreading in a small stream: a case study in Mala Nitra Canal in Slovakia. Environ Process 1 (3):265–276. ISSN 2198-7491 (Print) 2198-7505 (Online)
28. Parker FL (1961) Eddy diffusion in reservoirs and pipelines. J Hydraul Div ASCE 87(3):151–171
29. Yotsukura N, Fiering MB (1964) Numerical solution to a dispersion equation. J Hydraul Div 90 (5):83–104
30. Thackston EL, Krenkel PA (1969) Reaeration prediction in natural streams. J Sanitary Eng Div ASCE 95(SA1):65–94
31. McQuivey RS, Keefer TN (1974) Simple method for predicting dispersion in streams. J Environ Eng Div ASCE 100(4):997–1011
32. Kashefipour MS, Falconer RA (2002) Longitudinal dispersion coefficients in natural channels. Water Resour Res 36(6):1596–1608
33. Sahay RR, Dutta S (2009) Prediction of longitudinal dispersion coefficients in natural rivers using genetic algorithm. Hydrol Res 40(6):544
34. Brady JA, Johnson P (1980) Predicting times of travel, dispersion and peak concentrations of pollution incidents in streams. J Hydrol 53(1–2):135–150
35. Glover RE (1964) Dispersion of dissolved and suspended matarials in flowing streams. Geological survey Proffesional paper 433B, United States Government Printing Office, Washington, p 32
36. Cunge JA, Holly FM, Verwey A (1985) Practical aspects of computational river hydraulics. Energoatomizdat, Moscow. (in Russian)

Assessment of Heavy Metal Pollution of Water Resources in Eastern Slovakia

E. Singovszká and M. Bálintová

Contents

Abstract Sediment quality monitoring is amongst the highest priorities of environmental protection policy. Their main objective is to control and minimise the incidence of pollutant-oriented problems and to provide for water of appropriate quality to serve various purposes such as drinking water supply, irrigation water, etc.

   The quality of sediments is identified in terms of their physical, chemical and biological parameters. The particular problem regarding sediment quality monitoring is the complexity associated with analysing a large number of measured variables. This research was realised in order to determine and analyse selected heavy metals present in sediment samples from six river basins on East of Slovakia, represented by the rivers Hornád, Laborec, Torysa, Ondava, Topla and Poprad. Sampling points were selected based on the current surface water quality monitoring network. The investigation was focused on heavy metals (Zn, Cu, Pb, Cd, Ni, Hg,

E. Singovszká (✉) and M. Bálintová
Faculty of Civil Engineering, Institute of Environmental Engineering, Technical University of Košice, Košice, Slovakia
e-mail: eva.singovszka@tuke.sk

A. M. Negm and M. Zeleňáková (eds.), *Water Resources in Slovakia:*
Part I - Assessment and Development, Hdb Env Chem (2019) 69: 213–238,
DOI 10.1007/698_2017_216, © Springer International Publishing AG 2018,
Published online: 27 April 2018

As, Fe, Mn). The content of heavy metals reflected the scale of industrial and mining activities in a particular locality. The degree of sediment contamination in the rivers has been evaluated using an enrichment factor, pollution load index, geo-accumulation index and potential environmental risk index.

Keywords Heavy metals, Pollution indices, Sediments, Statistic methods

1 Introduction

The analysis of bottom sediment quality is an important yet sensitive issue. The anthropological influences (i.e. urban, industrial and agricultural activities) as well as the natural processes (i.e. changes in precipitation amounts, erosion and weathering of crustal materials) degrade surface water quality and impair its use for drinking, industrial, agricultural, recreational and other purposes. Based on spatial and temporal variations in water chemistry, a monitoring programme that provides a representative and reliable estimation of the quality of surface waters has become an important necessity. Heavy metals are usually present at low concentrations in aquatic environments; however, deposits of anthropogenic origin have raised their own concentrations, causing environmental problems in lakes [1, 2]. According to [3] the highest concentrations of heavy metals in sediment may be related to the terrigenous input and anthropogenic influence. The high content of trace metals in the sediments can be a good indication of man-induced pollution, and high levels of heavy metals can often be attributed to terrigenous input and anthropogenic influences, rather than the natural enrichment of the sediment by geological weathering [3]. An associated geochemical process plays an important role in the deposition of trace and heavy elements from the water column to the bottom sediments [1, 4, 5]. Heavy metals are non-biodegradable; they are not removed from the water as a result of self-purification. Once they are discharged into water bodies, they are adsorbed on sediment particles, accumulate in reservoirs and enter the food chain [6]. Consequently, comprehensive monitoring programmes include regular water sampling at numerous places and a whole analysis of a large number of physico-chemical parameters designed for the proper management of water quality in surface waters [7, 8]. Furthermore, they facilitate the identification of the possible factors/ sources influencing the system and provide not just a valuable tool for reliable management of water resources but also suitable solutions to pollution problems [9].

In the study of contaminated samples, the determination of the extent or degree of pollution by a given heavy metal requires that the pollutant metal concentration is compared with an unpolluted reference material. Such reference material should be an unpolluted or pristine substance that is comparable with the study samples. In assessing the impact of heavy metal pollution on environments, a number of different reference materials and enrichment calculation methods have been used by various publications [10–12]. There is thus a considerable variation in how the impact of anthropogenic pollution on a given site is quantified.

In the Slovak Republic, there are some localities with existing mining and industrial conditions. Overflows at the rivers in East of Slovakia produce flow with high metal concentrations and low values of pH (about 3–4) as a result of chemical oxidation of sulphides and other chemical processes. This was the reason for initiating the systematic monitoring of the geochemical development to prepare a prognosis in terms of environmental risk [13]. Till now, researchers have made some achievements on studies of heavy metal pollution. The degree of contamination in sediments is determined with the help of three parameters – enrichment factor (EF), pollution load index (PLI) and geo-accumulation index (I_{geo}). A common approach to estimate the degree to which sediment is impacted (naturally and anthropogenically) by heavy metals involves the calculation of the enrichment factor for metal concentrations above uncontaminated background levels [14]. The PLI is aimed at providing a measure of the degree of overall contamination at a sampling site. Sediment geo-accumulation index is the quantitative check of metal pollution in aquatic sediments [15]. Based on spatial and temporal variations in water and sediment chemistry, a monitoring programme which provides a representative and reliable estimation of the quality of surface waters and bottom sediments has become an important necessity [16]. The assessment model of heavy metal pollution in sediments can be used for environmental protection [17].

2   Materials and Methods

2.1   Study Area

Hornád River belongs to the river basin of Danube. Area of the Hornád River is 4,414 km^2. In the basin, 27.6% is arable land, 15.7% is agricultural land, 47.4% is of forests, 2.7% is shrubs and grasses and 6.6% is other lands. There are 165 surface water bodies, while 162 are in the category of the flowing waters/rivers and two are in the category of standing waters/reservoirs. Ten groundwater bodies exist in the basin, while one is in quaternary sediment, two are geothermal waters and seven are in pre-quaternary rocks. The Hornád River has 11 transverse structures without fishpass in operation. Significant industrial and other pollution sources are US Steel Kosice, Rudne bane š. p., Spišská Nová Ves, Kovohuty a.s., Krompachy and Solivary a.s. Prešov. From environmental loads, there are 11 high-risk localities which have been identified in the river basin. Diffuse pollution is from agriculture and municipalities without sewerage. The upper stretch of the Hornád River to Spišská Nová Ves is in good ecological status which gets worse to poor status or is potential for pollution and hydromorphological pressures. From the Ružín water reservoir, the Hornád River achieves moderate ecological status. According to chemical status assessment, the Hornád River is in good status. Fifty-six water bodies (34%) are failing to achieve good ecological status in Hornád river basin. The water body of intergranular groundwaters of quaternary alluviums of the Hornád river basin achieves poor chemical status (pollution from the point and diffuse sources) and poor quantitative status identified on the base of long-term decrease of groundwater levels. The water body of pre-quaternary rocks is in good status – quantitative and chemical [18].

Poprad River is in the river basin district of Vistula and is the only Slovak river that drains their waters into the Baltic Sea. Its source is in the High Tatras over Popradské Mountain Lake. It flows to the southeast direction up to Svit city. The river mouths into River Dunajec from the right side, in Poland, river km 117.00. It drains an area of 1,890 km^2. There are 83 surface water bodies all in the category of the flowing waters/rivers. Five groundwater bodies exist in the basin, while one is in quaternary sediment, one is geothermal waters and three are in pre-quaternary rocks. Poprad River has 27 transverse structures without fishpass in operation. Significant industrial and other pollution sources are Chemosvit Energochem, a.s., Svit, Whirlpool Slovakia, s.r.o., Poprad, screw factory Exim, Stará Ľubovňa and Východoslovenské stavebné hmoty a.s. (closed in 2013). From environmental loads, there are 17 high-risk localities which have been identified in the river basin. Diffuse pollution is from agriculture and municipalities without sewerage [19].

Ondava is a 146.5-km-long river in Slovakia, the northern source river of the Bodrog. It rises in the Low Beskids (Eastern Carpathian Mountains), next to Nižná Polianka village, close to the border with Poland. The Ondava flows south through the towns Svidník, Stropkov and Trhovište and through the Ondavská Highlands. Next to Cejkov village, the Ondava joins the Latorica and forms the Bodrog River, itself a tributary of the Tisza. The Ondava River is 44% regulated [18].

Torysa is a 129-km (80 mile)-long river in eastern Slovakia. It rises in the Levoča Mountains, and it flows through the towns of Lipany, Sabinov, Veľký Šariš, Prešov and into the Hornád River next to Nižná Hutka village, southeast from Košice [18].

Topla is a river in eastern Slovakia and a right tributary of the Ondava. It is 129.8 km long, and its basin covers an area of 1,544 km^2 (596 mile^2) [1, 22]. It rises in the Čergov mountains, flows through Ondava Highlands, Beskidian Piedmont, Eastern Slovak Hills and Eastern Slovak Flat and joins the Ondava River in the cadastral area of Parchovany. It flows through the towns of Bardejov, Giraltovce, Hanušovce nad Topľou and Vranov nad Topľou [18].

Laborec is a river in eastern Slovakia that flows through the districts of Medzilaborce, Humenné and Michalovce in the Košice Region and the Prešov Region. The river drains the Laborec Highlands. Tributaries of the Laborec River include River Uh which joins Laborec River near the city of Drahňov in Michalovce District and the River Cirocha. Laborec River itself is a tributary, flowing into the River Latorica. The catchment area of Ižkovce hydrometric profile at Laborec River is 4,364 km^2, and it is situated at 94.36 m a.s.l [18] (Fig. 1).

2.2   Sample and Preparation

Sediment was sampled according to ISO 5667-6 Water Quality, Sampling Part 6: Guidance on Sampling of Rivers and Streams [20]. This standard outlines the

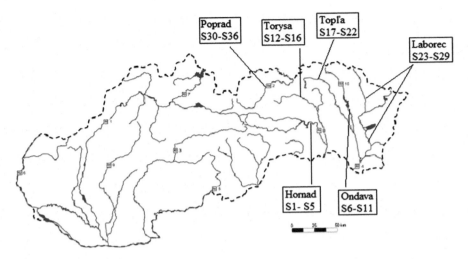

Fig. 1 Location of interested area: East of Slovakia

principles and design of sampling programmes and manipulation, as well as the preservation of samples. The samples of sediment were air-dried and ground using a planetary mill to a fraction of 0.063 mm. The chemical composition of sediments was determined using X-ray fluorescence (XRF) SPECTRO iQ II (Ametek, Germany). Sediment samples were prepared as pressed tablets with a diameter of 32 mm by mixing 5 g of sediment and 1 g of dilution material (Hoechst Wax C Micropowder – M – HWC – $C_{38}H_{76}N_2O_2$) and compressing them at a pressure of 0.1 MPa/m^2.

The mean total concentrations of 8 heavy metals in the sediment of 36 sediments samples are presented in Table 1.

Results of XRF analysis of sediments were compared with the limited values according to the Slovak Act. No. 188/2003 Coll of Laws on the application of treated sludge and bottom sediments to fields [21]; WHO standards (www.who.int); Canadian Sediment Quality Guidelines (CSQG) for protection of aquatic life 1999 [22], with the interim sediment quality values for Hong Kong [23]; Australian and New Zealand Environment and Conservation Council (ANZECC) [24]; and Egyptian drinking water quality standards [25] (Table 1).

The limit values were exceeding for Cu in all rivers excluding Topla River. Nickel and lead are exceeding limit values in all sediment samples according to WHO limit values. Cadmium exceeds the Hong Kong, CSQG, ANZECC and Egyptian limit values, but it is relevant because it depends on the extent of the XRF analysis.

Table 1 Concentration of heavy metals in sediment samples

		As	Cd	Cr	Cu	Hg	Ni	Pb	Zn
		mg/kg							
Hornád	S1	*14.9*	*<5.1*	35.8	110.3	<2	*59.4*	<2	*167*
	S2	<1	*<5.1*	24.3	27.4	<2	24.8	<2	38.7
	S3	**82.3**	*<5.1*	*141.2*	233	<2	*130.5*	37.9	*360.4*
	S4	<1	*<5.1*	*169.9*	108.4	<2	*45.2*	51.1	*177.4*
	S5	*12.6*	*<5.1*	*189.9*	188	<2	*64.6*	<2	202.7
Ondava	S6	<1	*<5.1*	*142*	46.3	<2	*88*	0	55.9
	S7	<1	*<5.1*	*110.2*	37.8	<2	*69.7*	<2	40.7
	S8	<1	*<5.1*	50.5	27.3	<2	*48.7*	<2	23.6
	S9	<1	*<5.1*	29.1	39.5	<2	*49.7*	<2	26.8
	S10	<1	*<5.1*	*125.9*	32.8	<2	*60.1*	<2	33.9
	S11	<1	*<5.1*	*200.4*	41	<2	*55.4*	<2	55.3
Torysa	S12	<1	*<5.1*	*94.1*	11.9	<2	32.5	<2	28
	S13	<1	*<5.1*	*73.5*	17.3	<2	34.8	<2	45.1
	S14	<1	*<5.1*	28.6	21	<2	38	<2	36.1
	S15	<1	*<5.1*	*70*	34.7	<2	*48.6*	<2	53.8
	S16	<1	*<5.1*	*141*	15.5	<2	3.4	<2	1
Topla	S17	<1	*<5.1*	23.7	15.3	<2	21.8	<2	25.8
	S18	<1	*<5.1*	*144.6*	0.3	<2	21.4	<2	1
	S19	<1	*<5.1*	*81.5*	13.1	<2	26.4	<2	22.5
	S20	<1	*<5.1*	49.6	27.3	<2	31.4	<2	24.7
	S21	<1	*<5.1*	62.7	19.2	<2	21.9	<2	30
	S22	<1	*<5.1*	68.2	25.5	<2	27.3	<2	30.1
Laborec	S23	<1	*<5.1*	52.6	18.4	<2	*51.7*	<2	36.3
	S24	<1	*<5.1*	21	33.5	<2	*46.2*	<2	31.7
	S25	<1	*<5.1*	28.1	30.1	<2	*66.5*	<2	51.7
	S26	<1	*<5.1*	36.6	35.8	<2	*54*	<2	33.7
	S27	<1	*<5.1*	5	8.7	<2	31.6	<2	30.2
	S28	1.3	*<5.1*	28	38	<2	*64.6*	<2	61.1
	S29	<1	*<5.1*	19	37.7	<2	*50.1*	<2	40.7
Poprad	S30	<1	*<5.1*	5	2.6	2.1	2	<2	1
	S31	<1	*<5.1*	*124.7*	51.6	<2	*65.7*	<2	*100.4*
	S32	<1	*<5.1*	28.7	24.7	<2	*50.3*	<2	58.1
	S33	<1	*<5.1*	5	6.3	<2	31.9	<2	148.2
	S34	<1	*<5.1*	56.9	2.9	<2	35.5	<2	118.6
	S35	<1	*<5.1*	38.5	5.6	<2	20	<2	105.6
	S36	<1	*<5.1*	16	1	<2	32.11	2.7	115.4
Limits	**SR**	**20**	**10**	**1,000**	**1,000**	**10**	**300**	**750**	**2,500**
	Hong Kong	12	1.5	–	65	–	40	200	75
	WHO		0.01	–	2	–	0.02	0.05	–
	CSQG	33	10	–	110	–	–	250	820
	ANZECC	20	1.2	–	34	–	–	47	200
	Egyptian	–	0.003	–	2	–	0.02	0.01	3

2.3   Pollution Indices

2.3.1   Enrichment Factor

Enrichment factor (EF) calculation is a common approach to estimate the anthropogenic impact on sediments [26]. It is mathematically expressed as [27]:

$$EF = \frac{[M_c/M_r]_s}{[M_c/M_r]_b} \tag{1}$$

where M_c is the content of contamination, M_r is the content of reference elements, s is the sample and b is the background. A reference element is often used as a conservative element [27]. The enrichment factor scale consists of six grades ranging, how indicate the Table 2.

2.3.2   Pollution Load Index

Pollution load index (PLI), for a particular site, has been evaluated using the following method proposed by Tomlinson et al. [28]. This parameter is expressed as:

$$PLI = (CF_1 \times CF_2 \times CF_3 \times \ldots \times CF_n)^{1/n} \tag{2}$$

where n is the number of the metals (11 in the present study) and CF is the contamination factor. The contamination factor can be calculated from the following relation:

$$CF = \frac{\text{Metal concentration in the sediment}}{\text{Reference value of the metal}} \tag{3}$$

The contamination factor scale and pollution load index scale are indicated in Tables 3 and 4.

Table 2 The enrichment factor scale

EF ≤ 1	Background concentration
EF 1–2	Deficiency to minimal enrichment
EF 2–5	Moderate enrichment
EF 5–20	Significant enrichment
EF 20–40	Very high enrichment
EF > 40	Extremely high enrichment

Table 3 The contamination factor scale

CF < 1	Low contamination
1 ≤ CF ≤ 3	Moderate contamination
3 ≤ CF ≤ 6	Considerable contamination
CF > 6	Very high contamination

Table 4 The pollution load
index scale

PLI < 1	Denote perfection
PLI = 1	Present that only baseline level of pollutants
PLI > 1	Deterioration of site quality

Table 5 Descriptive classes
for identifying sediment
contamination base on I_{geo}
values

I_{geo} values	I_{geo} class	Sediment quality
>5	6	Extremely polluted
4–5	5	Highly polluted
3–4	4	Moderately to highly polluted
2–3	3	Moderately polluted
1–2	2	Unpolluted to moderately polluted
0–1	1	Unpolluted
0	0	Background concentration

2.3.3 Geo-accumulation Index

Geo-accumulation index (I_{geo}), introduced by Muller [12] for determining the extent
of metal accumulation in sediments I_{geo}, is mathematically expressed as:

$$I_{geo} = \log_2 \frac{c_n}{1.5B_n} \tag{4}$$

where c_n is the concentration of element n and B_n is the geochemical background
value. The factor of 1.5 is incorporated in the relationship to account for possible
variation in background data due to lithogenic effect. The I_{geo} scale consists of six
grades ranging (Table 5) from unpolluted to very highly polluted.

2.3.4 Ecological Risk Assessment

For the assessment of sediment pollution, the contamination factor and contamination degree were used. In the version suggested by Hakanson, an assessment of
sediment contamination was conducted through references of contaminations in the
surface layer of bottom sediments:

$$C_f^i = \frac{C^i}{C_n^i} \tag{5}$$

where C_i is the mean concentration of an individual metal examined and $C_n{}^i$ is the
background concentration of the individual metal. In this work, as background
concentrations, the contents of selected elements in sediment unaffected by mining
activities in assessment area were used. C_f^i is the single-element index. The sum of

Table 6 Criteria for degree of contamination and classification

Contamination factor	Degree of contamination	Classification
$C_f < 1$	$C_d < 1$	Low
$1 \leq C_f < 3$	$1 \leq C_d < 3$	Moderate
$3 \leq C_f < 6$	$3 \leq C_d < 6$	Considerable
$C_f \geq 6$	$C_d \geq 6$	Very high

Table 7 Risk grade indexes and grades of potential ecological risk of heavy metal pollution

E_i^r	Risk grade	Risk level	R^i value	Risk grade
$E_i^r < 40$	Low risk	A	$R^i < 150$	Low risk
$40 \leq E_i^r < 80$	Moderate risk	B	$150 \leq R^i < 300$	Moderate risk
$80 \leq E_i^r < 160$	Considerable risk	C	$300 \leq R^i < 600$	Considerable risk
$160 \leq E_i^r < 320$	High risk	D	$R^i \geq 600$	Very high risk
$E_i^r \geq 320$	Very high risk	E		

contamination factors for all metals examined represents the contamination degree (C_d) of the environment:

$$C_d = \sum_{i=1}^{n} C_f^i \tag{6}$$

E_r^i is the potential ecological risk index of an individual metal. It can be calculated from

$$E_r^i = C_f^i \times T_r^i \tag{7}$$

where T_r^i is the toxic response factor provided by Hakanson (T_r^i for Cr, Cu, Cd, Zn, As, Pb, Ni and Hg are 2, 5, 30, 1, 10, 5, 5 and 40). R^i is the potential ecological risk index, which is the sum of E_r^i:

$$R^i = \sum_{i=1}^{n} E_r^i \tag{8}$$

Hakanson defined four categories of C_f^i, four categories of C_d, five categories of E_r^i and four categories of R^i, as indicated in Tables 6 and 7.

3   Results and Discussion

3.1   Hornád River

The enrichment factor was calculated from the concentrations of heavy metals in bottom sediments of four sampling sites in the study area. The heavy meal

Table 8 Enrichment factor values of heavy metals in Hornád River bed sediment

	As	Cd	Cr	Cu	Hg	Ni	Pb	Zn
S1	0.18	1.00	0.25	0.47	1.00	0.46	0.05	0.46
S3	0.01	1.00	0.17	0.11	1.00	0.19	0.05	0.11
S4	0.01	1.00	1.20	0.47	1.00	0.35	1.35	0.49
S5	0.15	1.00	1.35	0.81	1.00	0.49	0.05	0.56

concentration from sample site S2 was used as background concentration. EF calculation results for sediments are shown in Table 8. The EF values show a depletion trend for As, Cu and Zn ($<$1). The EF for Cr (S4, S5) and Pb (S4) show minimal enrichment (Fig. 2).

Table 9 shows very high values of PLI ($>$1) for all sampling sites, which means it is extremely polluted by heavy metals. High values of PLI indicated a deterioration of site quality. The results of the contamination factor for sediment are shown in Table 18. CF for As, Cr, Cu, Pb and Zn show very high contamination.

The calculated I_{geo} values are presented in Table 10. It is evident from the Table that the I_{geo} values for Cd and Hg fall in class "0", indicating that there is no pollution from these metals in the Hornád River sediments. The I_{geo} values for Ni fall within the range 0–2, indicating that it is unpolluted to moderately polluted. Cr and Cu indicated moderately polluted. Highly polluted shows concentration of Pb, which falls to class 5. The extremely polluted for Ondava River is presented by As.

All the values of R^i in the sediments were more than 250, which present moderate to very high risk. The E_r values of all parameters in all sampling locations were from 5 to 823, which reflects a very high ecological risk for the water body posed by these metals (Table 11).

3.2   Ondava River

EF calculation results for sediments are shown in Table 12. The enrichment factor was calculated from the concentrations of heavy metals in bottom sediments of five sampling sites in the study area. The heavy metal concentration from sample site S8 was used as background concentration. The highest enrichment shows chromium and zinc concentration (Fig. 3).

Table 13 shows considerable contamination for Cr and for other elements indicates moderate contamination by heavy metals. High values of PLI indicated a deterioration of site quality (PLI $>$ 1).

The calculated I_{geo} values are presented in Table 14. It is evident from Table 14 that the I_{geo} values for all elements expected Cr fall in class "1", indicating that there is no pollution from these metals in the Ondava River sediments. The I_{geo} values for Cr fall within the range 1–2, indicating that it is unpolluted to moderately polluted.

All the values of R^i in the sediments were less 150 which indicate a low risk for the water body posed by these metals (Table 15).

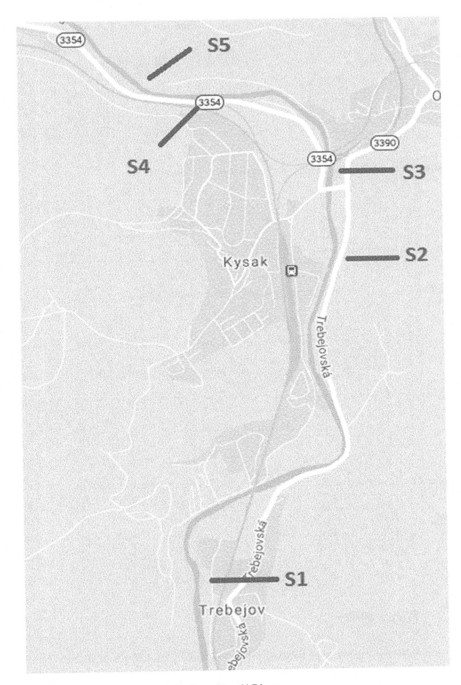

Fig. 2 Location of sediment samples from Hornád River

Table 9 Contamination factor (CF) values and pollution load index of heavy metals in the sediments of Hornád River

	As	Cd	Cr	Cu	Hg	Ni	Pb	Zn	
	CF								PLI
S1	14.90	1.00	1.47	4.03	1.00	2.39	1.00	4.31	2.35
S3	82.30	1.00	5.81	8.503	1.00	5.26	18.95	9.31	6.64
S4	1.00	1.00	6.99	3.96	1.00	1.82	25.55	4.58	2.96
S5	12.60	1.00	7.815	6.861	1.00	2.61	1.00	5.24	3.13

Table 10 Geo-accumulation indexes of heavy metals in Hornád River

	As	Cd	Cr	Cu	Hg	Ni	Pb	Zn
S1	3.31	−0.58	−0.03	1.42	−0.58	0.67	0.58	1.52
S3	5.78	−0.58	1.95	2.50	−0.58	1.81	3.65	2.63
S4	−0.59	−0.58	2.22	1.39	−0.58	0.28	4.09	1.61
S5	3.07	−0.58	2.38	2.19	−0.58	0.79	−0.58	1.80

Table 11 E_r and R^i of heavy metals in sediments from Hornád River

		E_r								R^i	Risk grade
		As	Cd	Cr	Cu	Hg	Ni	Pb	Zn		
Hornád	S1	149	30	2.95	20.13	40	11.98	5	4.32	263.36	Moderate risk
	S3	823	30	11.62	42.52	40	26.31	94.75	9.31	1,077.51	Very high risk
	S4	10	30	13.98	19.79	40	9.11	127.75	4.54	255.21	Moderate risk
	S5	126	30	15.63	34.31	40	13.02	5	5.24	269.19	Moderate risk

Table 12 Enrichment factor values of heavy metals in Ondava River bed sediment

	As	Cd	Cr	Cu	Hg	Ni	Pb	Zn
S6	1.00	1.00	2.81	1.70	1.00	1.81	1.00	2.37
S7	1.00	1.00	2.18	1.38	1.00	1.43	1.00	1.73
S9	1.00	1.00	0.57	1.45	1.00	1.02	1.00	1.13
S10	1.00	1.00	2.49	1.20	1.00	1.23	1.00	1.43
S11	1.00	1.00	3.96	1.50	1.00	1.14	1.00	2.34

3.3   Torysa River

The results for enrichment factor for Torysa River are shown in Table 16. The highest enrichment indicates zinc concentration. The pattern of the metal concentration at all the stations studied followed $Zn > Ni > Cu > As = Cd = Pb = Hg > Cr$ (Fig. 4).

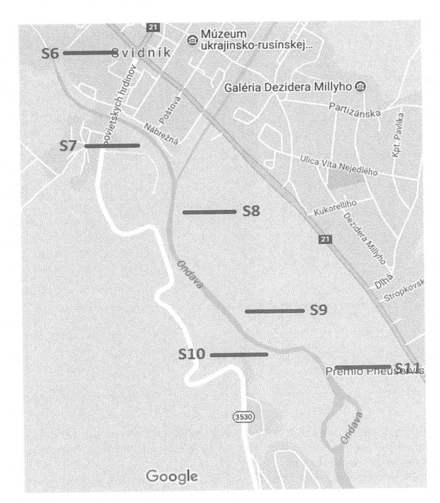

Fig. 3 Location of sediment samples from Ondava River

Table 13 Contamination factor (CF) values and pollution load index of heavy metals in the sediments of Ondava River

	As	Cd	Cr	Cu	Hg	Ni	Pb	Zn	
	CF								PLI
S6	1.96	1.00	2.82	1.70	1.00	1.81	1.00	2.37	1.59
S7	1.96	1.00	2.19	1.38	1.00	1.43	1.00	1.725	1.39
S9	1.96	1.00	0.57	1.44	1.00	1.02	1.00	1.14	1.08
S10	1.96	1.00	2.55	1.20	1.00	1.23	1.00	1.44	1.34
S11	1.96	1.00	3.98	1.50	1.00	1.13	1.00	2.34	1.54

Table 14 Geo-accumulation indexes of heavy metals in Ondava River

	As	Cd	Cr	Cu	Hg	Ni	Pb	Zn
S6	0.39	−0.58	0.91	0.18	−0.58	0.27	−0.58	0.66
S7	0.39	−0.58	0.55	−0.12	−0.58	−0.07	−0.58	0.20
S9	0.39	−0.58	−1.37	−0.05	−0.58	−0.56	−0.58	−0.40
S10	0.39	−0.58	−0.74	−0.32	−0.58	−0.28	−0.58	−0.06
S11	0.39	−0.58	1.41	0.001	−0.58	−0.39	−0.58	0.644

Table 15 E_r and R^i of heavy metals in sediments from Ondava River

	E_r								R^i	Risk grade
	As	Cd	Cr	Cu	Hg	Ni	Pb	Ni		
S6	19.6	30	5.64	8.45	40	9.05	5	2.36	120.1	Low risk
S7	19.6	30	4.38	6.92	40	7.15	5	1.72	114.77	Low risk
S9	19.6	30	1.16	7.20	40	5.1	5	1.15	109.19	Low risk
S10	19.6	30	5.01	6.01	40	6.17	5	1.44	113.22	Low risk
S11	19.6	30	7.96	7.51	40	5.65	5	2.3	118.06	Low risk

Table 16 Enrichment factor values of heavy metals in Torysa River bed sediment

	As	Cd	Cu	Cr	Ni	Pb	Zn	Hg
S12	1.00	1.00	0.76	0.67	9.56	1.00	28.00	1.00
S13	1.00	1.00	1.16	0.52	10.23	1.00	45.10	1.00
S14	1.00	1.00	1.35	0.20	11.17	1.00	36.10	1.00
S16	1.00	1.00	2.23	0.49	14.29	1.00	53.80	1.00

Table 17 shows very high contamination for Ni and Zn and for other elements indicates low to moderate contamination by heavy metals. High values of PLI indicated a deterioration of site quality (PLI > 1).

Table 18 presented values of I_{geo}. It is evident from the table that the I_{geo} values for As, Cd, Cu, Pb and Hg belong to class "1", indicating that there is no pollution from these metals in the Torysa River sediments. The I_{geo} values for Cr fall within the range 2–3, indicating that it is moderately polluted. Nickel belongs to class "4" and zinc falls into class "6" which indicates extremely polluted.

All the values of R^i in the sediments belong to range from 150 to 300 which indicate moderate risk for the water body posed by these metals (Table 19).

3.4   Topla River

Table 20 shows the results of enrichment factor for Topla River. As, Cd, Cr, Hg and Pb indicate background concentration. Nickel presents deficiency to minimal

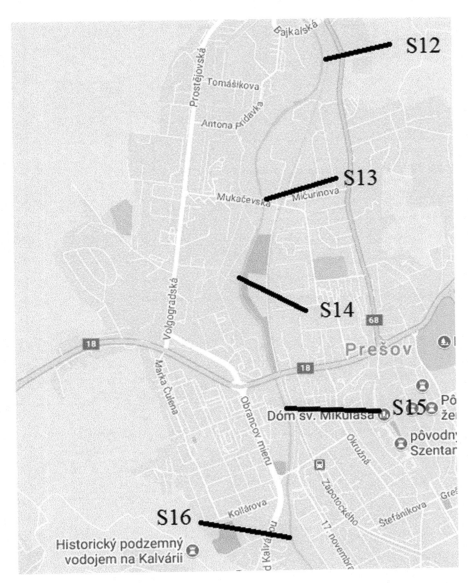

Fig. 4 Location of sediment samples from Torysa River

enrichment, and Zn and Cu indicate very high to extremely high enrichment. The heavy metal concentration from sample site S23 was used as background concentration (Fig. 5).

Table 21 shows very high values of PLI (>1) for all sampling sites, which means it is extremely polluted by heavy metals. High values of PLI indicated a deterioration

Table 17 Contamination factor (CF) values and pollution load index of heavy metals in the sediments of Torysa River

	As	Cd	Cu	Cr	Ni	Pb	Zn	Hg	
	CF								PLI
S12	1.00	1.00	0.76	0.67	9.56	1.00	28.00	1.00	1.05
S13	1.00	1.00	1.16	0.52	10.23	1.00	45.10	1.00	1.03
S14	1.00	1.00	1.35	0.20	11.17	1.00	36.10	1.00	1.04
S16	1.00	1.00	2.23	0.49	14.29	1.00	53.80	1.00	1.03

Table 18 Geo-accumulation indexes of heavy metals in Torysa River

	As	Cd	Cu	Cr	Ni	Pb	Zn	Hg
S12	−0.585	−0.585	−0.966	−1.168	2.672	−0.585	4.222	−0.585
S13	−0.585	−0.585	−0.426	−1.525	2.771	−0.585	4.91	−0.585
S14	−0.585	−0.585	−0.147	−2.887	2.897	−0.585	4.589	−0.585
S16	−0.585	−0.585	0.578	−1.595	3.252	−0.585	5.165	−0.585

Table 19 E_r and R^i of heavy metals in sediments from Torysa River

	E_r									
	As	Cd	Cu	Cr	Ni	Pb	Zn	Hg	R^i	Risk grade
S12	10.00	30.00	3.80	1.34	47.8	5.00	28.00	40.00	165.94	Moderate risk
S13	10.00	30.00	5.80	1.04	51.15	5.00	45.10	40.00	188.09	Moderate risk
S14	10.00	30.00	6.75	0.40	55.85	5.00	36.10	40.00	184.1	Moderate risk
S16	10.00	30.00	11.15	0.98	71.45	5.00	53.80	40.00	222.38	Moderate risk

Table 20 Enrichment factor values of heavy metals in Topla River bed sediment

	As	Cd	Cr	Cu	Hg	Ni	Pb	Zn
S17	1.00	1.00	0.164	51.00	1.00	1.019	1.00	25.80
S19	1.00	1.00	0.564	43.67	1.00	1.234	1.00	22.50
S20	1.00	1.00	0.343	91.00	1.00	1.467	1.00	24.70
S21	1.00	1.00	0.434	64.00	1.00	1.023	1.00	30.00
S22	1.00	1.00	0.472	85.00	1.00	1.276	1.00	30.10

of site quality. The results of the contamination factor for sediment are shown in Table 18. Contamination factor for Cu, Cr and Zn shows very high contamination by these metals.

The calculated I_{geo} values are presented in Table 22. It is evident from the Table that the I_{geo} values for As, Cd, Hg, Ni and Pb fall in class "1", indicating that there is no pollution from these metals in the Topla River sediments. The I_{geo} values for Cr fall within the range 2–3, indicating that it is moderately polluted. Zinc belongs to class "5" presenting highly polluted. Copper falls to class "6", indicating extremely polluted.

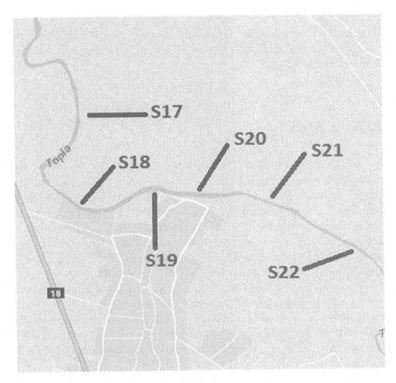

Fig. 5 Location of sediment samples from Topla River

Table 21 Contamination factor (CF) values and pollution load index of heavy metals in the sediments of Topla River

	As	Cd	Cr	Cu	Hg	Ni	Pb	Zn	
	CF								PLI
S17	1.00	1.00	1.00	51.00	1.00	1.02	1.00	25.80	167.60
S19	1.00	1.00	6.1	1.00	1.00	1.00	1.00	1.00	0.77
S20	1.00	1.00	3.44	43.67	1.00	1.23	1.00	22.50	521.37
S21	1.00	1.00	2.09	91.00	1.00	1.47	1.00	24.70	861.44
S22	1.00	1.00	2.65	64.00	1.00	1.02	1.00	30.00	650.63

Table 22 Geo-accumulation indexes of heavy metals in Topla River

	As	Cd	Cr	Cu	Hg	Ni	Pb	Zn
S17	−0.58	−0.58	−0.58	5.09	−0.58	−0.56	−0.58	4.10
S19	−0.58	−0.58	2.02	−0.58	−0.58	−0.58	−0.58	−0.58
S20	−0.58	−0.58	1.19	4.86	−0.58	−0.28	−0.58	3.90
S21	−0.58	−0.58	0.48	5.92	−0.58	−0.03	−0.58	4.04
S22	−0.58	−0.58	0.82	5.42	−0.58	−0.55	−0.58	4.32

The values of R^i in the sediment samples S17, S19, S20 and S21 present a considerable risk. The value for sediment site S22 indicates moderate risk ($R^i = 169.79$). The E_r reflects a very high ecological risk for the water body posed by these metals (Table 23).

3.5 Laborec River

The enrichment factor was calculated from the concentrations of heavy metals in bottom sediments of six sampling sites in the study area. EF calculation results for sediments are shown in Table 24. The EF for Cu indicates moderate enrichment. The EF values show a depletion trend for As, Cd, Cr, Pb and Zn (≤ 1). The heavy metal concentration from sample site S29 was used as background concentration (Fig. 6).

Table 25 shows very high values of PLI (>1) for all sampling sites which means it is extremely polluted by heavy metals. High values of PLI indicated a deterioration of site quality. The results of the contamination factor for sediment are shown in Table 25. Contamination factor for copper shows considerable contamination. CF for other elements indicates low to moderate contamination.

Table 26 shows the results for the geo-accumulation index for Laborec River. As, Cd, Pb, Zn and Hg indicate 0–1 which presents class "1" – unpolluted. Nickel and copper fall to class "2" – unpolluted to moderately polluted. Chromium belongs to class "4" which presents moderately to highly polluted.

On the base of R^i (Table 27) for Laborec River, it can be said that the river presents considerable risk for the water body posed by these metals.

Table 23 E_r and R^i of heavy metals in sediments from Topla River

	E_r								R^i	Risk grade
	As	Cd	Cr	Cu	Hg	Ni	Pb	Zn		
S17	10	30	2	255	40	5.05	5	25.8	372.85	Considerable risk
S19	10	30	6.86	218.35	40	6.15	5	22.5	338.86	Considerable risk
S20	10	30	4.18	455	40	7.45	5	24.7	576.33	Considerable risk
S21	10	30	5.28	320	40	5.10	5	30.0	445.38	Considerable risk
S22	10	30	5.79	425	40	6.40	5	30.1	169.79	Moderate risk

Table 24 Enrichment factor values of heavy metals in Laborec River bed sediment

	As	Cd	Cu	Cr	Ni	Pb	Zn	Hg
S23	1.00	1.00	1.82	0.39	0.89	1.00	0.87	1.00
S24	1.00	1.00	1.63	0.53	1.28	1.00	1.42	1.00
S25	1.00	1.00	1.94	0.69	1.04	1.00	0.92	1.00
S26	1.00	1.00	0.47	0.09	0.61	1.00	0.83	1.00
S27	1.30	1.00	2.06	0.53	1.24	1.00	1.68	1.00
S28	1.00	1.00	2.04	0.36	0.96	1.00	1.12	1.00

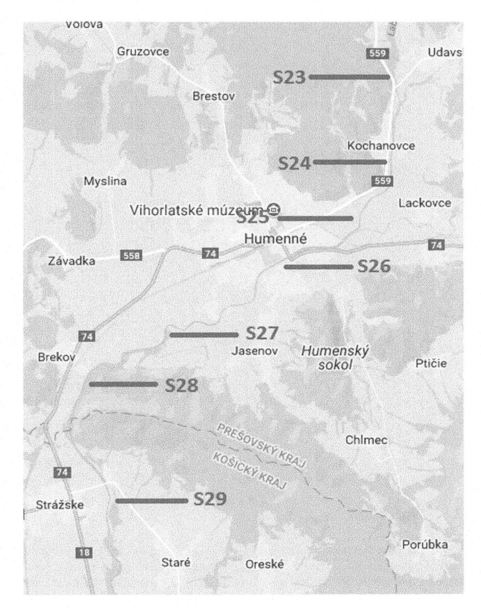

Fig. 6 Location of sediment samples from Laborec River

Table 25 Contamination factor (CF) values and pollution load index of heavy metals in the sediments of Laborec River

	As	Cd	Cu	Cr	Ni	Pb	Zn	Hg	
	CF								PLI
S23	1.00	1.00	1.82	0.39	0.89	1.00	0.87	1.00	0.552
S24	1.00	1.00	1.63	0.53	1.28	1.00	1.42	1.00	2.75
S25	1.00	1.00	1.94	0.69	1.04	1.00	0.92	1.00	1.01
S26	1.00	1.00	0.47	0.09	0.61	1.00	0.83	1.00	0.01
S27	1.30	1.00	2.06	0.53	1.24	1.00	1.68	1.00	11.17
S28	1.00	1.00	2.04	0.36	0.96	1.00	1.12	1.00	0.72

Table 26 Geo-accumulation indexes of heavy metals in Laborec River

	As	Cd	Cu	Cr	Ni	Pb	Zn	Hg
S23	−0.58	−0.58	0.27	−1.91	−0.74	−0.58	−0.78	−0.58
S24	−0.58	−0.58	0.12	−1.48	−0.22	−0.58	−0.07	−0.58
S25	−0.58	−0.58	0.37	−1.11	−0.52	−0.58	−0.69	−0.58
S26	−0.58	−0.58	−1.66	−3.98	−1.29	−0.58	−0.85	−0.58
S27	−0.58	−0.58	0.46	−1.49	−0.26	−0.44	0.16	−0.58
S28	−0.58	−0.58	0.44	−2.05	−0.63	−0.58	−0.41	−0.58

Table 27 E_r and R^i of heavy metals in sediments from Laborec River

	E_r								R^i	Risk grade
	As	Cd	Cr	Cu	Hg	Ni	Pb	Ni		
S23	10	30	2	255	40	5.05	5	25.8	372.85	Considerable risk
S24	10	30	12.2	5	40	5.00	5	1.00	108.2	Low risk
S25	10	30	6.86	218.35	40	6.15	5	22.5	338.86	Considerable risk
S26	10	30	4.18	455	40	7.45	5	24.7	576.33	Considerable risk
S27	10	30	5.28	320	40	5.10	5	30.0	445.38	Considerable risk
S28	10	30	5.79	425	40	6.40	5	30.1	169.79	Moderate risk

3.6 Poprad River

The enrichment factor was calculated from the concentrations of heavy metals in bottom sediments of six sampling sites in the study area. The heavy metal concentration from sample site S1 was used as background concentration. EF calculation results for sediments are shown in Table 28. The EF values show a depletion trend for As, Cu and Hg (≤ 1). The EF for Cr and Ni shows very high enrichment and for Zn indicates extreme enrichment (Fig. 7).

Table 29 shows very high values of PLI (>1) for all sampling sites, which means it is extremely polluted by heavy metals. High values of PLI indicated a deterioration of site quality. The results of the contamination factor for sediment are shown in Table 29. Contamination factor for Cu, Cr, Ni and Zn shows very high contamination.

Table 28 Enrichment factor values of heavy metals in Poprad River bed sediment

	As	Cd	Cu	Cr	Ni	Pb	Zn	Hg
S31	1.00	1.00	19.95	24.94	32.85	1.00	100.40	0.95
S32	1.00	1.00	9.50	5.74	25.15	1.00	58.10	0.95
S33	1.00	1.00	2.42	1.00	15.95	1.00	148.20	0.95
S34	1.00	1.00	1.12	11.38	17.75	1.00	118.60	0.95
S35	1.00	1.00	2.15	7.70	10.00	1.00	105.60	0.95
S36	1.00	1.00	0.39	3.20	16.05	1.35	113.40	0.95

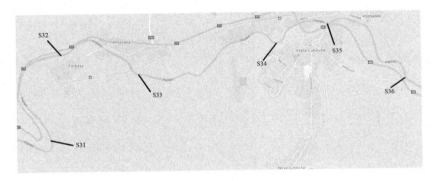

Fig. 7 Location of sediment samples from Poprad River

Table 29 Contamination factor (CF) values and pollution load index of heavy metals in the sediments of Poprad River

	As	Cd	Cu	Cr	Ni	Pb	Zn	Hg	
	CF								PLI
S31	1.00	1.00	19.95	24.94	32.85	1.00	100.40	0.95	5.55
S32	1.00	1.00	9.50	5.74	25.15	1.00	58.10	0.95	3.94
S33	1.00	1.00	2.42	1.00	15.95	1.00	148.20	0.95	2.95
S34	1.00	1.00	1.12	11.38	17.75	1.00	118.60	0.95	3.51
S35	1.00	1.00	2.15	7.70	10.00	1.00	105.60	0.95	3.26
S36	1.00	1.00	0.39	3.20	16.05	1.35	113.40	0.95	2.70

The calculated I_{geo} values are presented in Table 30. It is evident from the Table that the I_{geo} values for As, Cd, Hg and Pb fall in class "0", indicating that there is no pollution from these metals in the Poprad River sediments. Copper falls to class "4", indicating moderately to highly polluted. The I_{geo} values for Cr and Ni fall within the range 4–5, indicating that it is highly polluted. Zinc belongs to class "5" presenting extremely polluted.

Values of R^i (Table 31) in the sediments were from 150 to 600 which indicate considerable risk for the Poprad River posed by these metals.

Table 30 Geo-accumulation indexes of heavy metals in Poprad River

	As	Cd	Cu	Cr	Ni	Pb	Zn	Hg
S31	−0.59	−0.59	3.73	4.06	4.45	−0.59	6.07	−0.66
S32	−0.59	−0.59	2.66	1.94	4.07	−0.59	5.28	−0.66
S33	−0.59	−0.59	0.69	−0.59	3.41	−0.59	6.63	−0.66
S34	−0.59	−0.59	−0.43	2.92	3.57	−0.59	6.31	−0.66
S35	−0.59	−0.59	0.52	2.36	2.74	−0.59	6.14	−0.66
S36	−0.59	−0.59	−1.96	1.09	3.42	−0.15	6.27	−0.66

Table 31 E_r and R^i of heavy metals in sediments from Poprad River

	E_r								R^i	Risk grade
	As	Cd	Cr	Cu	Hg	Ni	Pb	Zn		
S31	10.00	30.00	49.88	99.73	164.25	5.00	100.40	38.08	497.34	Considerable risk
S32	10.00	30.00	11.48	47.50	125.75	5.00	58.10	38.08	325.91	Considerable risk
S33	10.00	30.00	2.00	12.12	79.75	5.00	148.20	38.08	335.27	Considerable risk
S34	10.00	30.00	22.76	5.58	88.75	5.00	118.60	38.08	301.59	Considerable risk
S35	10.00	30.00	15.4	10.77	50.00	5.00	105.60	38.08	260.22	Moderate risk
S36	10.00	30.00	6.4	1.93	80.25	5.00	113.40	38.08	282.31	Moderate risk

4 Conclusions

Environmental risk in the water catchments is closely related to the quality and quantity of water flows in the catchment, and quality is one of the most important indicators of risk in the river basin. The monitoring and evaluation of water quality have a permanent place in the process of risk management. The possibility of minimising the negative impact on the environment presents the assessment and management of environmental risks by using different methodologies. Methodology for assessing environmental risks in the basin presents a risk characterisation for the particular conditions of water flows. The results represent the basis for risk management in the river basin, whose task is to ensure the sustainability of water bodies.

Different calculation methods on the basis of different algorithms might lead to a discrepancy of the pollution assessment when they are used to assess the quality of sediment ecological chemistry. So it is of great importance to select a suitable method to assess sediment quality for decision-making and spatial planning. Pollution indices are a powerful tool for processing, analysing and conveying raw environmental information to decision-makers, managers, technicians and the public.

Ecological risk management provides policy makers and resource managers as well as the public with systematic methods that can inform decision-making. The results provide a comprehensive sediment contamination status of heavy metals and potential origin of contamination in the rivers, giving insight into decision-making for water source security.

The above analysis demonstrates the use of pollution index techniques to study the source of chemical parameters in sediments. The heavy metals of sediments were monitored in the six rivers on East of Slovakia. The data obtained in this study has presented consistency in metal pollution indexes of the sediment stations of the study area. This may be due to the continuous dilution of the water body from lower and upper reaches of the river; the similarity of the physical conditions of the sediments, particle composition and organic matter of the sediments may have also played a major role. Hárnad River indicated deficiency to minimal enrichment. The potential ecological risk index indicates moderate to high risk for water basin Hornád. Hornád River on the base of geo-accumulation index belongs to class "5", which indicates highly polluted.

Ondava River presents minimal to moderate enrichment. The highest enrichment shows chromium and zinc concentration. The I_{geo} values for this water basin fall within the range 1–2, indicating that it is unpolluted to moderately polluted. All the values of R^i in the sediments were less 150 which indicate a low risk for the water body posed by these metals.

The pattern of the metal concentration at all the stations studied in Torysa River followed Zn > Ni > Cu > As = Cd = Pb = Hg > Cr. The I_{geo} values for this water basin belong to class "6", which indicate extremely polluted. All the values of R^i in the sediments belong to range from 150 to 300 which indicate moderate risk for the Torysa River posed by these metals.

Topla River indicates very high to extremely high enrichment. The I_{geo} values for this water basin fall to class "5", which indicate extremely polluted. The potential ecological risk index presents a moderate risk.

The EF for Laborec River indicates moderate enrichment (Cu). The EF values show a depletion trend for As, Cd, Cr, Pb and Zn (≤ 1). The I_{geo} values for Laborec fall to class "4", which indicate moderate to highly polluted. On the base of Ri for Laborec River, it can be said that the river presents considerable risk for the water body posed by these metals.

The EF values show extremely enrichment for Poprad River. The I_{geo} values for this water basin fall to class "5", presenting extremely polluted. The potential ecological risk index presents considerable risk for the Porpad River posed by these metals.

Pollution load index for all water basins indicates a deterioration of site quality (PLI > 1).

Different calculation methods on the basis of different algorithms might lead to a discrepancy in pollution assessments when they are used to assess the quality of sediment ecological chemistry. Thus it is of great importance to select a suitable method to assess sediment quality for decision-making and spatial planning.

Ecological risk management provides policy makers and resource managers as well as the public with systematic methods that can facilitate informed decision-making. The results provide comprehensive sediment contamination status of heavy metals and potential origin of contamination in the creek, giving insight into decision – ensuring water source security.

5   Recommendations

Environmental risk management provides policy makers and resource managers as well as the public with systematic methods that can facilitate informed decision-making. The results provide comprehensive sediment contamination status of heavy metals and potential origin of contamination in the rivers, giving insight into decision – ensuring water source security.

There have been numerous sediment quality guidelines developed to monitor the sediments. Sediment quality guidelines are very useful to screen sediment contamination by comparing sediment contaminant concentration with the corresponding quality guidelines, provide useful tools for screening sediment chemical data to identify pollutants of concern and prioritise problem sites and relatively good predictors of contaminations. However, these guidelines are chemical specific and do not include biological parameters. Aquatic ecosystems, including sediments, must be assessed in multiple components (biological data, toxicity, physicochemistry) by using integrated approaches in order to establish a complete and comprehensive set of sediment quality guidelines.

The overview of existing sediment quality criteria enables us to state the worldwide harmonisation is missing. Such different outcome assessments occur because in different countries have been set for individual indicators various occupational exposure and also have different numbers of monitored indicators. These limit values were influenced by the background values as the concentration of the indicator depends on the geological conditions and so on. It should be properly used for the evaluation of indicators in the first place, and our laws and regulations in foreign countries should be used only as a supplementary assessment.

The present study suggests that these indices are useful tools for the identification of different sources of contamination of the bottom sediment. This paper will hopefully contribute to the development of a water and sediment pollution prevention strategy. The main topics that may need to be investigated are the control of industrial and domestic discharge, regular observation of pollutants, evaluation of the effects of pollutants on the ecosystem over the long term, coordination of the pollution source and prevention of inflow of pollutants to the water and sediment.

References

1. Goher ME, Farhat HI, Abdo MH, Salem GS (2014) Metal pollution assessment in the surface sediment of Lake Nasser, Egypt. Egypt J Aquat Res 40(3):213–224
2. Ntakirutimana T, Du G, Guo JS, Gao X, Huang L (2013) Pollution and potential ecological risk assessment of heavy metals in a lake. Pol J Environ Stud 22(4):1129–1134
3. Ahmed W, Mohamed AW (2005) Geochemistry and sedimentology of core sediments and the influence of human activities; Qusier, Safaga, and Hurghada harbors, Red Sea Coast, Egypt, Egyptian Mediterranean Coast. Egypt J Aquat Res 31(1):92–103
4. Ali MA, Dzombak DA (1996) Interactions of copper, organic acids, and sulfate in goethite suspensions. Geochim Cosmochim Acta 60:5045–5053
5. Tessier A, Fortin D, Belzile N, DeVitre RR, Leppard GG (1996) Metal sorption to diagenetic iron and manganese oxyhydroxides and associated organic matter: narrowing the gap between field and laboratory measurements. Geochim Cosmochim Acta 60:387–404
6. Loska K, Wiechula D (2003) Application of principal component analysis for the estimation of source heavy metal contamination in surface sediments from Rybnik Reservoir. Chemosphere 51:723–733
7. Lambrakis N, Antonakos A, Panagopoulos G (2004) The use of multicomponent statistical analysis in hydrogeological environmental research. Water Res 38:1862–1872
8. Kumru M, Bakac M (2003) R-mode factor analysis applied to the distribution of elements in soil from Aydin Basin, Turkey. J Geochem Explor 77:81–91
9. Simeonov V, Stratis J, Samara C, Zachariadis G, Voutsa D, Anthemidis A, Sofoniou M, Kouimtzis T (2003) Assessment of the surface water quality in northern Greece. Water Res 37:4119–4124
10. Salomon W, Forstner U (1984) Metals in the hydrocycle. Springer, Berlin, p 349
11. Hakanson L (1980) An ecological risk index for aquatic pollution control, a sedimentological approach. Water Res 14:975–1001
12. Muller G (1979) Heavy metals in the sediment of the Rhine, Veranderungem Seit 1971. Umschau, pp 778–783. (in German)
13. Slesarova A, Kusnierova M, Luptakova A, Zeman J (2007) An overview of occurrence and evolution of acid mine drainage in the Slovak Republic. In: Proceedings of the 22nd annual international conference on contaminated soils, sediments and water 2006, Curran Associates, Edt Amherst, pp 11–19. ISBN: 9781604239515
14. Mmolawa KB, Likuku AS, Gaboutloeloe GK (2011) Assessment of heavy metal pollution in soils along major roadside areas in Botswana. Afr J Environ Sci Technol 5(3):186–196
15. Jumbe AS, Nandini N (2009) Heavy metals analysis and sediment quality values in urban lakes. Am J Environ Sci 5(3):678–687
16. Angelovicova L, Fazekasova D (2013) The effect of heavy metal contamination to the biological and chemical soil properties in mining region of middle Spis (Slovakia). Int J Ecosyst Ecol Sci 3(4):807–812
17. Hong-Gui D, Teng-Feng G, Ming-Hui L, Xu D (2012) Comprehensive assessment model on heavy metal pollution in soil. Int J Electrochem Sci 7(6):5286–5296
18. Slovak Environmental Agency 1 Introduction. Pilot Project PiP1: Hornád/Hernád, Integrated Revitalisation of the Hornád/Hernád River Valley. TICAD, p 5
19. Ondruš Š (1991) Ešte raz o pôvode tatranskej rieky Poprad. In: Slovenská reč, vol 4. Veda, Vydavateľstvo Slovenskej akadémie vied, Bratislava. (In Slovak)
20. ISO 5667-6-2005 Water quality – sampling – Part 6: guidance on sampling of rivers and streams
21. Slovak Act. No. 188/2003 Coll of Laws on the application of treated sludge and bottom sediments to fields
22. Canadian Sediment Quality Guideline for protection of aquatic life, Canada (1999)
23. Chapman PM, Allard PJ, Vigers GA (1999) Development of sediment quality values for Hong Kong special administrative region: a possible model for other jurisdictions. Mar Pollut Bull 38(3):161–169

24. ANZECC (Australian and New Zealand Environment and Conservation Council) (1997) ANZECC interim sediment quality guidelines. Report for the Environmental Research Institute of the Supervising Scientist, Sydney
25. Egyptian Drinking Water Quality Standards (2007) Ministry of Health, Population Decision number 458
26. Huu HH, Rudy S, Van Damme A (2010) Distribution and contamination status of heavy metals in estuarine sediments near Cau Ong Harbor, Ha Long Bay Vietnam. Geol Belg 13(1–2):37–47
27. Sutherland RA (2000) Bed sediment – associated trace metals in an urban stream. Environ Geol, Oahu, pp 611–637
28. Tomlinson DC, Wilson JG, Harris CR, Jeffrey DW (1980) Problems in assessment of heavy metals in estuaries and the formation of pollution index. Helgoländer Meeresun 33(1–4):566–575

Influence of Acid Mine Drainage on Surface Water Quality

M. Bálintová, E. Singovszká, M. Holub, and Š. Demčák

Contents

Abstract Acid mine drainage (AMD) has been a detrimental by-product of sulphidic ores mining for many years. In most cases, this acid comes primarily from oxidation of iron sulphide, which is often found in conjunction with valuable metals. AMD is a worldwide problem, leading to ecological destruction in watersheds and the contamination of human water sources by sulfuric acid and heavy metals, including arsenic, copper and lead.

The Slovak Republic belongs to the countries with long mining tradition, especially in connection with the mining of iron, copper, gold, silver and another polymetallic ores. The abandoned mine Smolnik is one of these mines where AMD is produced.

Acid mine drainage from an abandoned sulphide mine in Smolnik, with the flow rates of 5–10 L s^{-1} and a pH of 3.7–4.1, flows into Smolnik creek and adversely affects the stream's water quality and ecology. High rainfall events increase the flow of Smolnik Creek, which ranges from 0.3 to 2.0 m^3 s^{-1} (monitored 2006–2016).

M. Bálintová (✉), E. Singovszká, M. Holub, and Š. Demčák
Faculty of Civil Engineering, Institute of Environmental Engineering, Technical University of Košice, Košice, Slovakia
e-mail: magdalena.balintova@tuke.sk

A. M. Negm and M. Zeleňáková (eds.), *Water Resources in Slovakia:*
Part I - Assessment and Development, Hdb Env Chem (2019) 69: 239–258,
DOI 10.1007/698_2017_220, © Springer International Publishing AG 2018,
Published online: 27 April 2018

Increased flow is associated also with a pH increase and precipitation of metals (Fe, Al, Cu and Zn) and their accumulation in sediment. The dependence of pH on flow in Smolnik Creek was evaluated using regression analysis.

The study also deals with the metal distribution between water and sediment in the Smolnik creek depending on pH and the metal concentrations.

Keywords Acid mine drainage, Heavy metals, pH, Surface water

1 Introduction

Mine waters origin during the exploitation, mainly after closing down the exploitation of mineral deposits running in the contact zones of water and geological environment [1]. Mine waters contaminate the ground and surface waters by a wide range of elements. Besides that, a part of heavy metals accumulates in both the inorganic part of the soil profile and the organic matter, thence inducing major deformations of their macro and microbiological structures. Acid mine drainage negatively affects the plants, animals, fish and aquatic insects (zoobentos) [2]. The pH of mine water is determined by the quality/quantity of present minerals in the deposit. Generally speaking, the mine water from deposits containing mainly acidic (sulphide) minerals produce acid mine water (pH < 6); deposits containing mainly alkaline (carbonate) minerals (also in case of the significant content of sulphide amounts) produce alkaline mine water (pH $\geq$ 6). In the deposits with sulphide content occurs specific type of mine water, called acid mine drainage (AMD) with pH values <4.5. Their formation is also determined by the existence of autochthonous chemolithotrophic iron- and sulphur-oxidizing bacteria of the genus *Acidithiobacillus*. AMD transport dissolved substances up to the surface, where oxidation of Fe occurs after their contact with air or surface water, producing the ochre precipitates (mainly goethite, jarosite, schwertmannite and ferrihydrite) [3, 4]. Various technologies have been developed and applied for treatment of AMDs, usually divided to passive and active approaches [5]. In the recent 30 years, the facilities of passive and active treatment of mine drainage waters have shifted from experimental testing in laboratory conditions, through the semi-pilot and pilot plants, to the implementation in large scale in numerous deposits throughout the world. Many research projects confirmed that AMDs treated by both passive and active systems do not negatively affect the environment.

The selection and application of the approaches depend on geochemical, technological, natural, financial and other factors. The virtue of passive treatment of AMDs resides in the use of naturally occurring chemical, biochemical and biological processes. Examples of passive AMD treatment include natural wetlands, constructed wetlands, anoxic limestone drains, systems gradually increasing the environment alkalinity, lime lagoons, open lime canals and bioremediation [1]. Passive systems produce a major disadvantage – production of large amounts of sludge requiring further treatment (as they are composed of a heterogeneous mixture of various compounds with metal content), or final disposal, which is quite finances consuming approach. Active systems of AMD treatment require a continuous presence of personnel, facilities and monitoring systems based on external energy power; however, they provide selective metals recovery from AMDs [6].

Active systems involve methods of chemical neutralization by addition of neutralization agents ($Ca(OH)_2$, CaO, NaOH, etc.), which induces pH increase and subsequent precipitation of metals in the form of hydroxides [5]. Other active systems for metal removal from AMDs use precipitation of metals in the form of weak soluble sulphides using the precipitation agents (sodium sulphide, ammonium sulphide or hydrogen sulphide) prepared either by chemical means [7] or biologically using the sulphate-reducing bacteria [6]. Active systems also involve aeration, neutralization (with precipitation of metals and sulphates), chemical precipitation of metals and sulphates, membrane processes, ion exchange, adsorption and biological-chemical methods.

Environmental technologies, specifically bioremediation gain the higher level of topicality by a solution of AMD problematic. The ground of the bioremediation is the controlled intensifying of the biogeochemical cycles of metals, routinely running in the natural waters under the influence of microorganisms (MO), which participate on the basis of their fundamental metabolic processes in the solubilization and immobilization of metals in AMD [8, 9]. Bioremediation is the economic and ecological option of conventional physical-chemical processes of metals elimination in waters and sediments. It makes use the genetic diversity and metabolic versatility of MO. Metals immobilization under the MO impression can be the result of biosorption, bioaccumulation, or precipitation [10].

1.1 The Sources of Mine Waters in Slovakia

The main sources of mine waters stem from remnants of mining activities (flooded shafts, dumps and sludge lagoons) representing the old mine loads belonging to the group of environmental loads [11]. The issues of elimination of environmental loads concerning the legislation are encompassed in various strategic documents of the Government of Slovak Republic, such as National Programme of Remediation of Environmental Loads (2010–2015), Regulation No. 153/2010, Act 409/2011 Coll. on Certain Measures Concerning Environmental Load, etc.

Typical examples of old mining loads are abandoned deposits Smolník, Poproč, Čučma, Pezinok, waste storage in Šobov, etc. [11–13]. The main sources of environmental risks in mentioned deposits are water discharges with limit exceeding concentrations of metals and metalloids in comparison with SR Government Regulation 269/2010 Coll [14, 15]. AMD production with the occurrence of genus *Acidithiobacillus* bacteria and limit exceeding concentrations of metals and sulphates is documented in bearings Smolník, Šobov, Pezinok, Slovinky, Rožňava and Rudňany [16]. Discharges of highly mineralized and mild alkaline/alkaline mine water, containing limit exceeding concentrations of metals/metalloids, are located in bearings Poproč, Čučma and Dúbrava [15]. Deposit Smolník belongs to historically most important and richest Cu-Fe ores deposits of Slovakia. Mining was carried out with pauses for several centuries, and the main raw materials were sulphidic pyrite-chalcopyrite ores, from which mainly copper was obtained. Besides classical mining of copper ore, there was also extracted copper by cementation at the site for many centuries.

In view of the spreading rate, pH 3.5–3.9, limit exceeding metal contents (Fe, Zn, Cu, Al and Mn) and sulphates content, as well as presence of genus *Acidithiobacillus* bacteria in acid mine drainage in the effluent from the former shaft Pech (Smolnik), is the object considered to be the most important source of contamination of that site. Based on the results of chemical analysis and flow rate of AMD (cca 10 L/s), it is possible to assume that from the shaft Pech, without spending any costs of mining, leak out 280 t of S, 90 t of Fe, 22 t of Al, 7 t of Mn, 2.5 t of Zn and 370 kg of Cu per year [17]. Given that in the flooded mine remains a large amount of pyrite (approximately 6 miles tonnes) and pyrite is additionally dispersed in the surrounding rock complexes, it is assumed that this process can continue for a very long time [18]. For the purpose of the mentioned AMD remediation, there were studied processes of water dilution, neutralization, application of sorptive/bio-sorptive, precipitating/bio-precipitating and testing pilot project of passive (in situ) treatment system for these mine water [18–20]. The research results have provided a number of positive experiences but also pointed out some negatives. They have contributed to the intention of further research, especially in the field of selective metal removal possibilities [21, 22].

Abandoned deposit Poproč belongs among important, historically mined stibnite ore deposit in Spis Gemer Ore Mountains. The antimony ore mining began probably in seventeenth century. In 1939 there was built flotation plant. Mining finally ended in 1965. After mining and mineral processing, activities remained at the site Poproč piles of mine tailings and ponds with deposited material from the treatment plant, which cause significant pollution of surface water, soils and stream sediments in the river basin Olšava. Another significant source of pollution is especially mine water leaking out from shafts Agnes and Anna. The main contaminants are As and Sb. The highest concentration of As (2,400 μg L^{-1}) was detected in mine water from the shaft Agnes and in seepage water from the tailing pond (1,950 μg L^{-1}). The highest concentration of Sb was detected in water from the shaft Anna, which flows through the heap material (840 μg L^{-1}), in mine water from the shaft Agnes (380 μg L^{-1}) and in seepage water from the tailing pond (400 μg L^{-1}) [23].

Mineralogical, hydrological, pedological and environmental-geological studies of Poproč deposit have been examined especially from the Faculty of Natural Sciences, Comenius University in Bratislava [24]. They obtained results pointing to the possibility of remediation of these mine waters by sorption using Fe0, which was applied in the form of granules, fragments, powder and waste Fe shavings.

Spontaneous self-improvement of the water quality is not feasible; hence it is necessary to monitor the condition of presented mine waters. Simultaneously, the attention should be focused on development of methods for their treatment with the aim to valorise their as potential resources of beneficial metals/metalloids in the form of useful products for practice [10, 25].

2   Influence of AMD on the Smolnik Creek

2.1   Characterization of Study Area

The stratiform deposit Smolnik belongs to the historically best known and richest Cu-Fe ore deposits in Slovakia. In 1990 the mining activity at the locality was stopped. The mine was flooded till 1994. In 1994 an ecological collapse occurred, which caused the death of fish and a negative influence on the environment. The mine system represents a partly opened geochemical system into which rain and surface water drain [26, 27]. More than 6 million tons of pyrite ores of various qualities have been abandoned in this mine. The analysis of water in the deserted mine and in the broader area surrounding this mine was made after the ecological accident in the Smolnik creek in 1995. Waters from the earth surface penetrated the mine, and they were enriched with metals, and their pH values decreased [12]. Acidity is caused mainly by the oxidation of sulphide minerals. The Pech shaft receives the majority of waters draining from the flooded Smolnik mine area and discharges them in the form of acid mine drainage (pH = 3–4, Fe 500–400 mg L^{-1}; Cu 3–1 mg L^{-1}; Zn 13–8 mg L^{-1} and Al 110–70 mg L^{-1}). "This water acidifies and contaminates not only the Smolnik creek water but transports pollution into the Hnilec River catchment" [28].

2.2   Water and Sediment Quality Monitoring in Smolnik Creek

Water and sediment sampling sites are located at 48° south latitude and 20° east longitude (Fig. 1). The first two sampling sites were situated in the upper part of the Smolnik creek not contaminated by acid mine water from the Pech shaft (1, outside Smolnik village; 2, small bridge, crossing to the Pech shaft). Another two sampling localities were located under the shaft (4, 200 m downstream of the Pech shaft; 5, inflow to the Hnilec river). The outflow of AMD from Pech shaft (Smolnik mine) is numbered as 3. Water and sediment samples were collected from the Smolnik creek during the years 2006–2016. GPS coordinates of sampling sites are in Table 1. Samples were collected according to ISO 5667-6-2005 Water quality – Sampling – Part 6: Guidance on sampling of rivers and streams. Samples were collected once a year (15 sediments and water per year), triplicate sampling from each sample sites.

To determine the pH of water samples, a multifunction device, MX 300 X mate pro (METLER TOLLEDO) was used. The concentrations of metals in water samples were determined by ICP-AES (Varian Vista-MPX, Australia). The samples of sediment were air-dried and ground by using a planetary mill and sieved to a fraction of 0.063 mm. The chemical composition of sediments was determined by means of X-ray fluorescence (XRF) method using SPECTRO iQ II (Ametek, Germany). The sediment samples were prepared as pressed tablets with a diameter of 32 mm by

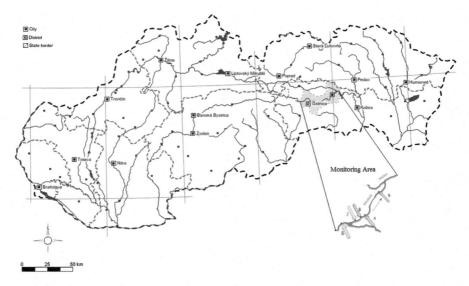

Fig. 1 Location of the Smolnik creek on the map of Slovak Republic and sampling sites No. 1–5 in the study area

Table 1 GPS coordinates of sampling sites

Sample site	GPS coordinates	Description
1	48° 43′ 27.6965658″ N 20° 42′ 59.2164803″ E	Site 1: Uncontaminated part
2	48° 44′ 21.9978463″ N 20° 45′ 37.2264862″ E	Site 2: Uncontaminated part
3	48° 44′ 18.0496747″ N 20° 45′ 44.9512482″ E	Site 3: Source of AMD – shaft Pech
4	48° 44′ 46.1817014″ N 20° 46′ 28.4995937″ E	Site 4: Contaminated part
5	48° 45′ 02.2642765″ N 20° 46′ 39.4108200″ E	Site 5: Contaminated part

mixing 5 g of sediment and 1 g of dilution material (M – HWC) and pressed at a pressure of 0.1 MPa m^{-2}.

For infrared spectroscopy in this study, the spectrum of 4,000–600 cm^{-1} (Alpha FTIR Spectrometer, BRUKER OPTICS) was used.

The crystal structure of sediments was identified with diffractometer Bruker D2 Phaser (Bruker AXS, GmbH, Germany).

Flow data of Smolnik creek were provided by the Slovak Hydro-meteorological Institute (SHI) in Kosice, and the corresponding pH of the surface water was provided by the Slovak Water Management Enterprise (SWME), Kosice. The results were compared to the limit values according to the Regulation of the Government of the Slovak Republic No. 269/2010 Coll. stipulating the requirements for a good water stage achievement. Results of chemical analyses of the sediment were compared with the limited values according to Slovak Act No. 188/2003 Coll. of Laws on the application of treated sludge and bottom sediments to fields.

3   Monitoring of Water Quality in the Smolnik Creek

The average values of chemical analysis of water samples (samples 1, 2, 4 and 5 in the Smolnik creek as well as AMD from shaft Pech (sample 3) in 2006–2016) are presented in Table 2 and pH of samples in Table 3.

The results in Tables 2 and 3 were compared to limited values in accordance with the Regulation of the Government of the Slovak Republic No. 269/2010 Coll. Stipulating requirements for the quality and qualitative goals of surface water and limit values of indicators of pollution of water wastes and separate waters.

Based on this comparison, we can state that acid mine drainage flowing from the Pech shaft has a permanent adverse effect on the surface water quality in Smolnik creek and produces values exceeding the limits values of the Regulation of the Government of the Slovak Republic No. 269/2010 Coll. Due to increased flows of Smolnik creek in 2011, 2012, 2013 and 2014, pH value of samples No. 1, 2 and 5 (2012, 2013, 2014) was in compliance with limits. From the chemical analysis, given in Table 1, follows, that AMD exceeds each of the evaluated indicators, except Ca, Pb, Mg and As. After AMD dilution with surface water in the Smolnik creek, the concentrations of sulphates, Fe, Mn, Al, Cu and Zn, also exceeded defined limits.

3.1   Study of Sediment Quality in the Smolnik Creek

Results of chemical analyses of the sediments (Table 4) were compared with the limit values according to the Slovak Act No. 188/2003 Coll. of Laws on the application of treated sludge and bottom sediments to fields. The results showed that rated sediments did not meet the limit values for arsenic and concentrations of lead was also exceeded in two samples.

From the results of chemical analysis of sediments (Table 4), the increase in the concentration of Fe, Cu and Zn in samples S4 and S5 is evident compared to sediment samples S1 and S2. The results are in accordance to literature [29, 30] where iron is precipitated at pH 3.5–4.5, copper at pH 5.5–6.5, Zn at pH 5.5–7.0 and Al at pH 4.5–5.5. The impact of flow on the pH of surface waters in the Smolnik creek was the subject of the next research.

Table 2 Results of chemical analyses of water from Smolnik in 2006–2014

Metals	Unit	Sampling stations					Limits
		1	2	3	4	5	
Ca	mg L^{-1}	9.99 ± 1.34	12.08 ± 2.00	**158.09 ± 19.07**	20.56 ± 5.92	20.03 ± 5.86	**100**
Mg		3.54 ± 0.28	4.74 ± 1.30	**249.64 ± 51.80**	15.90 ± 7.93	23.36 ± 31.76	**200**
Fe		0.12 ± 0.19	0.83 ± 0.88	**322.73 ± 87.58**	**12.74 ± 8.85**	**5.46 ± 5.64**	**2**
Mn		0.01 ± 0.01	0.12 ± 0.11	**25.34 ± 6.27**	**1.17 ± 0.74**	**0.91 ± 0.62**	**0.3**
Al		0.03 ± 0.03	0.15 ± 0.21	**65.11 ± 19.44**	**1.17 ± 1.60**	**0.35 ± 0.72**	**0.2**
Cu	µg L^{-1}	4.09 ± 4.35	12.55 ± 7.45	**1,512.0 ± 736.27**	**96.91 ± 116.09**	**43.00 ± 63.75**	**20**
Zn		4.27 ± 1.35	40.64 ± 42.67	**7,233.5 ± 2,268.8**	**348.82 ± 250.10**	**254.27 ± 213.64**	**100**
As		1.73 ± 1.27	1.36 ± 0.92	**37.36 ± 18.68**	1.55 ± 1.04	1.18 ± 0.60	**20**
Cd		0.30^a	0.31 ± 0.03	**12.19 ± 8.77**	0.73 ± 0.58	0.41 ± 0.24	**1.5**
Pb		5.00^a	5.00^a	**50.18 ± 16.91**	5.27 ± 0.91	5.00^a	**20**

Within row (for each medium and metal), mean ± standard deviation are significantly different at $p < 0.05$ ($n = 11$)

Within column (Total mean metals within a medium), mean ± standard deviation are significantly different at $p < 0.05$ ($n = 11$)

Bold values are the values that exceed limits according to the Slovak legislative: The Regulation of the Government of the Slovak Republic No. 269/2010 Coll.

Stipulating requirements for the quality and qualitative goals of surface water and limit values of indicators of pollution of water wastes and separate waters

^{a}Values under detection limits

	Sample site	pH	Limit
Table 3 Results of pH of water sampled from Smolnik in 2006–2016	1	6.29 ± 0.77	6–8.5
	2	6.44 ± 0.86	
	3	4.02 ± 0.13	
	4	5.82 ± 1.07	
	5	6.06 ± 1.05	

The sediment quality influenced by AMD was evaluated using FTIR and XRD methods. The infrared spectrum of sample S3 confirmed the presence of schwertmannite [31] which is dominated by a broad, OH-stretching band centred at $3,100 \text{ cm}^{-1}$ (Fig. 2). Another prominent absorption feature related to H_2O deformation is expressed at $1,634 \text{ cm}^{-1}$. Intense bands at 1,124 and $1,038 \text{ cm}^{-1}$ reflect a strong splitting of the $\nu_3(SO_4)$ fundamental due to the formation of a bidentate bridging complex between SO_4 and Fe. This complex may result from the replacement of OH groups by SO_4 at the mineral surface through ligand exchange or by the formation of linkages within the structure during nucleation and subsequent growth of the crystal. Related features due to the presence of structural SO_4 include bands at 981 and 602 cm^{-1} that can be assigned to $\nu_1 (SO_4)$ and $\nu_4 (SO_4)$, respectively. Vibrations at 753 and 424 cm^{-1} are attributed to Fe-O stretch; however, assignment of the former is tentative because similar bands in the iron oxyhydroxides usually occur at lower frequencies. A broad absorption shoulder in the $800–880 \text{ cm}^{-1}$ range is apparent in some specimens and is related to OH deformation ($\delta(OH)$) [32]. These results are in accordance with work [33] where was determined the presence of $Fe_{16}O_{16}(SO_4)_3(OH)_{10} \cdot 10H_2O$ by XRD method in sediment from AMD Smolnik.

FTIR spectra of all homogenized sediment samples (S2 and S4) showed similar features. Based on the concentration of silicon in Table 2 and data from the literature [34] IR spectrum (see Fig. 3), it can be said that the main part of compounds are silicates including quartz (982, 825, 753, 695, 518 cm^{-1}), but hydroxides ($3,600–3,650 \text{ cm}^{-1}$; $1,652 \text{ cm}^{-1}$) are present, too. The sample S4 has a bigger portion of hydroxides than sample 2. It is influenced by the metal concentration in surface water influenced by AMD.

The XRD patterns of sediments (S2, S3, S4) are shown together in Fig. 4. The spectra of S2 and S4 sediments are almost identical and contain the phases: Q, quartz SiO_2 (PDF 01 – 075 – 8322): M, muscovite 2M1, ferrian K $Al_{1.65}$ $Fe_{0.35}$ $Mn_{0.02}$ $(Al_{0.7}$ $Si_{3.3}$ $O_{10})$ $(OH)_{1.78}$ $F_{0.22}$ (PDF 01 – 073 – 9857); and C, clinochlore 1MIIb, ferroan (Mg, Fe)$_6$ (Si, Al)$_4O_{10}$ $(OH)_8$ (PDF 00 – 029 – 0701). The most dominant component is quartz with six broad peaks (the strongest line at $26.623°$ 2Θ).

The spectrum of sediment S3 points to a small part of the crystalline phase. It contains only three weak peaks of clinochlore and one peak of quartz. According to the literature [35], AMD precipitates from shaft Pech contains minerals such as ferrihydrite, goethite, jarosite or schwertmannite. Fresh precipitates are weakly crystallized; formed crystals are very small (tens to hundreds of nm), which is typical for all studied precipitates. Due to their weak crystallinity, it is hard to

Table 4 Results of chemical analyses of sediment from Smolník in 2006–2014

Elements	Unit	Sampling stations					Limits
		1	2	3	4	5	
$(SO_4)^{2-}$	%	0.03 ± 0.03	0.38 ± 0.28	11.98 ± 3.08	0.74 ± 0.70	0.73 ± 1.47	–
Ca		0.29 ± 0.08	0.21 ± 0.09	0.87 ± 2.61	0.25 ± 0.14	0.22 ± 0.08	
Mg		0.78 ± 0.08	0.73 ± 0.05	0.47 ± 0.40	0.74 ± 0.09	0.68 ± 0.17	
Fe		4.15 ± 0.51	5.05 ± 1.59	34.86 ± 5.49	7.58 ± 3.66	9.40 ± 7.76	
Mn		0.09 ± 0.03	0.05 ± 0.01	0.02 ± 0.03	0.06 ± 0.02	0.06 ± 0.02	
Al		7.39 ± 0.59	6.80 ± 0.38	2.16 ± 1.31	6.49 ± 0.40	6.28 ± 1.25	
Cu	mg kg^{-1}	131 ± 42	272 ± 103	478 ± 192	378 ± 231	509 ± 154	**1,000**
Zn		149 ± 25	173 ± 48	114 ± 80	189 ± 68	239 ± 47	**2,500**
As		**44 ± 12**	**77 ± 20**	**2,259 ± 812**	**134 ± 75**	**97 ± 25**	**20**
Cd		0.50^a	0.50^a	1.10 ± 1.96	0.50^a	0.50^a	**10**
Pb		43 ± 10	94 ± 33	**642 ± 837**	160 ± 91	111 ± 42	**750**

Within row (for each medium and metal), mean ± standard deviation are significantly different at $p < 0.05$ ($n = 11$)

Within column (Total mean metals within a medium), mean ± standard deviation are significantly different at $p < 0.05$ ($n = 11$)

Bold values are the values that exceed limits according to the Slovak legislative: The Slovak Act No. 188/2003 Coll. of Laws on the application of treated sludge and bottom sediments to fields

^{a}Values under detection limits

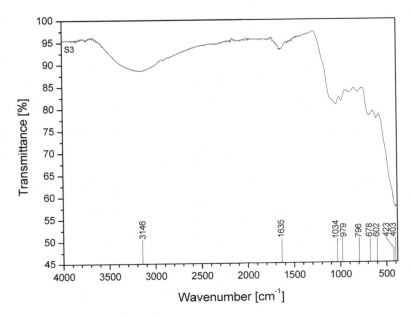

Fig. 2 FTIR spectrum of sediment S3

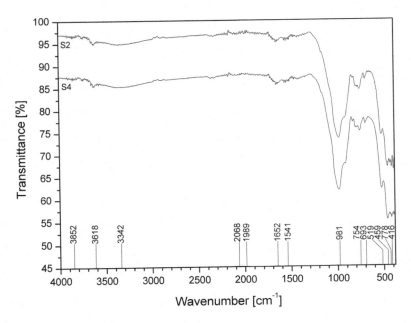

Fig. 3 FTIR spectrum of sediments S2 and S4

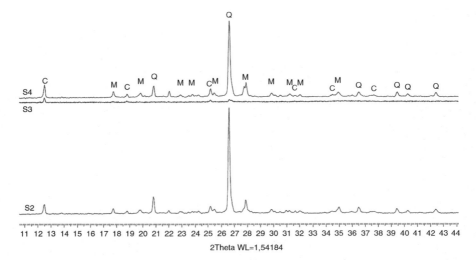

Fig. 4 XRD patterns of sediments S2, S3, S4 (identified compounds: Q, quartz; M, muscovite 2M1, ferrian; C, clinochlore1MIIb, ferroan)

identify only by X-ray diffractometry (XRD) [36], what is evident from XRD pattern of sample S3. Just by a combination of XRD, Mössbauer and infrared spectroscopy, a characterization of their structure is possible.

4 Influence of AMD on pH in Surface Water of the Smolnik Creek

The interdependence between the flow rate and pH in the study area was determined by Gnuplot software and MS Excel [37] (www.gnuplot.info). According to the nature of the data, a logarithmic relation between the values of u (pH) and the flow rate Q, expressed in m^3 s, was considered. This logarithmic relation can be explained by the mixing of creek water and mine drainage, considering process time to be an independent parameter and using the mixing equation:

$$10^{-u_S}Q + 10^{-u_A}Q_A = 10^{-u}(Q + Q_A),$$ (1)

in the form

$$u = u_S + \log(Q + Q_A \cdot 10^{u_S - u_A}),$$ (2)

where u_S denotes the pH of the water upstream of the pollution source and u_A is the pH of the mine drainage and Q_A the mine drainage flow. The typical values of Q_A can be neglected with respect to Q, and, if the creek flow is relatively small, then Q can be considered to be relatively small with respect to $10^{uS-uA} Q_A$ as $10^{uS-uA} \approx 10^3$. Thus, using the known approximate logarithm relationship:

$$log(1 + \varepsilon) \doteq \varepsilon, \text{ for any sufficiently small } \varepsilon, \tag{3}$$

provides the logarithms in the formula (2) with $\varepsilon = Q_A/Q$ and $\varepsilon = 10^{uA-uS} Q/Q_A$ in an approximate form:

$$u \doteq u_A - logQ + \frac{Q_A}{Q} - \frac{Q}{Q_A} 10^{uA-uS} \tag{4}$$

Then, only one logarithmic function remains. Additionally, the flow Q_A is small compared to Q within the given range of water flow, so the term Q_A/Q can be neglected. The regression model then consists of linear dependencies on Q and $log(Q)$. Hence, it can be considered in the form:

$$R1 : u = b_1 + a_1 logQ - c_1 Q \tag{5}$$

This was applied as the principal regression model. It required us to estimate $a_1{}^e$, $b_1{}^e$ and $c_1{}^e$ of unknown parameters a_1, b_1 and c_1 in the numerical analysis by the nonlinear least square method [38]. The computer program Gnuplot (www.gnuplot. info) and, in particular, the command *fit*, which fits a user-defined function to a set of data points, were used.

Given the resulting evaluation and relevance of the model, the calculation was supplemented by statistical analysis [39]. First, a normal distribution of values u with the constant standard deviation σ was assumed. The estimate of the standard deviation S, also calculated by the *fit* command, was calculated by the weighted sum of the squared residuals (WSSR), i.e.:

$$WSSR_r = \min_{a_r, b_r, c_r} \sum_i (u_{ri} - u_{ri}^e)^2, \tag{6}$$

where u_{ri}^e was obtained by using a particular regression model R. The *fit* command also provided asymptotic standard errors as a criterion for the qualitative assessment of the *fit* parameter estimates a_r^e, b_r^e and c_r^e.

The average flow rates Q and pH of the AMD from the Pech shaft is presented in Table 5. It can be observed that both flow rate and pH are in a very narrow interval of values. Due to this fact, this data were not further analysed.

Table 5 The average annual values of pH and water flow rates of acid mine drainage from the Pech shaft in 2002–2012

Year	2002	2003	2004	2005	2006	2007	2008	2009	2010	2011	2012	Average
Q (L/s)	5.65	6.57	7.3	8.89	7.01	5.25	5.89	6.29	8.24	6.13	4.19	6.49
pH	3.92	3.94	4.01	3.86	3.88	4.11	3.99	3.94	3.81	3.97	3.99	3.95

4.1 Regression Analysis of the Flow Rate and pH of Surface Water in Smolnik Creek

A regression analysis [38, 39] was made in order to find the dependence that pH, denoted as u, had on the flow of water (Q) in Smolnik Creek. For this analysis, 102 values of Smolnik Creek flow rate (SHI) and corresponding pH (SWME) data, collected from 2000 to 2012, were used. For a better interpretation of correlation, extreme flow rates (during the flood in June 2010) exceeding 2 m^3 s were excluded for, as mentioned earlier, Eq. (4) was formulated assuming that the values of Q_A were small with respect to Q. The typical values of Q_A in Table 5 correspond to this assumption.

A nonlinear least squares regression was used to estimate parameters a_1, b_1 and c_1 of Eq. (5). The results obtained by the command fit of the Gnuplot software are shown (Fig. 3). Nevertheless, it seems that the estimates of the parameters (see also the results below for the regression analysis relative to Eq. (8)) do not correspond to the proposed model Eq. (4). The best correspondence is achieved for the parameter b_1, which should correspond to the value $u_A - \log Q_A = 6.138$, using the average values from Table 5. It also reflects the expectation of the pH being a bit less than 7 for higher flow rates of Q. Because the other parameter's estimates are far from the expected values of Eq. (4). Another regression relationship was considered in addition to $R1$ in Eq. (5) to cope with the data obtained and the nature of u distribution. The second chosen model is based on a natural exponential relationship in the form:

$$R2 : u = b_2 - a_2 \exp(-c_2 Q). \tag{7}$$

From this model, the nonlinear least square method determines the estimates a_2^e, b_2^e and c_2^e of the unknown parameters a_2, b_2 and c_2. The results are shown in Fig. 5.

It can be observed that the estimate $b_2{}^e$ in the exponential model is approximately equal to 7 because increasing the flow rate neutralizes the acidic nature of the surface water (pH = 7). The other two parameters reflect the chosen exponential dependency, though there is no way to guess their expected values. In the calculation, we assumed a normal distribution. Such an assumption should confirm 95% of the measured data ranged in the interval ($u_r{}^e - 2\sigma$, $u_r{}^e + 2\sigma$). This interval is also shown in Fig. 2, where the standard deviation σ is estimated by S or WSSR from Eq. (6). Using the asymptotic error, the parameter estimates obtained from both regression models $R1$ and $R2$, Eqs. (5) and (7), can be written as:

$$\begin{aligned} R1 : a_1^e &= 4.451 \pm 0.618, b_1^e = 7.763 \pm 0.435, c_1^e = 1.088 \pm 0.437, \\ R2 : a_2^e &= 4.256 \pm 0.379, b_2^e = 6.985 \pm 0.178, c_2^e = -2.673 \pm 0.484. \end{aligned} \tag{8}$$

Although the measured data are rather scattered and affected by factors not included in the experiment such as metal precipitation, which lowers the pH [40], the results of the exponential model can be used to predict the values of pH (u), depending on the flow rate Q for a relatively wide range of flow rates. While the

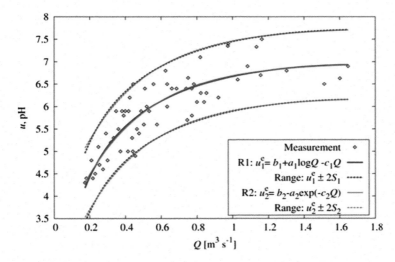

Fig. 5 Dependence of pH and water flow – regression analysis with two models: $R1$ from (Eq. 5) and $R2$ from (Eq. 6)

graphs in Fig. 2 for both proposed regression models are rather coincidental, the asymptotic errors in the parameters are smaller for the exponential model than for the logarithmic one.

5 Study of Metals Distribution Between Water and Sediment in the Smolnik Creek

To study surface water and sediment quality, two sampling localities along the Smolnik creek were chosen (4, approx. 200 m under the shaft Pech; 5, inflow into the Hnilec river). The influence of AMD on surface water and sediment quality and redistribution of the selected metals was studied in the water samples (data from Table 1, sampling stations 4 and 5) and sediments from the Smolnik creek in 2006–2011 (data from Table 4, sampling stations 4 and 5).

Based on laboratory results oriented to the selected metals precipitation from AMD Smolnik [41, 42] and the data from the literature, the redistribution of metals Cu, Fe, Mn, Zn and Al between water sediment in the Smolnik creek was evaluated.

The results of metal concentration decrease in surface water and an increase of metal concentration in sediment were compared with the results of the experimental study focused on pH influence on iron, copper, aluminium, zinc and manganese precipitation from raw AMD from mine Smolnik [21, 41, 42]. It was determined that aluminium is precipitated (98.5%) in the pH range from 4 to 5.5. Precipitation of copper was carried out in accordance with the literature, where copper begins to precipitate at pH > 4 and total precipitation occurs at pH 6 with the efficiency 92.3%.

In spite of iron occurrence in AMD mainly as Fe^{2+}, which precipitates at pH < 8.5, the experimental results confirmed the iron precipitation across studied pH range (4–8) by the progressive oxidation of Fe^{2+} to Fe^{3+} by oxygen from air and its precipitation in the form of $Fe(OH)_3$, which starts at pH 3.5. Zinc is precipitated in the range of pH 5.5–7, and 84% of total Zn was precipitated in this interval.

Precipitation of copper begins at pH > 4 and total precipitation occurs at pH 6. Figure 6 presented dependence of immediate Cu concentration in surface water on its concentration in sediment independence of the pH. As it is seen in Fig. 3, in spite the concentration of Cu in AMD, the decreasing of Cu concentration in surface water with the increasing of pH is connected with its increase in sediment. This is in accordance with literary data and our results [43, 44].

It was determined [41] that iron is in AMD present mainly as Fe^{2+}, which should be precipitated at pH < 8.5 [30, 45, 46]. The reason of the iron precipitation across the range of studied pH may be progressive oxidation of Fe^{2+} to Fe^{3+} in the presence of oxygen and its precipitation in the form of $Fe(OH)_3$, which starts at pH 3.5. From the study of the dependence of Fe concentration in water and sediment resulted, that Fe concentration in sediment varies in the slightest measure in comparison to its concentration in water (Fig. 7).

The interaction among the metals can influence the reaction rate and oxidation state of the metals in the precipitate. For example, manganese will be simultaneously precipitated with iron (II) in water solution at pH 8, only if the concentration of iron in the water is much greater than the manganese content (about four times more). If

Fig. 6 Influence of pH on Cu concentration in water and sediment in the Smolnik creek

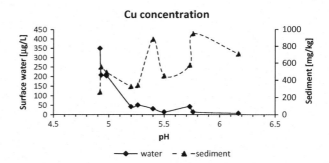

Fig. 7 Influence of pH on Fe concentration in water and sediment in the Smolnik creek

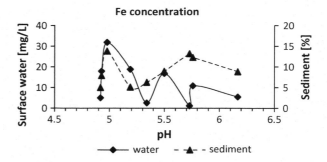

the concentration of iron in AMD is less than four times of the manganese content, then the manganese can be removed from the solution at pH > 9 [47].

The fact that in the presence of a large excess of iron the manganese is precipitated at pH 8 was not confirmed. At pH 8.2 was precipitated only 15.9% of total Mn in AMD. Only at pH 11 was precipitated 93.0% of Mn [42].

In Fig. 8 the dependence of the pH on Mn concentration in water and sediment is presented. As it results from Fig. 3, the variation of Mn concentration in water has minimal influence on its concentration in sediment. The result is in accordance with literature and our research [48, 49].

According to Xinchao et al. [30], Balintova and Kovalikova [45] and Balintova et al. [46], zinc is precipitated in the range of pH 5.5–7. In this interval was precipitated 84% of total Zn [41]. This effect was confirmed by rapid decreasing of Zn concentration in water and its simultaneous increasing in sediment at pH 5.8 (Fig. 9).

Aluminium hydroxide usually precipitates at pH > 5.0 but again dissolves at pH 9.0 [30, 47]. According to Balintova and Petrilakova [42], 98.5% of total aluminium is precipitated from AMD Smolnik in the pH range from 4 to 5.5. The similar tendency can be observed for aluminium, where at the pH > 5.0, the content of Al is decreasing in water and increases in sediment (Fig. 10).

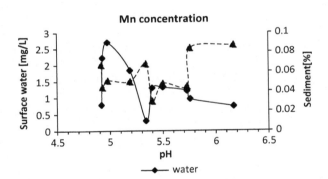

Fig. 8 Influence of pH on Mn concentration in water and sediment in the Smolnik creek

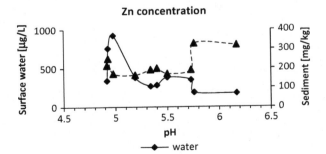

Fig. 9 Influence of pH on Zn concentration in water and sediment in the Smolnik creek

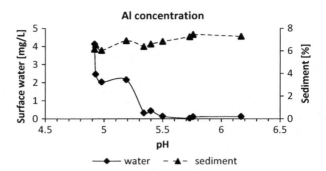

Fig. 10 Influence of pH on Al concentration in water and sediment in the Smolnik creek

6   Conclusions and Recommendations

Smolnik deposit belongs to many localities in Slovakia, where the unfavourable influence of acid mines drainage on the surface water can be observed. Acid mine drainage discharged from abandoned mine Smolnik (shaft Pech) contaminates the downstream from the Smolnik mine works to confluence of the stream with the Hnilec river, because of decreasing pH and heavy metal production. This fact was confirmed by exceeding the limited values of followed physical and chemical parameters in water and sediments in Smolnik creek according to Slovak legislation.

The effect of pH and water flow was studied using regression analysis. The statistical analysis confirmed the significance of the exponential relationship between pH and flow rate. Though both of the models were statistically relevant, the exponential relationship is preferred due to its asymptotic behaviour for increasing flow rate. The obtained numerical results also provide expected values of parameters in the proposed exponential model. The confidence is limited by the scattered character of the experimental data caused by phenomena not considered in the test.

The variability of pH also influences the sediment-water partitioning of heavy metals (e.g. Fe, Cu, Zn, Al, Mn) in Smolnik creek polluted by acid mine drainage, that has been confirmed by presented results. Because AMD generation at the Smolnik locality is not possible to stop and there is no chance of self-improvement of this area, it is necessary to accept this situation, monitor the quality of these waters and develop treatment methods.

References

1. Wolkersdorfer C (2008) Water management at abandoned flooded underground mines: funda-
   mentals, tracer tests, modelling, water treatment. Springer, Heidelberg, p 1580
2. Jennings SR, Neuman DR, Blicker PS (2008) Acid mine drainage and effects on fish health and
   ecology: a review. Reclamation Research Group, Bozeman, p 29

3. Johnson DB, Hallberg KB (2005) Acid mine drainage remediation options: a review. Sci Total Environ 338:3–14
4. Andráš P, Nagyová I, Samešová D, Melichová Z (2012) Study of environmental risks at an old spoil dump field. Pol J Environ Stud 21:1529–1538
5. Skousen JG, Ziemkiewicz PF (1995) Acid mine drainage control and treatment. West Virginia University, Morgantown, p 243
6. Kaksonen AH, Puhakka JA (2007) Sulfate reduction based bioprocesses for the treatment of acid mine drainage and the recovery of metals. Eng Life Sci 7:541–564
7. Veeken AHM, Rulkens WH (2003) Innovative developments in the selective removal and reuse of heavy metals from wastewaters. Water Sci Technol 47:9–16
8. Gadd GM (2000) Bioremedial potential of microbial mechanisms of metal mobilization and immobilization. Curr Opin Biotechnol 11:271–279
9. Sheoran AS, Sheoran V, Choudhary RP (2010) Bioremediation of acid-rock drainage by sulphate-reducing prokaryotes: a review. Miner Eng 23:1073–1100
10. Hennebel T, Boon N, Maes S, Lenz M (2015) Biotechnologies for critical raw material recovery from primary and secondary sources: R&D priorities and future perspectives. New Biotechnol 32:121–127
11. Andráš P, Turisová I, Marino A, Buccheri G (2012) Environmental hazards associated with heavy metals at Ľubietová Cu-deposit (Slovakia). Chem Eng 28:259–264
12. Šottník P, Dubikova M, Lintnerova O, Rojkovič I, Šucha V, Uhlik P (2002) The links between the physico-chemical character of different mining waste in Slovakia and their environmental impacts. Geol Carpath 53:227–228
13. Majzlan J, Chovan M, Andráš P, Newville M, Wiedenbeck M (2010) The nanoparticulate nature of invisible gold in arsenopyrite from Pezinok (Slovakia). Neues Jb Mineral Abh J Mineral Geochem 187:1–9
14. Hiller E, Lalinská B, Chovan M, Jurkovič Ľ, Klimko T, Jankulár M, Hovorič R, Šottník P, Fľaková R, Ženišová Z, Ondrejková I (2012) Arsenic and antimony contamination of waters, stream sediments and soils in the vicinity of abandoned antimony mines in the Western Carpathians, Slovakia. Appl Geochem 27:598–614
15. Chovan M, Lalinská B (2009) Evaluation of contaminated areas affected by Sb mining (Slovakia). Contaminated areas. Ekotoxikologické centrum, Bratislava, pp 177–182
16. Luptakova A, Prascakova M, Kotulicova I (2012) Occurrence of Acidithiobacillus ferrooxidans bacteria in sulfide mineral deposits of Slovak Republic. Chem Eng 28:31–36
17. Kupka D, Pállová Z, Horňáková A, Achimovičová M, Kavečanský V (2012) Effluent water quality and the ochre deposit characteristics of the abandoned Smolnik mine, East Slovakia. Acta Montan Slovaca 17:56–64
18. Lintnerová O, Šoltés S, Šottník P (2003) Stream sediment and suspended solids in the Smolník mining area (Slovakia). Slovak Geol Mag 9:201–203
19. Lintnerova O, Sottnik P, Soltes S (2006) Dissolved matter and suspended solids in the Smolník Creek polluted by acid mine drainage (Slovakia). Geol Carpath 54:311–324
20. Luptakova A, Kusnierova M (2005) Bioremediation of acid mine drainage contaminated by SRB. Hydrometallurgy 77:97–102
21. Bálintová M, Petriláková A (2011) Study of pH influence on selective precipitation of heavy metals from acid mine drainage. Chem Eng Trans 25:1–6
22. Luptakova A, Ubaldini S, Macingova E, Fornari P, Giuliano V (2012) Application of physical–chemical and biological–chemical methods for heavy metals removal from acid mine drainage. Process Biochem 47:1633–1639
23. Jurkovič Ľ, Šottník P (2010) Abandoned Sb-deposit Poproč: source of contamination of natural constituents in Olšava River Catchment. Mineral Slov 42:109–120
24. Klimko T, Lalinska B, Majzlan J, Chovan M, Kucerova G, Paul C (2011) Chemical composition of weathering products in neutral and acidic mine tailings from stibnite exploitation in Slovakia. J Geosci 56:327–340
25. Ehinger S, Janneck E (2010) ProMine – nano-particle products from new mineral resources in Europe: Schwertmannite – raw material and valuable resource from mine water treatment processes. In: Freiberger Forschungsforum research conference, Freiberg

26. Spaldon T, Brehuv J, Bobro M, Hanculak J, Sestinova O (2006) Mining development the Spis-Gemer ore-location. Acta Montan Slovaca 11:375–379
27. Balintova M, Petrilakova A, Singovszka E (2012) Study of metal ion sorption from acidic solutions. Theor Found Chem Eng 46:727–731
28. Luptakova A, Macingova E, Apiariova K (2008) The selective precipitation of metals by bacterially produced hydrogen sulphide. Acta Metall Slovaca 1:149–154
29. Matlock MM, Howerton BS, Atwood DA (2002) Chemical precipitation of heavy metals from acid mine drainage. Water Res 36:4757–4764
30. Xinchao W, Roger C, Viadero J, Karen M (2005) Recovery of iron and aluminum from acid mine drainage by selective precipitation. Environ Eng Sci 22:745–755
31. Bigham JM, Carlson L, Murad E (1994) Schwertmannite, a new iron oxyhydroxysulphate from Pyhasalmi, Finland, and other localities. Mineral Mag 58:641–648
32. Balintova M, Singovszka E, Holub M (2012) Qualitative characterization of sediment from the Smolnik Creek influenced by acid mine drainage. Proc Eng 42:1654–1661
33. Pállová Z, Kupka D, Achimovičová M (2010) Metal mobilization from AMD sediments in connection with bacterial iron reduction. Mineral Slov 42:343–347
34. Guihua L, Xiaobin L, Chuanfu Z, Zhihong P (1998) Formation and solubility of potassium aluminosilicate. Trans Nonferrous Met Soc Chin 8:120
35. Lintnerová O (1996) Mineralogy of Fe-ochre deposits formed from acid mine water in the Smolnik mine (Slovakia). Geol Carpath 5:55–63
36. Bigham JM (1994) Mineralogy of ochre deposits formed by sulfide oxidation. In: Blowes DW, Jamdor JL (eds) The environmental geochemistry of sulfide mineral-wastes, vol 22. Mineralogical Association of Canada, Waterloo, pp 103–132
37. Drever JI (1997) The geochemistry of natural waters.3rd edn. Prentice Hall, Upper Saddle River, p 436
38. Bates DM, Watts DG (1988) Nonlinear regression and its applications. Wiley, New York
39. Balintova M, Singovszka E, Vodicka R, Purcz P (2016) Statistical evaluation of dependence between pH, metal contaminants, and flow rate in the AMD-affected Smolnik Creek. Mine Water Environ 35:10–17
40. Bhattacharyya GK, Johnson RA (1977) Statistical concepts and methods. Wiley, New York
41. Calmano W, Hong J, Forstner U (1993) Binding and mobilization of heavy metals in contaminated sediments affected by pH and redox potential. Water Sci Technol 28:223–235
42. Balintova M, Petrilakova A (2011) Study of pH influence on selective precipitation of heavy metals from acid mine drainage. Chem Eng Trans 25:345–350
43. Balintova M, Petrilakova A, Singovszka E (2012) Study of metals distribution between water and sediment in the Smolnik Creek (Slovakia) contaminated by acid mine drainage. Chem Eng Trans 28:73–78
44. Loska K, Wiechula D (2000) Effects of pH and aeration on copper migration in above-sediment. Pol J Environ Stud 9:433–437
45. Balintova M, Kovalikova N (2008) Testing of various sorbents for copper removal from acid mine drainage. Chem Lett 102:343
46. Balintova M, Luptakova A, Junakova N, Macingova E (2009) The possibilities of metal concentration decrease in acid mine drainage. Zeszyty Naukowe Politechniki Rzeszowskiej Budownictwo i Inżynieria Środowiska 266:9–17
47. Luptakova A, Spaldon T, Balintova M (2007) Remediation of acid mine drainage by means of biological and chemical methods. Adv Mater Res 20:283–286
48. Gerringa LJA (1990) Aerobic degradation of organic matter and the mobility of Cu, Cd, Ni, Pb, Zn, Fe and Mn in marine sediment slurries. Mar Chem 29:355–374
49. Sheremata T, Kuyucak N (1996) Value recovery from acid mine drainage, metals removal from acid mine drainage-chemical methods, MEND project 3.21.2a. Noranda Technology Center, Pointe-Claire

Formation of Acid Mine Drainage in Sulphide Ore Deposits

A. Luptáková and P. Andráš

Contents

Abstract Acid mine drainage (AMD) is the product of the natural oxidation of sulphide minerals. The simultaneous influence of water, oxygen and indigenous microorganisms represents the necessary conditions for AMD formation. The occurrence of AMD is associated mainly with the presence of sulphide minerals in the polymetallic, coal and lignite deposits. AMD contaminates the groundwaters and soils because it contains mainly sulphuric acid, heavy metals and metalloids. During the exploitation, and mostly after the mine closure, the produced AMD pollutes the environment. The continuance of AMD generation is difficult to halt. Self-improvement situation is not possible. It is necessary to monitor the quality of AMD and develop the methods of their treatment. Slovakia belongs to the countries

A. Luptáková (✉)
Department of Mineral Biotechnology, Institute of Geotechnics, Slovak Academy of Sciences, Košice, Slovakia
e-mail: luptakal@saske.sk

P. Andráš
Department of Environmental Management, Faculty of Natural Sciences, Matej Bel University, Banska Bystrica, Slovakia

A. M. Negm and M. Zeleňáková (eds.), *Water Resources in Slovakia: Part I - Assessment and Development*, Hdb Env Chem (2019) 69: 259–276, DOI 10.1007/698_2018_313, © Springer International Publishing AG, part of Springer Nature 2018, Published online: 6 September 2018

with significant mining tradition, especially with regard to the exploitation of iron, copper, gold and silver. Currently, only one deposit is being exploited, namely, Au-ore deposit in Hodruša. The other deposits are mostly flooded. They present the suitable conditions for creation and intensification of chemical and biological-chemical oxidation of the sulphide minerals, i.e. formation of AMD. In Slovakia, Smolník and Pezinok deposits, as well as the Šobov dump, are typical examples of the old mining loads with production of AMD.

Keywords Acid mine drainage, *Acidithiobacillus ferrooxidans*, Sulphide minerals

1 Introduction

The acid mine drainage (AMD) refers to the drainage of acidic water from a mining site caused by the natural oxidation of sulphide minerals. This process can be accelerated by autochthonous microorganisms [1]. The high acidity, elevated levels of toxic metals/metalloids and sulphates in AMD effected a negative impact on the environment. The phenomenon of AMD is related to the presence of sulphides in the deposit. It is typically for the polymetallic as well as coal and lignite deposits. However, it can occur in the case when sulphide-bearing rocks are exposed to air and water [2].

The existence of AMD has been known since the early civilizations of Sumeria, Assyria and Egypt, as well as Greek and Roman scholars, were familiar with the salts formed from the oxidation of pyrite [3]. The first records of the negative impact of AMD on the environment come from the sixteenth century. In 1556, Agricola described to the criticism raised against mining activities [4]. At the some time, Diego Delgada noticed to contamination of river Rio Tinto by sulphuric acid [5].

Polymetallic sulphide deposits represent the resources of metals (Cu, Zn, Pb, Au, Ni, Fe). The mining activities induce input of the sulphides into the unstable conditions. Besides in consequence of the mining activities and the following processing processes come in on the formation of the mining wastes and processing tailings with residual contents of sulphides, when are the unstable conditions for sulphides, too. In these conditions, the contact of sulphides with atmospheric oxygen and water as well as the presence of indigenous microorganisms leads to their oxidation and subsequent production of AMD. For all that, the sources of AMD are underground as well as open-pit mining works, mining waste rock, and over-burden dumps, processing tailings, temporary and permanent stockpiles of sulphide concentrate (especially containing pyrites), flooded tunnels and shafts, heaps, sludge lagoons, etc. These sources are active during and mainly after closing the exploitation of mineral deposits. The remains of mining activities can cause the production of AMD decades or even centuries [6, 7].

2 Creation of Acid Mine Drainage

Understanding the creation of AMD requires knowledge of the geological environment of the deposit; the content of mineral phases (especially sulphide minerals, i.e. acid-forming minerals, carbonates and aluminosilicates, i.e. acid-consuming minerals); climatic, topographical and hydrological conditions; presence of autochthonous microorganisms; methods of extraction and treatment of ores and disposal of mining waste, etc. In the course of AMD genesis, many physical, physicochemical, chemical and biological-chemical processes are taking place [6]. The most important process of them is the complex series of chemical reactions involving the generation of sulphuric acid and the consumption of the generated acid [2]. The basis of the sulphuric acid generation is the oxidation of sulphides (mainly pyrite) by the simultaneous influence of water and oxygen. These reactions are autocatalytic, and their rate can be increased by microbial activity particularly of iron-oxidizing bacteria [7]. The main of the consumption of the generated acid is the acid neutralization by the reactions of acid-consuming minerals (carbonates and aluminosilicates) with the generated acid. These reactions induce the precipitation of gypsum, metal hydroxides, oxyhydroxides and other complex compounds (e.g. iron-oxyhydroxysulfate such as schwertmannite, jarosite, etc.) [6, 8]. The formation of AMD depends on the relative representation of individual minerals in the deposit. Deposits with higher sulphide content (>3%) produce the acidic effluents and deposits with a sulphide content of about 1% the neutral to slightly alkaline effluents [9, 10]. AMD is formed when the buffering capacity of accompanying minerals, especially carbonates, is not sufficient to neutralize acidic sulphide oxidation products.

2.1 Oxidation of Pyrite

The most common sulphide minerals in polymetallic, noblemetalic and coal deposits are pyrite (FeS_2), chalcopyrite ($CuFeS_2$), arsenopyrite (FeAsS) and the like [2, 7]. Their oxidation is under way through a complex series of reactions involving direct, indirect and microbial-influenced mechanisms [11]. Some oxidation reactions result in the acid formation, and others result in the dissolution and mobilization of heavy metals or metalloids. In the AMD process, the oxidation of pyrite has the greatest importance. Under natural conditions, the pyrite is closed in the rocks. Its oxidation is slow, and the slight production of acidity is in most cases either immediately diluted or neutralized by the influence of surrounding alkaline rocks. If the pyrite is released from the rock by mining and exposed to oxidation conditions, the completely different situation occurs. Under these conditions pyrite begins to react, atmospheric oxygen and water attack it and the multistage process of chemical and biological-chemical reactions producing AMD is started [2, 10]. Biological processes are caused by the presence of the indigenous microorganisms. The most extensively studied are iron- and/or sulphur-oxidizing bacteria *Acidithiobacillus* spp. and iron-oxidizing bacteria *Leptospirillum* spp. [12]. The development of molecular approaches in the studies of the AMD autochthonous microorganisms has resulted in remarkable insights into the diversity and function of these extraordinary organisms. At present between the representative prokaryotic microorganisms detected in AMD ecosystems are associated bacteria such as *Acidiphilium multivorum*, *Ferrovum myxofaciens*, *Sulfobacillus acidophilus*, etc. and archaea such as *Acidianus brierleyi*, *Ferroplasma acidiphilum*, *Sulfolobus acidocaldarius*, etc. [12, 13].

The most important chemical and biological-chemical reactions of the pyrite oxidation in connection with the formation of AMD are as follows [2]:

- Chemical oxidation of pyrite (so-called initiation reaction), resulting in the formation of the acidity and subsequent release of iron (and other metals) into solution and providing suitable conditions in terms of pH on the growth of the iron- and sulphur-oxidizing bacteria *Acidithiobacillus ferrooxidans* (Fig. 1):

$$2FeS_2 + 7O_2 + 2H_2O \rightarrow 2Fe^{2+} + 4H^+ + 4SO_4^{2-} \tag{1}$$

- Biological-chemical oxidation of Fe^{2+} to Fe^{3+} under the influence of *Acidithiobacillus ferrooxidans*:

$$4Fe^{2+} + 4H^+ + O_2 \xrightarrow{Acidithiobacillus\ ferrooxidans} 4Fe^{3+} + 2H_2O \tag{2}$$

At the abiotic conditions, the reaction rate of the Fe^{2+} chemical oxidation (i.e. only by oxygen) is slow. However, in the presence of *Acidithiobacillus ferrooxidans* bacteria, the reaction rate is multiply faster. According to Singer and Stumm (1970), these bacteria can accelerate the rate of reaction (2) by a factor of 10^6 [15]. Produced Fe^{3+} is a strong chemical sulphide-oxidizing agent. For all these

Fig. 1 Bacterial cells of *Acidithiobacillus ferrooxidans* adhered to the pyrite surface [14]

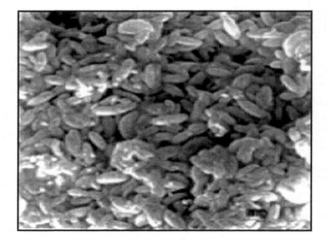

reasons, Eq. (2) is considered to be the rate-determining step in the overall AMD-generating sequence. The biological-chemical oxidation of the dissolved Fe^{2+} by bacteria depends on the oxygen concentration, the bacterial activity and pH values. The bacterial metabolism will begin to fully manifest at pH <3.

- Chemical hydrolysis of Fe^{3+} at pH >3.0 resulting in subsequent acidification of the environment and precipitation of iron in the form of the yellow-orange precipitates (commonly referred to as "ochre" or yellow boy):

$$Fe^{3+} + 3H_2O \leftrightarrow Fe(OH)_3 + 3H^+ \tag{3}$$

- Chemical oxidation of pyrite by the bacterially produced Fe^{3+} (Eq. 2) to form sulphates and ferrous ions:

$$FeS_2 + 14Fe^{3+} + 8H_2O \rightarrow 15Fe^{2+} + 2SO_4{}^{2-} + 16H^+ \tag{4}$$

Next, the Fe^{2+} will be oxidized to Fe^{3+} according to the reaction (2) and will become again disposable to oxidize additional pyrite (autocatalysis). Pyrite oxidation (Eqs. 1–4) products are Fe^{2+}, Fe^{3+}, $SO_4{}^{2-}$ and H^+ ions, and results are the formation of acidic solutions.

Dominant position in the AMD process has pyrite. Other sulphide minerals (galena, chalcocite, chalcopyrite, arsenopyrite, etc.) are oxidizing directly or indirectly by the action of Fe^{3+} and contribute to the heavy metal/metalloids dissolution. Oxidation of bivalent metal sulphides (MeS, where Me = Fe, Zn, Cd, Pb, Cu, Ni) describes reactions (5) and (6):

$$MeS + O^2 \rightarrow Me^{2+} + 4SO_4{}^{2-} \tag{5}$$

$$MeS + 2Fe^{3+} + 3/2O_2 + H_2O \rightarrow Me^{2+} + 2Fe^{2+} + 2H^+ + SO_4{}^{2-} \tag{6}$$

Therefore AMD contains increased concentrations of sulphates, heavy metals (Fe, Cu, Pb, Zn, Cd, Co, Cr, Ni, Hg), metalloids (As, Sb) and other elements (Al, Mn, Si, Ca, Na, K, Mg, Ba, F).

2.2 Acid Neutralization

In polymetallic, noblemetalic and coal deposits, the sulphide minerals coexist with the acid-consuming minerals. Typical representatives are carbonates (calcite, aragonite, siderite, magnesite, etc.) and aluminosilicates (olivine, pyroxenes, feldspar, micas, etc.). [2, 10]. These minerals react with sulphuric acid produced by the oxidation of sulphides and neutralize it. Neutralization reactions with calcite (Eqs. 7 and 8) and K-feldspar (Eqs. 9 and 10) are as follows:

$$CaCO_3 + 2H^+ \rightarrow Ca^{2+} + H_2O + CO_2 \qquad (pH < 6.4) \qquad (7)$$

$$CaCO_3 + H^+ \rightarrow Ca^{2+} + HCO_3^- \qquad (pH > 6.4) \qquad (8)$$

$$4H^+ + KAlSI_3O_8 + 4H_2O \rightarrow K^+ + Al^{3+} + 3H_4SiO_4 \qquad (pH < 4.5) \qquad (9)$$
$$H^+ + KAlSI_3O_8 + 7H_2O \rightarrow K^+ + Al(OH)_3 + 3H_4SiO_4 \qquad (pH < 4.5) \qquad (10)$$

In the course of the neutralization reaction sequence evolution, and as the pH rises, heavy metals precipitate in the form of hydroxides (such as $Fe(OH)_3$, $Al(OH)_3$, $Cu(OH)_2$, etc.), producing acidity. Neutralization reactions result in the production of gypsum (Eq. 11), too:

$$Ca^{2+} + SO_4^{2-} + 2H_2O \rightarrow CaSO_4 \cdot 2H_2O \qquad (11)$$

At low pH be in progress, other precipitating reactions mainly involving Fe^{3+} ions may also form oxyhydroxides (FeOOH) and other complex compounds, e.g. iron-oxyhydroxysulfate such as schwertmannite, jarosite, etc.

The intensity of pyrite oxidation and hence the intensity of production of acidic mining waters and their properties depend on the following factors: reactive pyrite surface, oxygen concentration, pyrite sulphur form, pH value, temperature, oxidative activity of bacteria, frequency of precipitation, quality and quantity of bacterial culture, occurrence and amount of accompanying minerals (mainly acid-consuming minerals) in the parent rock, etc.

3   Influence of Acid Mine Drainage on the Environment

AMD causes the decomposition of other minerals; the devastation of the surrounding environment; the contamination of underground water and water streams by a wide range of elements, including the toxic ones; and the penetration of metals into the food chain. The fish kill belongs to the first observable negative impacts of the AMD inflow into the streams. Fish are exposed directly to metals and H^+ ions through their gills (the damage of the respiration) or indirectly through their ingestion of contaminated sediments and food items. Effluents of AMD have characteristic yellow-orange colour associated to the ferric hydroxide formation (the yellow-orange precipitates, known as "ochre" or "yellow boy"); sometimes the blue-green colour (if it contains iron in the ferrous state), which convert to brown-red by the oxidation of ferrous iron to ferric iron; and occasionally other colours in consequence of the other precipitation reactions products (e.g. white colour upon contents of aluminium hydroxide). The precipitates mainly in the form of the iron precipitates of different types (oxides, oxyhydroxides, sulphates, oxyhydroxysulfates) are carried by water into the stream, river, lake, water basin, etc. They coat the surface of stream/river sediments and become the part of the bottom sediments of lake and basin. Gradually they induce the colourations of waterbody, diminish the availability of gravels for the fish spawning and reduce the benthic macro-invertebrates hence the fish food items, etc. Effluents of AMD are accountable to for physical, chemical and biological degradation of aquatic life [16, 17]. Influences of AMD on the environment are shown in Table 1. Differed colouring of AMD samples coming from typical localities with the occurrence of sulphide minerals in Slovakia is described in Fig. 2.

4 Acid Mine Drainage in Slovakia

Slovakia has the significant mining tradition in the main exploitation of iron, copper, gold and silver. At present, there is only one deposit being exploited, namely, Au-ore deposit in Hodruša. The other deposits are mostly flooded within the exploitation-attenuating process and present the suitable environment for gradual generation and intensification of chemical and biological-chemical oxidation resulting in the formation of AMD. Smolník and Pezinok deposits, as well as the Šobov dump, are the typical examples of AMD production under the influence of *Acidithiobacillus ferrooxidans* bacteria. All three localities can be currently considered as natural biogeoreactors producing AMD with pH 2.0–3.5 and high content of heavy metals and sulphates [18]. On other deposits of sulphide minerals in Slovakia, the occurrence study of autochthonous Fe- and S-oxidizing microbial cultures with the catalytic effect of AMD formation is not given sufficient attention [19]. From

Table 1 Influences of AMD on the environment [17]

Factor	Form	Concentration/ value	Environmental impact
Acidity	H^+	pH < 4.5	Mobilization of metals/metalloids, animal death, damage to vegetation, a decline of the drinking water quality, corrosion pipe
Precipitation of Fe	Fe^{2+}, Fe^{3+}, Fe $(OH)_3$	100–1,000 mg l^{-1}	The colouration of water, clogging of the fish gills, damage of macroinvertebrates, a coating of the surface of stream sediments
Dissolved heavy metals and metalloids	Cu, Pb, Zn, Cd, Co, Cr, Ni, Hg, As, Sb	0.01–1,000 mg l^{-1}	Degradation and death of organisms and vegetable, bioaccumulation, a decline of the drinking water quality, contamination of soils and sediments
Total dissolved solids	Ca, Mg, K, Na, Fe, Al, Si, Mn, SO_4^{2-}	100–10,000 mg l^{-1}	Decline of the drinking water quality, decline of the technical/utility water quality, formation of sheet and scale, contamination of soils and sediments

Fig. 2 Samples of AMD from Smolník deposit and Šobov dump (Slovakia) [18]. 1, Šobov dump (effluent from heap); 2, Smolník deposit, Karitas shaft; 3, Smolník deposit, Pech shaft; 4, Smolník deposit, setting-pit

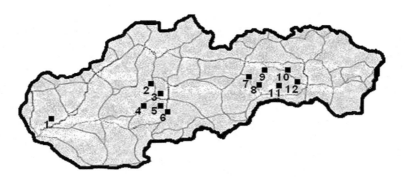

Fig. 3 The localities with the occurrence of *Acidithiobacillus ferrooxidans* bacteria in Slovakia. 1, Pezinok; 2, Kremnička; 3, Horná Ves; 4. Voznica; 5, Šobov; 6, Odkalisko sedem žien; 7, Nižná Slaná; 8, Rožňava; 9, Rudňany; 10, Slovinky; 11, Smolník; 12, Fichtenhűbel [19]

2006 to present time, our research is focused on the study of the occurrence of autochthonous chemolithotrophic Fe- and S-oxidizing bacteria of *Acidithiobacillus ferrooxidans* species in the mine water from the selected sulphide mineral deposits on the Slovakia territory. Silverman's and Lundgren's selective nutrient medium 9 K [20] was used for the isolation and following cultivation of the studied bacteria. Their identification by the investigation of morphological, physiological and cultivation properties was realized. The samples collection was carried out from 60 sampling points from 24 sulphide mineral deposits. The orange precipitates occurrence is the typical attribute of the *Acidithiobacillus ferrooxidans* bacteria positive growth (according to the Eqs. 2–3). This effect was detected at 32 sampling points in 12 sulphide deposits (Fig. 3).

Predominantly occurrence of the studied bacteria was detected in Smolník deposits and Šobov dump. The occurrence of Acidithiobacillus ferrooxidans bacteria was also determined in a mine water effluent with pH >5.0 (e.g. Pernek–Pezinok tunnel and Slovinky deposit). These results mention on the possibility of the presence of *Acidithiobacillus ferrooxidans* also in slightly acidic or slightly neutral mining waters with increased iron and sulphate concentrations, too [19]. The primary isolation of *Acidithiobacillus ferrooxidans* bacteria was carried out in the following sampling points of the sulphide mineral deposits: Rožňava, Maria mine; Nižná Slaná, Jozef shaft; Slovinky, Alžbeta shaft; Rudňany, New tunnel; Kremnička, Dedičná tunnel; Voznica, Voznická dedičná tunnel; Voznica, New drain tunnel; and Fichtenhűbel, Raky tunnel. The obtained results demonstrate a real possibility of the acid mine drainage generation in Slovinky, Rožňava and Rudňany deposits [19].

4.1 Smolník Deposit

The Smolník deposit is located between Smolnická Huta and Smolník villages in the eastern side of the Smolník brook valley. It belongs to the most critical historical deposits of Slovakia. It was in the past well-known in the worldwide scale [21]. The mineralization consists of stratiform massive pyrite mineralization, disseminated pyrite-chalcopyrite and pyrite-pyrrhotite mineralization, keratofyres and their pyroclastics. The deposit was exploited 725 years since thirteenth century by German colonists [21, 22]. Originally was exploited rich Fe-ore, later, in the Medieval times also Cu-Ag-Au ore. During the period 1326–1990, 19 Mt of ore from the deposit were produced (150 kt copper). For many centuries besides the traditional mining, copper was exploited by cementation. Therefore the mining fields, heaps and water percolation were adapted, so there was maximum volume of mine waters with the copper contents produced. These activities represent today's serious environmental problem. The mining activity stopped only in 1990. The mine was flooded till 1994. In the same year occurred the ecological collapse which caused the fishkill. In 1996 the companies Aquipur a.s. Bratislava and Geological Service of Slovakia, Spišská Nová Ves, realized the technological measures concerning the mine hydrogeological regime for purpose of decreasing the acidity of mine drainage. However, due to complicated situation in the area, the problem was not treated thoroughly, only delivered a partial elimination of the unfavourable state. The estimated resources of pyrite ore (approximately 6 miles tonnes) [23], water and the occurrence of autochthonous chemolithotrophic Fe- and S-oxidizing bacteria of *Acidithiobacillus ferrooxidans* present the basic conditions for AMD production [19]. Average monthly values of the observed indicators of water quality showed that the situation is still unfavourable due to over-limit concentrations of toxic metals (Fe, Al, Zn, Cu and Pb) and sulphates, according to the Regulation of the Slovakia 269/2010 Coll (Table 2). The outflow of AMD from the galleries is in the surface conditions mixed with surface water of Smolník stream, which flows to the Hnilec river and

Table 2 Values of pH and concentration of chosen heavy metals/metalloid and sulphates of AMD samples from the Smolník deposit in 2006–2016 [27]

AMD	pH	Concentration (mg l^{-1})					SO$_4^{2-}$(mg l^{-1})
		Cu	Al	Zn	Pb	Fe	
Pech shaft	3.94	1.40	68	7.25	0.06	351	4,421
Charitas shaft	2.96	15.50	158	16.20	0.02	680	6,775
Limits[a]	6.0–8.5	0.02	0.2	0.1	0.02	2.0	250

[a]Limits according to the Regulation of the Slovakia 269/2010 Coll

(a)   (b)   (c)

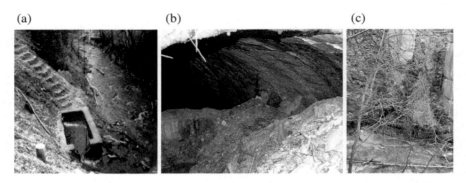

Fig. 4 The effluent of AMD in Smolník deposit to Smolník stream (Slovakia) [27]. (**a**) Pech shaft, (**b**) Charitas shaft, (**c**) percolation of AMD from Charitas shaft

consecutively to the Ružín dam reservoir [24]. AMD is flowing out mainly from the Pech shaft (Fig. 4a), which collects most of drainage waters of the flooded mine [25]. The Pech shaft is the most important source of contamination of that site. The second source of AMD is Charitas shaft (Fig. 4b, c).

For the purpose of the AMD from Pech shaft, was studied remediation processes of water dilution, neutralization, sorption/biosorption, precipitation/bioprecipitation [18, 26]. The research results have provided a number of positive experiences but also pointed out some negatives. They have contributed to the intention of further research, especially in the field of selective metal removal possibilities [25, 26].

4.2 Pezinok Deposit

At the Pezinok deposit, two types of ore mineralization were described: (1) metamorphosed, primarily exhalation-sedimentary pyrite-pyrrhotite mineralization genetically related to Devonian basic volcano-sedimentary cycle which was subsequently metamorphosed and (2) hydrothermal Sb-Au-As mineralization of epigenetic character which is most frequently localized in beds of tectonically deformed black schists [28].

The exhalation-sedimentary pyrite-pyrrhotite was exploited from 1848 to 1896. The ore was used for production of sulphuric acid. About 20,000 tons of antimony was exploited from this deposit in the period from 1939 to 1992. The published Sb content vary from 1% to 4% and the As content from 0.5% to 1.5%, and the average Au content is 3.60 ppm [29]. The mine was closed in 1992.

The mining waste is deposited in several heaps and two sludge lagoons containing 380,000 m^3 of material [30]. As and Fe minerals (predominantly arsenopyrite and pyrite) were during the ore dressing process suppressed and moved to the waste. The gangue minerals are represented mainly by carbonates and quartz. The schist fragments occur only rarely. The dominant clay mineral is illite. Chlorite is abundant, but kaolinite is very rare [31]. Also, Fe oxyhydroxides and Sb oxides are formed in the oxidation zone of the sludge lagoons [30]. The high residual concentrations of metals Sb, Fe and As in the deposited solid wastes and contaminated soils are currently the permanent source of in situ pollution and due to the activity of autochthonous microflora, the source of AMD generation. Surface and underground waters are also polluted with elements from the floatation agents used in the ore processing. The released metals and other chemical agents may enter to the food chain of animals and humans through plants and water.

In the mining area of Pezinok deposit, it is possible to distinguish two types of mining waters: the first acid type (pH 3.5–5.2) is derived from the exhalation-sedimentary pyrite-pyrrhotite ore and the second type is connected with the hydrothermal Sb-Au-As mineralization (pH 5.5–7.2), containing carbonate gangue minerals [32].

Sludge lagoons and setting-pits contain a lot of waste sulphide minerals which represent the main substrates necessary for the metabolic activity of autochthonous, bacteria *Acidithiobacillus ferrooxidans*, *Acidithiobacillus thiooxidans* and *Leptospirillum ferrooxidans* catalysing the sulphide minerals oxidation processes [14]. Mine waters discharging from the Pyritová (Fig. 5) and Augustín galleries drain mainly the massive pyrite-pyrrhotite ore body. The oxidation of sulphides

Fig. 5 Abandoned Pyritová gallery

causes the formation of AMD (Table 3). At about 6 l of water per second is discharging from the sludge lagoons. This water is partially cleaned up in the sludge lagoons by the natural mechanical settling and sorption. The mineralization of draining water (Table 4) is about 1,250 mg l^{-1}, and the water contains an increased amount of sulphates and metals like Fe, Sb, As, Mn, Pb and Cd [30]. This water is collected in great marshland under the dam of the setting-pit (Fig. 6).

The occurrence of autochthonous chemolithotrophic Fe- and S-oxidizing bacteria of *Acidithiobacillus ferrooxidans* in the mine galleries, mining water and surface water percolating the dump sediments at the Pezinok deposit and the relic content of sulphide minerals in the deposited wastes makes us to assume the biogenic catalysis of the oxidation processes.

The influence of the bacterial leaching at the pH <3.5 can substantially accelerate the ore minerals degradation. Despite the present favourable pH stage of the AMD, it is necessary to monitor the development of the conditions at the sludge lagoons. The neutralizing potential of the carbonates could be exhausted, and the character of

Table 3 pH of the AMD samples from the Pezinok deposit

Water source	pH
Pyrite gallery	3.5–5.1
Augustín gallery	3.7–5.2
Percolation water from the sludge lagoons	3.3–7.8
Stream water from the area of the mining plant	5.5–6.6

Table 4 Values of pH and concentration of chosen heavy metals/metalloids and sulphates of the AMD samples from the Pezinok deposit

Sample	pH	As	Fe	Sb	Zn	SO_4^{2-}
		mg l^{-1}				
P1	5.54	<0.005	24.50	0.002	0.16	530
P2	4.50	0.01	0.92	0.002	0.11	450
P3	6.63	<0.005	8.36	0.003	0.12	230

Explanatory notes: P1, mine water from Pyritová gallery; P2, water from the sludge lagoon; P3, mine water from the creek near the mining plant

Fig. 6 Acid drainage water in the marshland under the dam of the setting-pit

the AMD pH could change to the markedly acid values. Such a change could activate the catastrophic biodegradation process and cause the substantial contamination of the surrounding landscape [33]. The kinetics of the ore minerals degradation decrease in the range: löllingite, FeAs; arsenopyrite, FeAsS; native Sb; stibnite, Sb_2S_3; gudmundite, FeSbS; berthierite, $FeS.Sb_2S_3$; sphalerite, ZnS; pyrite, FeS_2; and chalcopyrite, $CuFeS_2$. Despite the relatively favourable pH values of the mixed two types of mining water, it is necessary to monitor the acidity of the water.

4.3 Šobov Dump

The Banská Štiavnica-Hodruša ore field is developed in central part of the statovolcano caldera [34]. On the north-eastern top part of the caldera rim is nearby the Nová shaft situated the Šobov hydroquartzite quarry (Fig. 7) and underneath its dump and the setting-pit Sedem Žien [35]. The setting-pit dam is 44 m tall, and its volume is 2.5 million m^3 of dump material from Pb-Zn mineralization near Nová shaft. It was used from 1963 to 1994. The secondary hydroquartzite in the Šobov quarry is rich in fine-grained pyrite of two generations. The Šobov dump is active from 1956 up today [36].

The dump, as well as the setting-pit, is percolated by acid drainage water from the spring beginning the nearby quarry. The acid waters are of red colour and contain high Fe, Mn and Zn contents (Table 5, sample V-1). It is collected in a great retention impound (Fig. 8, sample V-2) from which it flows through artificial anoxic and oxic wetland system, where it is neutralized and partly cleaned from the potentially toxic elements (samples V-3 and V-4). This water percolates sediments of the Sedem Žien setting-pit and outflow at the bottom of the setting-pit dam (Fig. 9).

Both in acid drainage water (Table 6) and soil/technogenous sediments of the setting-pit (Table 7) was described the presence of various bacteria, which accelerate decomposition of sulphide minerals (mainly of fine-grained pyrite from hydroquartzite) and cause the formation of acidity [35].

Fig. 7 The Šobov quarry

Table 5 Atom absorption spectrometric analysis of the AMD

Sample	pH	Fe mg l^{-1}	Mn	Zn	Mg	Pb	Cu	Co	Ni
V-1	2.17	290.12	9.88	1,438	90.01	97	765	184	112
V-1	2.20	311.00	10.74	1,720	98.35	126	890	270	160
V-3	8.81	0.75	0.00	400	64.82	3,773	160	0	0
V-4	7.60	9.63	4.37	1,690	59.11	73	140	0	0

Explanatory notes: sample 1, acid drainage water from the spring nearby the hydroquartzite quarry; sample 2, acid water from the retention impounds; samples 3 and 4, outflow of the percolating water from the setting-pit

Fig. 8 Retention impound collecting AMD (sample V-2) from the spring

Fig. 9 Outflow of the Sedem Žien setting-pit percolating water from the bottom of the dam (samples V-3 and V-4)

Table 6 Occurrence of bacteria in water percolating the dump material [35]

Sample	Locality	Medium	pH	Bacteria
Š – 1	Dump of Šobov quarry	Acid drainage water	2.3	*At.f., At.t.*
Š – 2			2.3	*At.f.*
Š – 3			2.4	*At.f.*
Š – 4			2.1	*At.f., L.f.*
Š – 5			2.3	*At.f., At.t., L.f.*
Š – 6			2.0	*At.f., At.t., L.f.*
ŠQ – 1	Šobov quarry – spring under the quarry	Acid water spring	3.0	*ATF, ATT*
ŠQ – 2			2.7	*At.f., At.t., L.f.*
NŠ – 1	Dump of Nová shaft	Acid drainage water	5.6	*At.f., L.f.*
NŠ – 2			6.8	*ATF*
NŠ – 3			6.3	*At.f., At.t., L.f.*
SŽ – 1	Spoil-dump Sedem Žien	Water lixivium from the soil	5.7	*At.f.*

Explanations: *At.f. Acidithiobacillus ferrooxidans*, *At.t. Acidithiobacillus thiooxidans*, *L.f. Leptrospirillum ferroxidans*

Table 7 Occurrence of bacteria in soil samples of the setting-pit material [37]

Sample	Bacteria
S-1	*Staphylococcus haemolyticus*
	Bacillus megaterium
	Bacillus simplex
S-2	*Bacillus cereus*
S-3	*Lysinibacillus fusiformis*
	Bacillus cereus
	Bacillus weihenstephanensis

The soil reaction (pH) in the setting-pit material is very variable (2.07–7.26). It depends both on the content of the sulphide minerals (mainly of pyrite) and on the bacteria activity. The drainage water percolates the sediments (Fig. 9) of the setting-pit flow to the near creek. The pH values of the water at the place of inflow to the brook vary from 7.55 to 7.71.

5 Conclusions and Recommendations

Formation and treatment of AMD belong to the most topical worldwide environmental problems concerning the mining and processing activities. However, according to the current expert studies, the AMD may be considered as a substantial atypical resource of metals [38]. Present trends of AMD treatment are focused not only on metal elimination down to/under the legislation requirements, nevertheless on the efficient recovery of metals [39]. Various procedures and approaches have been and still are developed for selective metals recovery. The industry is mainly interested in such procedures, which enable the removal of the metals in the form suitable for further practical application, e.g. pigments for dyes production. Environmental technologies specifically bioremediation gain the higher level of topicality by a solution of AMD problematic. The ground of the bioremediation is the controlled intensifying of the biogeochemical cycles of metals, routinely running in the natural waters under the influence of microorganisms (MO), which participate on the basis of their fundamental metabolic processes in the solubilization and immobilization of metals in AMD [40]. Bioremediation is the economic and ecological option of conventional physical-chemical processes of metals elimination from AMD. The combination of the metal precipitation using the sodium hydroxide (chemical methods) with the metal precipitation using the bacterially produced hydrogen sulphide (biological-chemical method) presents the base of the selective sequential precipitation (SSP). This method constitutes the possibility of the recovery metals in a suitable form for commercial or industrial utilization [26, 41].

References

1. Chen LX, Huang LN, Mendez-Garcia C, Kuang JL, Hua ZS, Liu J, Shu WS (2016) Microbial communities, processes and functions in acid mine drainage ecosystems. Curr Opin Biotechnol 38:150–158
2. Kontopoulos A (1998) Acid mine drainage control. In: Castro HF, Vergara F, Sanchez MA (eds) Effluent treatment in the mining industry. University of Concepcion, Chile, pp 57–118
3. Karpenko V, Norris JA (2002) Vitriol in the history of chemistry. Chem List 96:997–1005
4. Agricola G (1556) De Re Metallica. Book I, (trans: Hoover HC, Hoover LH). Dover, New York 1950, p 8
5. Salkield LU (1987) A technical history of the Rio Tinto mines: some notes on exploitation from pre-Phoenician times to the 1950s. The Institution of Mining and Metallurgy, London. 114 pp
6. Lottermoser BG (2010) Mine wastes: characterization, treatment and environmental impacts.3rd edn. Springer, Berlin, Heidelberg. 400 pp
7. Hallberg KB, Johnson DB (2001) Biodiversity of acidophilic prokaryotes. Adv Appl Microbiol 49:37–84
8. Paikaray S, Schröder C, Peiffer S (2017) Schwertmannite stability in anoxic Fe(II)-rich aqueous solution. Geochim Cosmochim Acta 217:292–305
9. White WW, Jeffers TH (1994) Chemical predictive modelling of acid-mine drainage from metallic sulphide-bearing waste rock. In: ACS symposium series, vol 550, pp 608–630
10. Karavaiko GI, Rossi G, Agate AD, Groudev SN, Avakyan ZA (1988) Biotechnology of metals - manual. Centre for International Projects GKNT, Moscow. 350 pp
11. Sand W, Jozsa PG, Gehrke T, Schippers A (2001) (Bio)chemistry of bacterial leaching direct vs. indirect bioleaching. Hydrometallurgy 59:159–175
12. Johnson DB, Rolfe S, Hallberg KB, Iversen E (2001) Isolation and phylogenetic characterization of acidophilic microorganisms indigenous to acidic drainage waters at an abandoned Norwegian copper mine. Environ Microbiol 3:630–637
13. Kuang JL, Huang LN, Chen LX, Hua ZS, Li SJ, Hu M, Li JT, Shu WS (2013) Contemporary environmental variation determines microbial diversity patterns in acid mine drainage. ISME J 201(7):1038–1050

14. Kušnierová M, Fečko P (2001) Mineral biotechnology I. VŠB-TU, Ostrava. 143 pp
15. Singer PC, Stumm W (1970) Acidic mine drainage, the rate-determining step. Science 167:1121–1123
16. Jennings SR, Neuman DR, Blicker PS (2008) Acid mine drainage and effects on fish health and ecology: a review. Reclamation Research Group, Bozeman, p 29
17. Andráš P, Dirner V, Turisová I, Vojtková H (2014) Remnants of old activity at abandoned Cu-deposits. Vodní zdroje Ekomonitor, Chrudim. 439 pp
18. Luptakova A, Kusnierova M, Fecko P (2002) Mineral biotechnology II. – Sulfuretum in nature and industry. Ostrava, VŠB-TU. 152 pp
19. Luptakova A, Prascakova M, Kotulicova I (2012) Occurrence of Acidithiobacillus ferrooxidans bacteria in sulfide mineral deposits of Slovakia. Chem Eng 28:31–36
20. Silverman MP, Lundgren DC (1959) Studies on the chemoautotrophic iron bacterium Ferrobacillus ferrooxidans I. An improved medium and a harvesting procedure for securing high cell yields. J. Bacteriol 77:642–647
21. Grecula P, Abonyi A, Abonyiová M, Antaš J, Bartalský B, Ďuďa R, Gargulák M, Gazdačko Ľ, Hudáček J, Kobulský J, Lőrinczz L, Macko J, Návesňák D, Németh Z, Novotný L, Radvanec M, Rojkovič I, Rozložník L, Rozložník O, Varček C, Zlocha J (1995) Mineral deposits of the Slovak Ore Mountains I. Geocomplex, Bratislava. 829 pp
22. Radvanec M, Bartalský B (1987) Geochemical zoning of stratiform sulphide ore mineralization in the Smolník – Štós – Medzev area. Mineralia Slovaca 19:443–445
23. Rojkovič I (2003) Rudné ložiská Slovenska. Univerzita Komenského, Bratislava. 180 pp
24. Balintova M, Petrilakova A, Singovszka E (2012) Study of metals distribution between water and sediment in the Smolnik Creek (Slovakia) contaminated by acid mine drainage. Chem Eng Trans 28:73–78
25. Balintova M, Petrilakova A (2011) Study of pH influence on selective precipitation of heavy metals from acid mine drainage. Chem Eng Trans 25:345–350
26. Luptakova A, Ubaldini S, Macingova E, Fornari P, Giuliano V (2012) Application of physical–chemical and biological–chemical methods for heavy metals removal from acid mine drainage. Process Biochem 47:1633–1639
27. Balintova M, Luptakova A (2012) Treatment of acid mine drainage. TU in Košice, Faculty of Civil Engineering, Košice. 131 pp
28. Chovan M, Rojkovič I, Andráš P, Hanas P (1992) Ore mineralisation of the Malé Karpaty Mts. (Western Carpathians). Geol Carpath 43:275–286
29. Uher P, Michal S, Vitáloš J (2000) The Pezinok antimony mine, Malé Karpaty Mts., Slovakia. Mineral Rec 31:153–162
30. Trtíková S, Chovan M, Kušnierová M (1999) Oxidation of pyrite and arsenopyrite in the mining wastes (Pezinok - Malé Karpaty Mts). Folia Fac. Sci. Nat., Univ. Mas. Brun. Geologia 39:225–231
31. Lukianenko Ľ, Čerňanský S, Štubňa J (2008) Mine tailing site Pezinok – Kolársky vrch (Slovakia) – an example of anthropogenic contaminated landscape. Acta Environmentalistica Universitatis Comenianae Bratislava 16(1):64–68
32. Andráš P, Adam M, Chovan M, Šlesárová A (2008) Environmental hazards of the bacterial leaching of the ore minerals from waste at the Pezinok deposit (Malé Karpaty Mts., Slovakia). Carpath J Earth Environ Sci 3(1):7–22
33. Andráš P, Kušnierová M, Adam M, Chovan M, Šlesárová A (2004) Bacterial leaching of ore minerals from waste at the Pezinok deposit (Western Slovakia). Slovak Geol Mag 12(2):79–90
34. Burian J, Slavkay M, Štohl J, Tőzsér J (1985) Metalogenéza neovulkanitov Slovenska. Alfa, Bratislava. 269 pp
35. Križáni I, Andráš P, Ladomerský J (2007) Banícke záťaže Štiavnických vrchov. Technická univerzita vo Zvolene, Zvolen. 100 pp
36. Križáni I, Andráš P, Šlesárová A (2009) Percolation modeling of the dump and settling pit sediments at the Banská Štiavnica ore-field (Western Carpathians, Slovakia). Carpath J Earth Environ Sci 4(1):109–126

37. Remešicová E (2016) Analysis of problems associated with the decontamination of mine water in ore mining. PhD thesis, VŠB-TU, Ostrava, 148 pp
38. Hennebel T, Boon N, Maes S, Lenz M (2015) Biotechnologies for critical raw material recovery from primary and secondary sources: R&D priorities and future perspectives. New Biotechnol 32:121–127
39. Ehinger S, Janneck E (2010) ProMine – nano-particle products from new mineral resources in Europe: Schwertmannite – raw material and valuable resource from mine water treatment processes. In: Freiberger Forschungsforum research conference, Freiberg
40. Gadd GM (2000) Bioremedial potential of microbial mechanisms of metal mobilization and immobilization. Curr Opin Biotechnol 11:271–279
41. Sheoran AS, Sheoran V, Choudhary RP (2010) Bioremediation of acid-rock drainage by sulphate-reducing prokaryotes: a review. Miner Eng 23:1073–1100

Groundwater: An Important Resource of Drinking Water in Slovakia

D. Barloková and J. Ilavský

Contents

Abstract Groundwater resources are mainly used as the supply of drinking water in Slovakia (87.3% of inhabitants are supplied with drinking water from underground resources), of which approximately 22% of this amount has to be treated. Water treatment is mostly needed for the removal of iron and/or manganese. Concentrations of dissolved iron and manganese are evaluated every year within the groundwater monitoring done by the Slovak Hydrometeorological Institute (SHMI) for the whole territory of Slovakia.

The presence of iron and manganese compounds in water creates technological problems, failures of water supply systems and deterioration of water quality with respect to sensory properties. Also, if these waters are slightly over-oxidized, unfavourable incrustations are formed.

The objective of the pilot plant tests in the water treatment plant Kúty was to verify the efficiency of manganese and iron removal from water with the use of different filtration materials with MnO_2 layer on the surface – Klinopur-Mn, Greensand and Cullsorb M.

D. Barloková (✉) and J. Ilavský
Department of Sanitary Engineering, Faculty of Civil Engineering, Slovak University of Technology, Bratislava, Slovak Republic
e-mail: danka.barlokova@stuba.sk

A. M. Negm and M. Zeleňáková (eds.), *Water Resources in Slovakia:*
Part I - Assessment and Development, Hdb Env Chem (2019) 69: 277–302,
DOI 10.1007/698_2017_215, © Springer International Publishing AG 2018,
Published online: 27 April 2018

Keywords Filtration materials, Groundwater quality, Removal of iron and manganese, Water treatment

1 Introduction

Groundwater is an irreplaceable component of the environment. It represents an invaluable, well-available and quantitatively and economically most suitable source of drinking water. The abundance of natural groundwater resources, their quality and the potentially less possibility of their contamination predict groundwater as the dominant source of drinking water in the Slovak Republic (Fig. 1).

Groundwater resources are mainly used as the supply of drinking water in Slovakia (87.3% of inhabitants are supplied with drinking water from underground resources), of which approximately 22% of this amount has to be treated.

The monitoring of groundwater quality and chemical status was divided in accordance with the Water Framework Directive 2000/60/EC (WFD) into the following groups:

– Surveillance monitoring
– Operational monitoring.

The groundwater quality is monitored approximately in 171 sites of the surveillance monitoring. These are sites included in the national monitoring network of the Slovak Hydrometeorological Institute or springs not affected by point sources of pollution. Operational monitoring is done in all groundwater bodies that were assessed as being at risk of failing to achieve good chemical status. There are approximately 295 sites

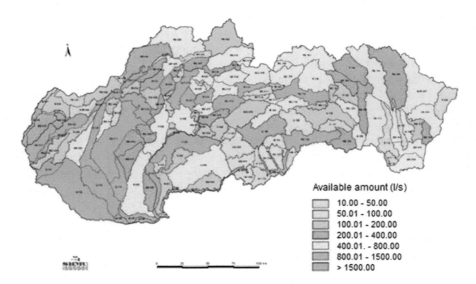

Fig. 1 Available amount of groundwater in Slovakia at 2012 [1, 2]

monitored within the operational monitoring programme (except the region of Žitný ostrov) where the potential input of pollution to the groundwater from potential source/ sources of pollution is expected. The sampling frequency is one to four times a year depending on the type of rock environment. The samples are taken in spring and autumn when the extreme condition of groundwater could be monitored. The region of Žitný ostrov represents a separate part of the SHMI monitoring network as it plays an important role in the process of monitoring the changes in water quality in Slovakia since this region is the most significant drinking water resource in our territory. The monitoring network of Žitný ostrov comprises 34 piezometric multilayer wells (84 layers) that are monitored two to four times a year.

In terms of quality of groundwater which is used for drinking purposes, the main indicators are the amount of iron, manganese, ammonium, heavy metals (e.g. arsenic, antimony, nickel, lead), etc. Furthermore, we can also classify here the content of CO_2, hydrogen sulphide and microbiological quality of water.

Concentrations of dissolved iron and manganese, arsenic, antimony and other parameters of water quality are evaluated every year within the groundwater monitoring done by the Slovak Hydrometeorological Institute (SHMI) for the whole territory of Slovakia. The results of laboratory analyses were assessed under the Regulation of the Slovak Government 496/2010 amending the Regulation 354/2006 defining the requirements for drinking water intended for human consumption and for drinking water quality monitoring. The assessment of results was done through a comparison between measured values and limit values for each of analysed parameters. The results were published in the annual report "Groundwater Quality in Slovakia" [1–4].

According to the 2015 Reports on the Environment in Slovakia, the concentration of iron exceeded the 0.2 mg/L limit in more than 13.4% of the samples, and the concentration of manganese exceeded the 0.05 mg/L limit in more than 15.8% of 373 groundwater samples. These samples represent 166 objects of surveillance monitoring at this year. The concentration of iron exceeded the 0.2 mg/L limit in more than 36.2% of the samples, and the concentration of manganese exceeded the 0.05 mg/L limit in more than 38.0% of 687 groundwater samples (which represents 220 objects of operational monitoring at this year). The limit values are defined under the Government Regulation of the Slovak Republic No. 496/2010 on Drinking Water [4].

Exceeded limits of parameters at surveillance and operational monitoring sites according to the GR 496/2010 in 2015 are present in Figs. 2 and 3.

In the case of groundwater used for drinking purposes, water treatment is mostly needed for the removal of iron and/or manganese and heavy metals (arsenic, antimony). The occurrence of iron and manganese in groundwater in Slovakia is shown in Figs. 4 and 5.

In Slovakia conditions, the treatment processes are usually mainly focused on the removal of iron and manganese from the raw water. These technological processes of treatment are demanding not just from the aspect of investment but also from the aspect of operating costs. "In searching for suitable water treatment technology, an emphasis is placed on new, more efficient and cost-effective methods and materials compared to the technology currently used" [6].

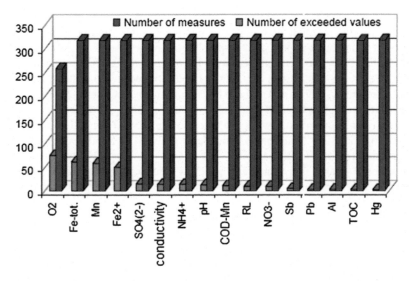

Fig. 2 Exceeded limits of parameters at surveillance monitoring sites according to the Government Regulation 496/2010 in 2015 [4]

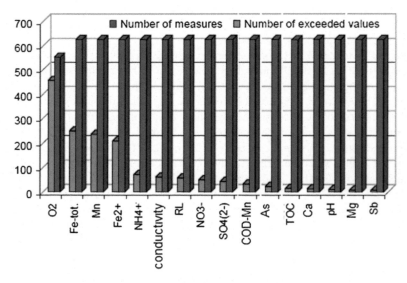

Fig. 3 Exceeded limits of parameters at operational monitoring sites according to the Government Regulation 496/2010 in 2015 [4]

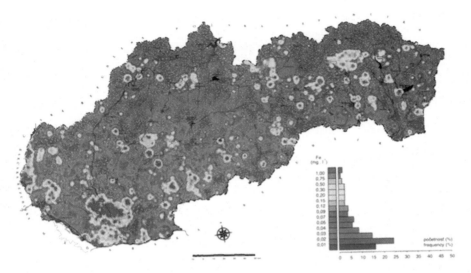

Fig. 4 Occurrence of iron in groundwater [5]

Fig. 5 Occurrence of manganese in groundwater [5]

2 Iron and Manganese: Occurrence in Water

Iron and manganese occur in dissolved forms as single ions (Fe^{2+}, Mn^{2+}) or in undissolved higher forms, mainly as $Fe(OH)_3$ or $MnO_2 \cdot xH_2O$, respectively. They can also be present in colloidal form (bound to humic substances). The form of their occurrence depends on the oxygen concentration, the solubility of Fe and Mn compounds in water, the pH value, the

redox potential, hydrolysis, the presence of complex-forming inorganic and organic substances and the water temperature and composition (e.g. CO_2 content) [7, 8].

The process of the oxidation of Fe^{2+} can be represented by the schematic equation:

$$4Fe^{2+} + O_2 + 10H_2O \rightarrow 4Fe(OH)_3 + 8H^+.$$

In waters containing bicarbonate, this reaction can also take place:

$$4Fe^{2+} + 8HCO_3^- + 2H_2O + O_2 \rightarrow 4Fe(OH)_3 + 8CO_2.$$

The rate of oxidation depends on the pH, the concentration of the iron, the dissolved
oxygen concentration and the redox potential. Since the reaction produces hydrogen ions,
the oxidation is accelerated in an alkaline medium [9–11].

The dependence of the oxidation on the pH is very strong. At a pH range of 5–8.2,
the oxidation rate is about 100 times higher when the pH level rises by one. Furthermore, it is affected by the effects of temperature and light. The positive or negative
impact of different anions or organics depends on the stability of their Fe^{2+} or Fe^{3+}
complexes. If they form stable complexes with Fe^{3+}, the rate of oxidation increases
and vice versa.

The stability of iron ion depends not only on pH but also on the activity of electrons
which is represented by a redox potential E [V] (Fig. 6). High positive value of pE
indicates oxidizing conditions where iron is insoluble, and the low values of pE
indicate reducing conditions where iron is soluble [12].

The occurrence and behaviour of manganese are not similar to iron. Manganese in
the oxidation state of Mn^{2+} in waters containing dissolved oxygen under certain
conditions is unstable. In alkaline conditions, manganese is rapidly oxidized and
hydrolysed to form the less soluble oxides of manganese in the higher oxidation state
Mn^{4+}:

$$Mn^{2+} + 2H_2O \rightarrow MnO_2 + 4H^+ + 2e^-.$$

The mechanism of the oxidation of Mn^{2+} in an actual rock environment is complicated. This is a set of the interconnected processes of oxidation, catalysis, sorption, ion
exchange and biological oxidation. The composition of the final products of oxidation,
which is partially secreted in a colloidal form, depends on factors such as the pH,
temperature, oxidation-reduction potential, reaction time and rocks. The general scheme
of Mn^{2+} oxidation by oxygen dissolved in water can be represented as follows:

$$Mn^{2+} \rightarrow Mn(OH)_2(s) \rightarrow Mn_2O_3 \cdot xH_2O \rightarrow MnO(OH) \rightarrow MnO_2 \cdot xH_2O.$$

The relationship between iron and manganese under increasing pH and redox
potential (pE) suggests that ferrous iron (Fe^{2+}) normally occurs in the area with lower
redox potential (<200 mV) and within the pH range of 5.5–8.2. This also means that
Fe^{2+} is more easily and rapidly oxidized than Mn^{2+}. The latter is often occurring with
Fe^{3+} under pH values larger than 8 and redox potentials between 420 and 790 mV.
Above this redox potential, the stable form of MnO_2 is found [9, 12–15].

Fig. 6 Eh-pH stability diagram of iron [10]

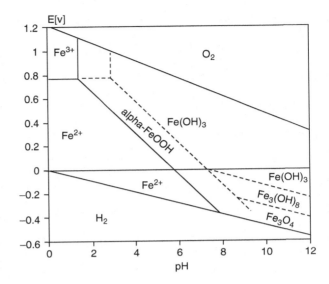

Diagram of the existence of dominant areas Mn-CO_2-H_2O-O_2 is in Fig. 7 [7].

Alkalinity and pH have a marked effect on the solubility of Mn(II). This solubility is governed by the formation of manganese carbonate. Manganese hydroxide has a much higher solubility. At pH values of 8 and higher, the calculated solubility of Mn (II) is very limited (1–2 mg/L or lower) even at low alkalinity (1.2 mmol/L) [16].

3   Effect of Iron and Manganese on Water Quality

Higher Fe and Mn concentrations in drinking water have adverse effects which can be summarized as follows [17]:

1. Iron (II) and manganese (II) ions are oxidized to higher forms in water distribution system, and this results in the formation of hydroxide suspensions causing undesirable turbidity and colour of water.
2. Presence of iron and manganese bacteria in water supply system causing a change in water quality (smell) and bacterial growth in pipes.
3. In case of occurrence of iron (II) and manganese (II) ions at the consumer's point, iron and manganese are oxidized and precipitated under suitable conditions (e.g. in washing machines, boilers).

Due to the facts mentioned above, higher concentrations of iron and manganese in water can cause technological problems, failures in the operation of water supply systems and deterioration in water quality. In water with slightly higher concentrations of oxygen, iron and manganese form undesirable incrustations, resulting in the reduction of the flow in a pipe's cross-section [17].

Fig. 7 Eh-pH diagram of
the existence of dominant
areas $Mn-CO_2-H_2O-O_2$. The
total concentration of
2 mmol/L CO_2 and the total
concentration of manganese
0.055 mg/L. Dashed lines
are marked area at a total
concentration of manganese
5.5 mg/L

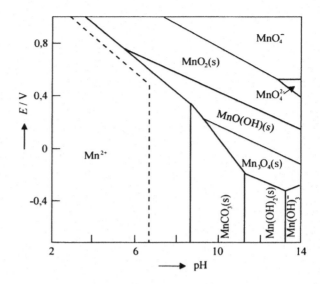

4   Iron and Manganese: Methods of Removal from Water

The principle of most methods used for iron and manganese removal is that originally
dissolved iron and manganese are transformed into undissolved compounds that can
be removed through single-stage or two-stage separation. Oxidation and hydrolysis of
these compounds are done under strict conditions with respect to water properties and
type of equipment for iron and manganese removal [17].

The single-stage water treatment (filtration) is designed for iron and manganese
concentrations to 5 mg/L, and the two-stage treatment (settling tanks or clarifiers and
filters) is used for water with iron and manganese concentrations higher than 5 mg/L.
In case water contains higher concentrations of Ca, Mg and CO_2 (eventually H_2S),
aeration is done before settling or filtration.

Removal of Fe and Mn from groundwater and surface water can be done by
several methods [17]:

• Oxidation by aeration
• Removal of Fe and Mn by oxidizing agents (O_2, Cl_2, O_3, $KMnO_4$)
• Removal of Fe and Mn by alkalinization (by adding the lime)
• Contact filtration for removal of Fe and Mn
• Removal of Fe and Mn by ion exchange
• Removal of Fe and Mn using membrane processes
• Removal of Fe and Mn using biological filtration
• Removal of Fe and Mn using in situ method

The Fe^{2+} and Mn^{2+} oxidation rates, as well as hydrolysis of emerging oxides of higher iron and manganese oxidation forms in groundwater, depend on the pH value. Various graphic dependencies of these relationships also with respect to oxidation time are listed in the literature. The pH value should be equal or greater than 7 in the removal of iron from groundwater. For removal of manganese without a catalyst, the pH value should be equal or greater than 8 [7, 8].

Removal by using the oxidized film on grains of the filter medium is one of the methods for elimination of dissolved manganese. The film is formed on the surface of filter medium by adding permanganate potassium (not only $KMnO_4$ but also other strong oxidizing agents). The MnO_x coating serves as a catalyst for the oxidation process. Grains of filter medium are covered by higher oxides of metals. In such a case, it is related to special filtration so-called contact filtration – filtration by using manganese filters. The oxidation state of the film of $MnO_x(s)$ filter medium is very important for removal of dissolved manganese. Manganese removal efficiency is a direct function of $MnO_x(s)$ concentration and its oxidation state. The films with different ability to remove dissolved manganese from water are formed on the surface of various filter media [18–22].

Natural or synthetic zeolite can be used as a filtration material for removal of iron and manganese from water [23–40]. Birm, Greensand, Pyrolusite, Pyrolox, Cullsorb M, MTM, Everzit Mn, Klinopur-Mn and Klinomangan are the most frequently used materials in filtration.

Contact with filter material results in oxidation of dissolved iron and manganese. Subsequently, precipitated Fe and Mn hydroxides (pH of 8–9 is required for Mn removal) are easily removed by filtration. The filter medium is cleaned by backwashing. There is no need of chemical regeneration. In cleaning process, the time of backwashing and wash water velocity are important factors. Long service life is also one of the advantages of this medium [17].

5   Materials Used for Contact Filtration in Removal of Fe and Mn

Birm® is a granulated filtration medium (imported from the USA) used mainly for removal of iron and manganese from water. It is a specially developed material with MnO_2 film on the surface (serves as a catalyst). Properties of Birm are listed in Table 1 (Fig. 8).

Birm is recommended to be used at lower concentrations of iron (up to Fe^{2+} 6.0 mg/L and Mn^{2+} 3.0 mg/L) and for household water treatment. It can be used in gravity or pressure filters. Treated water shall not contain oils, sulphates, organic substances and high concentration of chlorine. Water with low oxygen level has to be pretreated by using aeration.

Table 1 Conditions of Birm use

Filtration material	Operating range	Filtration material	Operating range
Manganese content	<3 mg/L	Alkalinity	>2× (SO_4^{2-} Cl^-)
Iron content	<6 mg/L	Organic matters	<5 mg/L
Temperature range	3–45°C	Chlorine (Cl_2)	<0.5 mg/L
pH	6.8–9.0	H_2S	=0 mg/L
Dissolved oxygen	>15% of Fe content	Oils	=0 mg/L

Fig. 8 Birm material [41]

The efficiency of Birm also depends on pH value. Water with pH < 6.8 should be treated by adding the alkaline agents. The most suitable pH value is in the range from 8.0 to 9.0.

Greensand® (imported from the USA) is a glauconitic mineral with a zeolite-type structure. It is produced from glauconitic sand, which is activated by potassium permanganate ($KMnO_4$). The resulting product is a granulated material covered by a MnO_2 film on its surface and other higher oxides of manganese. It is used for the removal of iron, manganese and hydrogen sulphide from water. Dissolved iron and manganese are oxidized and precipitated in contact with the higher oxides of manganese on the surface of Greensand. Undissolved iron and manganese are trapped in the "Greensand medium" and removed by backwash. After the exhaustion of its oxidizing capacity, the bed is regenerated using a $KMnO_4$ solution or chlorine. The regeneration frequency depends on the amount of iron, manganese and oxygen in the water as well as the filter size. We recognize two regeneration processes, i.e. with discontinuous or continuous regeneration.

The pH value of water is an important factor influencing the efficiency of filters. If the pH is lower than 6.8, the efficiency of Greensand is reduced. The operating conditions for Greensand are listed in Table 2 [6] (Fig. 9).

Greensand has been used for several decades for the removal of Fe and Mn from the water. A thin layer of manganese dioxide gives the dark sand a definite green colour and thus its name. The combination of a strong oxidant and Greensand filtration media for iron and manganese removal is commonly referred to as the "manganese Greensand process".

Table 2 Conditions of Greensand use

Parameter	Operating range	Parameter	Operating range
Manganese content	<5 mg/L	Alkalinity	$>2\times (SO_4^{2-}\ Cl^-)$
Iron content	<10 mg/L	Organic matters	<5 mg/L
Temperature	5–30°C	Chlorine concentration	1.2 times of Fe
pH	6.8–9.0	Oils	≈0 mg/L
Dissolved oxygen	>15% of Fe content	H_2S	<5 mg/L

Fig. 9 Manganese
Greensand material [41]

The advantage is that water with a low oxygen content does not have to be pre-oxidized. Greensand can be used in cases with a higher content of iron (over 10 mg/L) and manganese (over 3 mg/L) [6]. It can also be used in industry. A content of organic substances, oils and hydrogen sulphides has an adverse effect on its efficiency.

Pyrolusite is the common name for naturally occurring manganese dioxide and is available in the USA, the UK, South America and Australia. It is distributed under brand names such as Cullsorb M, Pyrolox, Filox-R and MetalEase. It is a mined ore consisting of 40–85% manganese dioxide by weight. The various configurations of pyrolusite provide extensive surface sites available for oxidation of soluble iron [41], manganese and hydrogen sulphide. Removal rates of iron in excess of 15 and 3 mg/L of manganese are achievable.

Pyrolusite is a coarse oxidizing media with a high specific gravity of about 4.0. Like silica sand, pyrolusite is a hard media with small attrition rates of 2–3% per year. Pyrolusite [41] may be used in the following two ways: (1) mixing with sand, typically at 10–50% by volume, to combine a filtering media with the oxidizing properties of pyrolusite and (2) installing 100% pyrolusite in a suitably graded filter to provide oxidation and filtration [41] (Fig. 10).

Cullsorb M (imported from the USA) is a natural, highly selected mineral, lacking in additives and impurities. It is based on manganese dioxide, which has an extremely high capacity for catalytic oxidation. It is specifically used in filtering plants for the removal of iron, manganese and hydrogen sulphide by oxidation in water. The filter cartridge requires periodic or continuous regeneration of the oxidized reagent, either potassium permanganate or air [6] (Table 3).

Fig. 10 Pyrolusite material

Table 3 Basic properties of Cullsorb Mn

Parameter	Operating range	Parameter	Properties
pH	6.8–9.0	Colour	Dark brown to black
Dissolved oxygen	>15% of Fe content	Aspect	Granular
Fe^{2+}/Mn^{2+}	<15 mg/L	Water content	<2%
H_2S	<5 mg/L	Iron (as Fe)	<3.5%
Alkalinity	>2× (SO_4^{2-} Cl^-)	Manganese (as Mn)	>53%
Organic matters	<5 mg/L	Manganese (as MnO_2)	>80%
Oils	≈0 mg/L	Temperature	5–30°C

Klinopur-Mn is produced in Slovakia from natural zeolites. It is an activated zeolite – clinoptilolite (rich deposits of clinoptilolite are in the East Slovakia Region). On the surface of clinoptilolite grains, there is a factory-made film consisting of manganese oxides (MnO_x) which enable this material to be used in the contact filtration. The filter material is produced by Zeocem Bystré, and it is much cheaper compared to materials imported from the USA [17].

Clinoptilolite $(NaK)_6(Al_6Si_{30}O_{72})\cdot20H_2O$ is one of the most frequently used natural zeolites. At present, it is also applied to water treatment process. Sufficient mechanical resistance, chemical stability and abrasion values, even if they categorize it among soft materials, enable clinoptilolite to be used as a filtration material.

The specific weight of clinoptilolite is lower than the weight of silica sand. Moreover, its porosity and sludge capacity are 1.5 times greater compared to filtration sand. Using the zeolite in slow sand filtration allows filtration rate to be increased by four times. Furthermore, it reduces also the amount of wash water and time needed for filter backwashing [17, 42, 43] (Fig. 11, Table 4).

The most important properties of zeolitic minerals are their ability to change cations, to adsorb inorganic and organic molecules of certain dimensions, catalytic properties, high content of Si causes the chemical and thermal stability thereof.

Fig. 11 Clinoptilolite from
Slovakia (grain size
1–2.5 mm)

Table 4 Basic properties of clinoptilolite

Parameter	Operating range	Parameter	Operating range
Colour	Grey-green	Effective diameter of pores	0.4 nm
Compressive strength	33 MPa	Absorbability	34–36%
Specific gravity	2.39 g/cm^3	Water solubility	0
Mass density	0.84 g/cm^3	Thermal stability	<450°C
pH	6.8–7.2	Stability against acids	79.50%

Klinomangan – activated zeolite – clinoptilolite $(K, Na, Mn)^{6+}[(AlO_2)_6(SiO_2)_{30}]\cdot$ $24H_2O$ from the bearing Rátka in Hungary. Superficial layers of manganese oxide as in the case of Klinopur-Mn allow using this material in contact filtration for removal of iron and manganese from water. Depending on the quality of the treated water, the filter cartridge is required after a certain time to regenerate with the solution of potassium or sodium permanganate [42, 43, 44].

Table 5 shows the content of essential minerals forming clinoptilolite deposit, in Table 6 compares the chemical composition of clinoptilolite from Nižný Hrabovec and clinoptilolite from the deposit Rátka.

Comparison of basic properties of some filtration materials used for the removal of Mn and Fe is listed in Table 7. Table 8 shows chemical composition of the most commonly used filtration materials with MnO_2 layers on the surface (on the basis of our results).

6 Removal of Iron and Manganese by Contact Filtration at WTP Kúty

The objective of the technological trials in the locality of Kúty (a water treatment plant) was to verify the efficiency of manganese and iron removal in water treatment using a filtration medium based on a chemically modified natural zeolite (Klinopur-Mn). At the same time, the efficiency of manganese and iron removal was compared with the imported Greensand and Cullsorb M (USA) materials, which are often used

Table 5 Mineralogical composition of the zeolite deposits in Slovakia and Hungary

Mineral	Nižný Hrabovec (Slovakia)	Rátka (Hungary)
	Content [%]	
Clinoptilolite	84	55
Cristobalite	8	15
Feldspar	3–4	10
Illite	4	–
Montmorillonite	–	10

Table 6 Chemical composition of the nature zeolites from Nižný Hrabovec and Rátka

Zeolite	Content [%]							
	SiO_2	Al_2O_3	K_2O	CaO	Fe_2O_3	MgO	Na_2O	TiO_2
Nižný Hrabovec	66.40	12.20	3.33	3.04	1.45	0.56	0.29	0.15
Rátka (Hungary)	72.15	12.86	3.72	1.84	1.22	0.53	0.26	0.10

Table 7 Filtration materials and some selected parameters

Material	Klinopur	Birm	Cullsorb	Greensand	Klinomangan
Grain size [mm]	0.3–2.5	0.4–2.0	0.42–1.6	0.25–0.8	00.5–1.2
Specific gravity [g/cm^3]	2.39	2.0	3.5–4.0	2.4	20.66
Apparent density [g/cm^3]	0.84	0.7–0.8	1.75–1.85	1.36	10.04
Porosity [%]	64.8	–	–	–	41.7
Abrasion [%]	8.2	–	2–3	–	00.57

Table 8 Chemical composition of the filtration materials for contact filtration

Material	Content [%]					
	SiO_2	Al_2O_3	K_2O	CaO	MnO_2	Fe_2O_3
Birm	50.32	10.55	1.59	19.85	**7.15**	6.78
Greensand	47.43	8.19	4.55	3.53	**14.95**	16.98
Cullsorb M	12.13	12.13	1.53	0.83	**69.25**	1.96
Everzit Mn	10.95	5.19	2.07	0.53	**75.80**	4.68
Klinopur-Mn	69.56	8.19	5.58	3.79	**6.92**	3.32
Klinomangan	64.68	8.29	4.77	3.51	**12.16**	2.87
Manganized sand	27.52	–	0.14	32.67	**33.58**	5.12

The layer of MnO_2 on the surface of grains is marked with bold numbers

abroad for dissolved manganese and iron removal from water in small-scale water treatment plants (small water resources) [6].

The water treatment plant in Kúty is a part of the Senica group of water supply systems. The water from two wells with a yield of 80 L/s does not meet the requirements of Regulation No. 496/2010 on Drinking Water for iron, manganese, ammonium ions and aggressive carbon dioxide. The technological water treatment process consists of aeration, a dosage of calcium hydrate, slow mixing, filtration and disinfection [6] (Figs. 12 and 13).

Fig. 12 WTP Kúty (view from the street)

Fig. 13 WTP Kúty (aeration units, pump and filtration halls)

The technological scheme of the WTP Kúty is shown in Fig. 14. The figure also indicates the location of the filter columns (sampling points) used in our experiments.

The methodology for the verification of suitable filtration materials for iron and manganese removal is based on their properties and possible technological applications in the water treatment process. The following technological water treatment method was proposed [6]:

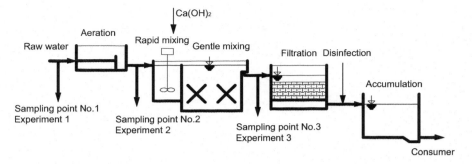

Fig. 14 Scheme of the technology of WTP Kúty and the location of the filter columns

Raw water → filtration and oxidation (backwashing and regeneration).

Raw water is passed through the filtration equipment, and the removal of the Fe^{2+} and Mn^{2+} ions is carried out directly in the filtration column beds (the media). The following were used as filtration materials:

- Greensand
- Cullsorb M
- Natural activated zeolite with MnO_2 (Klinopur-Mn)

The experiments were designed to optimize the filtration rate (contact time of the raw water with the filter media) and washing and regenerating the filter materials (filter length cycles).

The quality of the raw water (Fe and Mn content) and treated water at the outlet from the separate filtration columns was monitored during the experiments. At the same time, the amount of water (filtration rate) at the outlet from the columns was measured.

To verify the efficiency of iron and manganese removal from the water resources in the locality of WTP Kúty, three filtration columns containing Greensand, Cullsorb M and Klinopur-Mn were used. The adsorption columns were made of glass. The parameters of each adsorption column are as follows: diameter = 5.0 cm, height = 2 m, surface = 19.635 cm^2, filtration medium height 110 cm and volume of filtration medium 2,160 [cm^3]. The filtration equipment is shown in Fig. 15. The figure shows a simple device that allows splitting the incoming water either for washing or filtration through a valve system [6].

The water was supplied to filtration columns from three different sites for the technological water treatment process. The water for Experiment 1 (sampling point No. 1) was taken from the inlet of the raw water to the water treatment plant. The water for Experiment 2 (sampling point No. 2) was taken after aeration of the water,

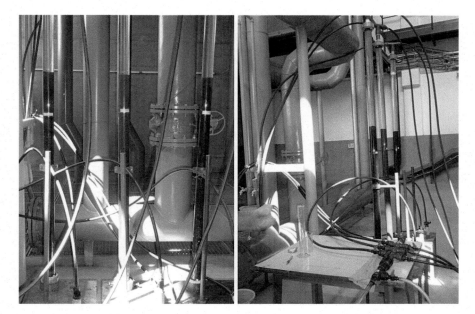

Fig. 15 Filtration equipment – filtration columns

where the content of the oxygen in the water has increased. The water for Experiment 3 (sampling point No. 3) was taken after aeration and lime dosing, where the optimal conditions for the removal of the iron and manganese (increased oxygen content and a pH of more than 8) were achieved [6]. Table 9 shows the values of the basic parameters during the experiments.

The model tests and the results of the experiments are divided into three parts:

1. For raw water from wells
2. For raw water after aeration
3. For raw water after aeration and the addition of lime

6.1 Experiment 1

Raw water passed through the filtration columns in a downward direction. The average filtration rate was 6.23 m/h for the Greensand, 5.98 m/h for the Klinopur-Mn and 5.83 m/ h for the Cullsorb M. The filtration conditions are shown in Table 10.

The results of removing the iron and manganese from the raw water are documented in Figs. 16 and 17 [6]. They show the concentration of manganese and iron in the raw water and the values measured after they passed through the monitored filter materials. The figures also show the manganese limit value (0.05 mg/L) and, respectively, the iron limit value (0.2 mg/L) for drinking water in the Regulation of the Government of the Slovak

Table 9 The basic parameters during the pilot test

Parameter	Sampling point		
	No. 1	No. 2	No. 3
Fe (mg/L)	2.28–5.16	0.90–3.87	1.96–4.22
Mn (mg/L)	0.82–1.12	0.816–1.092	0.198–0.524
pH	6.64–6.98	6.81–7.14	8.40–8.62
Oxygen (% saturation)	6–7	59–60	56–57

Table 10 Filtration conditions for the first sampling point

Parameter	Greensand	Klinopur	Cullsorb
Grain size [mm]	0.25–0.8	0.6–1.6	0.4–0.6
Height of filtration medium [cm]	110	110	110
Average discharge through column [mL/min]	204	196	191
Average filtration rate [m/h]	6.23	5.99	5.84
Filtration total time [h]	260	260	260
Total volume of water flown through [m^3]	3.182	3.058	2.980
Average residence time in column [min]	10.587	11.019	11.308

Republic No. 496/2010 on Drinking Water. The arrow represents the regeneration time of the filter media [6].

Figure 16 shows that the quality of the raw water (low pH – 6.6 to 6.9, low oxygen content – 6 to 7%) has an influence on the efficiency of the removal of the manganese and that the efficiency for the monitored materials was different. The best results were achieved with Cullsorb M, which even after 260 h of operation did not exceed the limit value for manganese (0.05 mg/L). It is a fact that Cullsorb M is the material with the highest MnO_2 content on its surface. The Greensand exceeded the limit of 0.05 mg/L after 20 h and the Klinopur-Mn after 42 h of operation. Those materials had to be regenerated (2.5% solution of $KMnO_4$). After the regeneration, the columns were operating again, but for this type of water, their efficiency was too low.

The Greensand and Klinopur-Mn released the manganese from their surfaces into the water because of the pure quality of the treated water, the low oxygen content and the low pH value [6].

Figure 17 shows that all the materials are effective for removing iron from the water during the operation of the filtration columns and did not exceed the limit value of 0.20 mg/L.

6.2 Experiment 2

For Experiment 2 (sampling point No. 2), water was taken after aeration, where the content of the oxygen in the water has increased. Raw water passed through the filtration columns in a downward direction. The average filtration rate was 5.84 m/h

Fig. 16 Course of removing the manganese from the water

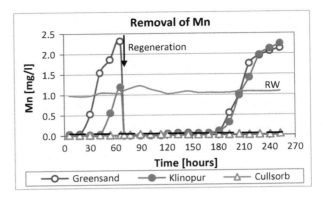

Fig. 17 Course of removing the iron from the water

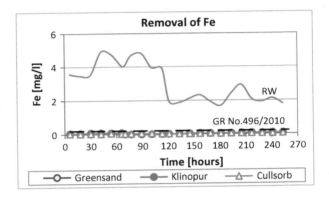

for the Greensand, 5.13 m/h for the Klinopur-Mn and 5.65 m/h for the Cullsorb M. The filtration conditions are shown in Table 11.

Figures 18 and 19 show the results of removing iron and manganese from the raw water after aeration (sampling point No. 2). The concentration of manganese and iron in the raw water and the values measured after they passed through the monitored filter materials, contrasted with the manganese limit value (0.05 mg/L) and, respectively, the iron limit value (0.2 mg/L) in the drinking water defined under Regulation of the Government of the Slovak Republic No. 496/2010 on Drinking Water, and the regeneration time for the filter media are illustrated.

Figure 18 shows that the influence of changes in the quality of the raw water (pH 6.8–7.2, the oxygen content from 56 to 57% saturation) for the efficiency of the manganese removal from the water through the filtration materials improved significantly. All three materials obtained a high level of efficiency of the manganese removal from the water. In the case of the Cullsorb M and Greensand materials during the 910 h of the operation of the filtration process, the limit value for manganese (0.05 mg/L) was not exceeded [6]. Klinopur-Mn is necessary to modify by the gradual backwashing and regeneration with a solution of $KMnO_4$. In the first filtration step, it exceeded the limit value of 0.05 mg/L after 162 h of operation; in the second filter

Table 11 Filtration conditions for the second sampling point

Parameter	Greensand	Klinopur	Cullsorb
Grain size [mm]	0.25–0.8	0.6–1.6	0.4–0.6
Height of filtration medium [cm]	110	110	110
Average discharge through column [mL/min]	191	168	185
Average filtration rate [m/h]	5.84	5.13	5.65
Filtration total time [h]	910	910	910
Total volume of water flown through [m³]	10.429	9.173	10.101
Average residence time in column [min]	11.308	12.856	11.675

Fig. 18 Course of removing the manganese from water

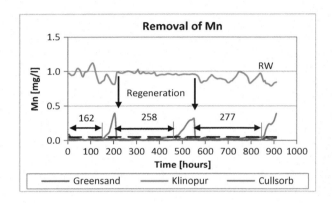

Fig. 19 Course of removing the iron from water

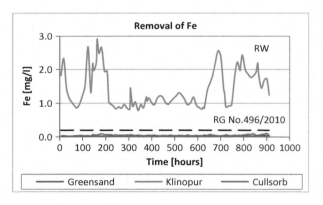

cycle, the limit value was exceeded after 258 h, and the limit value was exceeded in the third filter cycle in 277 h. The filtration time without regeneration was gradually extended. This means that the industrially activated clinoptilolite (Klinopur-Mn) should be modified on-site directly in water treatment plant. The filter cycles will be extended, and after some time, regeneration will not be needed.

The filter media were backwashed continuously (approximately every 2–3 days) (given the amount of precipitated ferric hydroxide collected). Over time, as shown in Fig. 18, the concentration of manganese in the treated water after passing through Klinopur-Mn exceeded the value of 0.05 mg/L; then the filter materials were regenerated with a solution of $KMnO_4$ (2.5% solution).

Figure 19 shows the course of removing the iron from the water for sampling point No. 2 (after the water aeration). The value of the iron in the raw water was quite changed, depending on which well was used for pumping or the production of precipitated $Fe(OH)_3$, which gradually clogged the system. In general, all three materials removed the iron effectively and, during the operation of the filtration columns, did not exceed the limit value of 0.20 mg/L as defined under Regulation No. 496/2010 on Drinking Water.

The filtration rates during the second experiment were lower compared to the first experiment [6], what was caused with precipitations of iron after the oxidation, ferric hydroxide clogged the columns.

6.3   Experiment 3

For Experiment 3 (sampling point No. 3), water was taken after aeration and the addition of lime, where the value of pH and content of the oxygen in the water have increased. Raw water passed through the filtration columns in a downward direction. The average filtration rate was 5.53 m/h for the Greensand, 5.38 m/h for the Klinopur-Mn and 5.47 m/h for the Cullsorb M. The filtration conditions are shown in Table 12.

The results from removing the iron and manganese from the raw water after aeration and the addition of lime are best documented by Figs. 20 and 21 in which the concentration of manganese and iron in the raw water and the values measured after passing through the monitored filter materials are shown. The figures also show the manganese limit value (0.05 mg/L) and, respectively, the iron limit value (0.2 mg/L) in the drinking water as defined under Regulation No. 496/2010 on Drinking Water. The arrow represents the regeneration time of the filter media.

Figure 20 shows that the changes in the quality of the raw water (pH 8.4–8.6; an oxygen content of 59–60% saturation) have an efficiency in the removal of manganese from the water. The high removal efficiency of the manganese was achieved by all three materials – Cullsorb, Greensand and Klinopur-Mn. Even after 802 h of

Table 12 Filtration conditions for sampling point No. 3

Parameter	Greensand	Klinopur	Cullsorb
Grain size [mm]	0.25–0.8	0.6–1.6	0.4–0.6
Height of filtration medium [cm]	110	110	110
Average discharge through column [mL/min]	181	176	179
Average filtration rate [m/h]	5.53	5.38	5.47
Filtration total time [h]	802	802	802
Total volume of water flown through [m³]	8.710	8.469	8.613
Average residence time in column [min]	11.933	12.272	12.066

Fig. 20 Course of
removing the manganese
from water

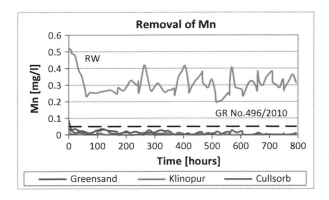

Fig. 21 Course of
removing the iron from
water

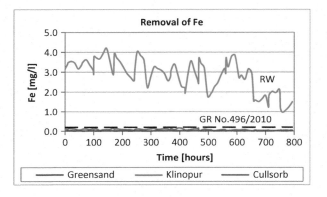

operation of the filtration system, they did not exceed the value of the manganese in the treated water limit value of 0.05 mg/L. The filter materials were backwashed continuously (approximately every 2–3 days). There was no need to regenerate these materials [6].

Figure 21 shows the progress made in removing the iron from the water for sampling point No. 3 (after the water aeration and the addition of lime). The value of the iron in the raw water is quite changed, depending on the production of precipitated $Fe(OH)_3$, which gradually clogged the system. In general, all three materials removed the iron effectively and, during the operation of the filtration columns, did not exceed the limit value of 0.20 mg/L as defined under Regulation No. 496/2010 on Drinking Water.

7   Conclusion

The results obtained proved the possibility of using Klinopur-Mn in removing iron and manganese from water (the so-called contact manganese removal) that is comparable to the Greensand and Cullsorb M materials imported from abroad.

The materials observed exhibit different efficiencies of manganese removal from water since the quality of the treated water plays a major role (oxygen content and pH value). In the case of the removal of the iron from the water, the quality of the raw water is a limiting factor; all the materials removed Fe from the water to below the limit value (0.20 mg/L).

The rate of filtration, backwashing time and intensity (self-carriage filter material during the washing) and the method of regeneration of the filter media with $KMnO_4$ (a 2.5% solution of $KMnO_4$) were also measured during the pilot plant experiments.

The insertion of aeration and the pH adjustment of the raw water before the filtration column to increase the efficiency of the filter media provide an effective treatment of the water as seen from the experimental results.

The technology of the removal of Fe and Mn with contact filtration is often used for small water resources (the water treatment is directly on the water resource). Based on our experiments, the most suitable material for the water quality and filtration conditions was Cullsorb M (contains over 80% of MnO_2 on its surface) [6].

Acknowledgements The technological trials were performed within the APVV-15-0379 grant project and VEGA 1/0400/15 project.

References

1. Ministry of Environment of the Slovak Republic (2012) Water management in the Slovak Republic in 2012 (Blue Report). Prepared by Water Research Institute, Bratislava 2013, p 23 (in Slovak). http://www.minzp.sk/files/sekcia-vod/modra-sprava-2012-slovenska.pdf
2. (2017)   www.enviroportal.sk   http://www1.enviroportal.sk/indikatory/detail.php?id_indikator=1624
3. Čaučík P, Leitmann Š, Možiešiková K, Sopková M, Molnár L, Bodácz B, Lehotová D, Mada I (2013) Balance of water management. Water balance of groundwater quality for 2012. SHMI, Bratislava, 318 p (in Slovak)
4. Lieskovská Z, Némethová M et al (eds) (2016) Report on the State of the Environment of the Slovak Republic in 2015. The Ministry of the Environment of the Slovak Republic, Slovak Environmental Agency, Bratislava, p 236. ISBN 978-80-89503-60-5
5. Rapant S, Vrana K, Bodiš D (1996) Geochemical atlas of Slovakia-part I. Groundwater. Ministry of the Environment of the Slovak Republic, Geological Survey of Slovak Republic, Bratislava, p 127. ISBN 80-85314-67-3
6. Barloková D, Ilavský J (2012) Modified clinoptilolite in the removal of iron and manganese from water. Slovak J Civil Eng 20(3):1–8. https://doi.org/10.2478/v10189-012-0012-9
7. Pitter P (2009) Hydrochemie, 4th edn. Institute of Chemical Technology Press, Prague, 568 pp, (in Czech)
8. Edzwald JK (2011) Water quality and treatment. A handbook of drinking water, 6th edn. AWWA, McGraw-Hill Companies, New York, 1696 pp
9. Stumm W, Morgan JJ (1996) Aquatic chemistry. Wiley, New York, 1022 pp
10. Scheffer F, Schachtschabel P (1989) Lehrbuch der Bodenkunde. Enke Verlag, Stuttgart, 491 pp
11. Brezonik PL, Arnold WA (2011) Water chemistry: an introduction to the chemistry of natural and engineered aquatic systems. Oxford University Press, New York, 809 pp
12. Mouchet P (1992) From conventional to biological removal of iron and manganese in France. J AWWA 84(4):158–167

13. Katsoyiannis IA, Zouboulis AI (2004) Biological treatment of Mn(II) and Fe(II) containing groundwater: kinetic considerations and product characterization. Water Res 38(7):1922–1932
14. Trace Inorganic Substances Committee (1987) Committee report: research needs for the treatment of iron and manganese. J AWWA 79:119–122
15. Sommerfeld EO (2007) Iron and manganese removal handbook. AWWA, Denver, 172 pp
16. Buamah R, Petrusevski B, Schippers JC (2008) Adsorptive removal of manganese(II) from the aqueous phase using iron oxide coated sand. J Water Supply Res Technol-AQUA 57:1–11
17. Barloková D, Ilavský J (2010) Removal of iron and manganese from water using filtration by natural materials. Polish J Environ Stud 19:1117–1122. http://www.pjoes.com/pdf/19.6/1117-1122.pdf, 2017
18. Doula KK (2006) Removal of Mn^{2+} ions from drinking water by using Clinoptilolite and a Clinoptilolite-Fe oxide system. Water Res 40:3167–3176
19. Knocke WR, Van Benschoten JE, Keanny MJ, Soborski AW, Reckhow DA (1991) Kinetics of manganese (II) and iron (II) oxidation by potassium permanganate and chlorine dioxide. J AWWA 83:80–87
20. Knocke WR, Occiano S, Hungate R (1991) Removal of soluble manganese by oxide-coated filter media: sorption rate and removal mechanism issues. J AWWA 83:64–69
21. Knocke WR, Hamon JR, Thompson CP (1988) Soluble manganese removal on oxide-coated filter media. J AWWA 80:65–70
22. Merkle PB, Knocke WR, Gallagher DL, Solberg T (1996) Characterizing filter media mineral coatings. J AWWA 88:62–73
23. Bruins JH (2016) Manganese removal from groundwater. Role of biological and physico-chemical autocatalytic processes. Dissertation of the Delft University of Technology and of the Academic Board of the UNESCO-IHE Institute for Water Education. CRC Press/Balkema, Leiden, The Netherlands, 163 p
24. Jeż-Walkowiak J, Dymaczewski Z, Weber L (2015) Iron and manganese removal from groundwater by filtration through a chalcedonite bed. Water Supply Res Technol-AQUA 64(1):19–34
25. Taffarel SR, Rubio J (2010) Removal of Mn^{2+} from aqueous solution by manganese oxide coated zeolite. Miner Eng 23:1131–1138
26. Olańczuk-Neyman K, Częścik P, Łasińska E, Bray R (2001) Evaluation of the effectivity of selected filter beds for iron and manganese removal. Water Sci Technol Water Supply 1(2):159–165
27. Bruins JH, Petrusevski B, Slokar YM, Kruithof JC, Kennedy MD (2015) Manganese removal from groundwater: characterization of filter media coating. Desalin Water Treat 55(7):1851–1863
28. Bruins JH, Petrusevski B, Slokar YM, Huysman K, Joris K, Kruithof JC, Kennedy MD (2015) Reduction of ripening time of full-scale manganese removal filters with manganese oxide-coated media. J Water Supply Res Technol-AQUA 64(4):434–441. https://doi.org/10.2166/aqua.2015.117
29. Coffey BM, Gallagher DL, Knocke WR (1993) Modeling soluble manganese removal by oxide-coated filter media. J Environ Eng 119:679–695
30. Hargette AC, Knocke WR (2001) Assessment of fate of manganese in oxide-coated filtration systems. J Environ Eng 127:1132–1138
31. Merkle PB, Knocke WR, Gallagher DL (1997) Method for coating filter media with synthetic manganese oxide. J Environ Eng 123:642–649
32. Merkle PB, Knocke WR, Gallagher DL, Little JC (1997) Dynamic model for soluble Mn^{2+} removal by oxide-coated filter media. J Environ Eng 123:650–658
33. Islam AA, Goodwill JE, Bouchard R, Tobiason JE, Knocke WR (2010) Characterization of filter media MnOx(s) surfaces and Mn removal capability. J AWWA 102:71–83
34. Kim J, Jung S (2008) Soluble manganese removal by porous media filtration. Environ Technol 29:1265–1273
35. Chen L, Zhang J, Zheng X (2016) Coupling technique for deep removal of manganese and iron from potable water. Environ Eng Sci 33(4):261–269

36. Tiwari D, Yu MR, Kim MN, Lee SM, Kwon OH, Choi KM, Lim GJ, Yang JK (2007) Potential application of manganese coated sand in the removal of Mn(II) from aqueous solutions. Water Sci Technol 56:153–160
37. Inglezakis VJ, Doula KK, Aggelatou V, Zorpas AA (2010) Removal of iron and manganese from underground water by use of natural minerals in batch mode treatment. Desalin Water Treat 18:341–346
38. Buamah R, Petrusevski B, de Ridder D, van de Wetering TSCM, Shippers JC (2009) Manganese removal in groundwater treatment: practice, problems and probable solutions. Water Sci Technol Water Supply 9:89–98
39. Rose P, Hager S, Glas K, Rehmann D, Hofmann T (2017) Coating techniques for glass beads as filter media for removal of manganese from water. Water Sci Technol Water Supply 17:95–106
40. Hamilton G, Chiswell B, Terry J, Dixon D, Sly L (2013) Filtration and manganese removal. J Water Supply Res Technol AQUA 62:417–425
41. Birm, Pyrolusite, Greensand – Osmo Sistemi (2017) www.osmosistemi.it. http://www.osmosistemi.it/index.php?option=com_content&view=article&id=14&Itemid=126&lang=en
42. Barloková D, Ilavský J, Sokáč M (2015) Modified zeolites in ground water treatment. GeoSci Eng LXI(1):10–17. ISSN 1802-5420
43. Barloková D (2008) Natural zeolites in the water treatment process. Slovak J Civil Eng 16(2): 8–12. https://www.svf.stuba.sk/buxus/docs/sjce/2008/2008_2/file1.pdf, 2017
44. (2017) http://www.zeoclay.eu/domain/zeoclay/files/klinomangan-1000-data-sheet.pdf

Influence of Mining Activities on Quality of Groundwater

J. Ilavský and D. Barloková

Contents

Abstract The increased pollution of water resources leads to a deterioration in the quality of surface water and groundwater, and it initiates the application of various methods for water treatment. The Slovak Technical Standards – STN 75 7111 Water – and the enactment of the Decree of the Ministry of Health of the Slovak Republic No. 151/2004 on requirements for drinking water and monitoring of the quality of drinking water quality have resulted in the reduction of heavy metal concentrations or, for the first time, in defining the limit concentrations for some heavy metals (As, Sb), respectively. Based on this fact, some water resources in Slovakia have become unsuitable for further use, and they require appropriate treatment.

The objective of the study was to verify the sorption properties of some new sorption materials for the removal of antimony (Bayoxide E33, GEH, CFH12). Technological tests were carried out at the facility of the Slovak Water Company

J. Ilavský (✉) and D. Barloková
Department of Sanitary Engineering, Faculty of Civil Engineering, Slovak University of Technology in Bratislava, Bratislava, Slovak Republic
e-mail: jan.ilavsky@stuba.sk

A. M. Negm and M. Zeleňáková (eds.), *Water Resources in Slovakia:*
Part I - Assessment and Development, Hdb Env Chem (2019) 69: 303–332,
DOI 10.1007/698_2017_213, © Springer International Publishing AG 2018,
Published online: 27 April 2018

Liptovský Mikuláš in the locality of Dúbrava. Technological tests have proved that the new sorption materials can be used for reduction of antimony concentration in water to meet the values set under the Decree of the Ministry of Health of the Slovak Republic No. 247/2017 on requirements for drinking water – 5 µg/L.

Keywords Groundwater quality, Removal of antimony, Sorption materials, Water treatment

1 Introduction

One of the primary goals of WHO and its member states is that "all people, whatever their stage of development and their social and economic conditions, have the right to have access to an adequate supply of safe drinking water." A major WHO function to achieve such goals is the responsibility "to propose . . . regulations, and to make recommendations with respect to international health matters . . ." [1].

Since 1998, an intensive attention has been paid to the presence of heavy metals in the water, when standard, STN 75 7111 Drinking Water, was introduced into the Slovak legislation. By transposition of European Directive 98/83/EC and WHO recommendation [2, 3] into our legislation, the limit concentrations of some of the heavy metals (e.g., As, Sb) were decreased, resp., determined for the first time which caused that some of the Slovak water sources has become nonconforming and they need to be adjusted properly for their next use. The risk of the heavy metal presence rests mainly in their tendencies of being accumulated in the tissue of plants and animals. Some metals are quite equally presented in the earth crust from where they may move into the groundwater. Heavy metal occurrence presents the same risks as the risks of industrial or agricultural contaminants. The knowledge about the health aspects of heavy metal presence in drinking water are included in the paper Water Quality and Treatment: A Handbook of Community Water Suppliers [4] and literature [5, 6].

On the map of Slovakia (Fig. 1) are marked places with a higher concentration of antimony in groundwater. In these places are deposits of antimony ore.

The deposits of antimony ore in Slovakia are in five metallogenetic areas – in the Little Carpathians, Low Tatras, Spišsko-Gemerské Rudohorie Mts., Banskoštiavnicko-kremnické Mountains, and Prešov Mountains. The basic minerals of these sites are antimony, which is often accompanied by gold and silver [7].

The Dúbrava deposit contains quartz veins with antimony mineralization – arsenopyrite, Pb–Sb–Bi sulfosalts, sphalerite, tetrahedrite, bournonite, chalkostibit, gold, scheelite, Fe dolomite, and barite. Sb contents in the ore range from 1.5 to 5.0%. From the middle of the eighteenth century until the beginning of the twentieth century, iron ore and antimony were extracted here. The mining was then restored only before the Second World War when it started to grow antimony. The Dúbrava deposit was a significant producer of Sb in the Czechoslovakia in the last four decades until the end of the mining in 1993. In the years 1753–1985, 1,046 kt of

Fig. 1 The occurrence of antimony in Slovakia (https://uvp.geonika.sk/teslo/images/archive/a/a3/20151013103415%21Sb.png)

antimony ore was used on the Dúbrava deposit, of which 1,033 kt after 1945, 27 kt of metal was extracted from the total extracted ore [8].

The abandoned Dúbrava Mine is situated in the northern part of the Low Tatras in the middle of Slovakia. Mine drainage from adits (containing up to 9,300 μg/L of Sb), mine waste dumps, and the leachate from mine tailings contribute Sb and arsenic (As) into nearby Paludzanka Creek and groundwater. Some drinking water resources have been closed due to excessive Sb concentrations; the concentration of Sb in one household well (126 μg/L) far exceeds the Sb drinking water limit of 5 μg/L [9–11].

The Pernek deposit contains antimony mineralization with Au-bearing arseno-pyrite and pyrite, bound to black slate, which lie in the environment of actinolitic sands and amphibolites. The mining of antimony and pyrite ores on the Pernek deposit began at the end of the eighteenth century (1790) and lasted with breaks until the early twentieth century (1922).

Pernek–Pezinok mining area is important Sb deposit in the Malé Karpaty Mts. Many dump piles and mine adits left abandoned when the mining activity had stopped. At the present time, these become sources of the surface, groundwater, soils, and stream sediment contamination. Arsenic and antimony are the trace elements transforming and accumulating in several natural components. Sulfide oxidation and silicate weathering are the main processes participating in surface and groundwater chemical composition. The antimony shows an elevated concentration ranging from 1 to 31 μg/L together with elevated concentrations of Ni, Zn, Fe, and sulfates. The stream situated above dump piles is considered to be the site with background values which is confirmed by a relatively low concentration of Fe, As, Sb, Ni, and Zn. The highest concentration of As (0.005 mg/L), Ni (189 μg/L), Zn (161 μg/L), Fe (6.94 mg/L), Mn (0.655 mg/L), and sulfates (488 mg/L) was detected in the mine adit outflow (Pavol). The concentration of Sb was 0.014 mg/L [12, 13].

The Medzibrod deposit is located in the area of the southern slopes of the Ďumbierske Tatry Mts. The monitoring sites are situated in Močiar valley, which is drained by Borovský potok creek. Drainage water from Murgaš mine adit represents a significant source of contamination, where elevated concentrations of sulfates, arsenic (500 µg/L) and antimony (180 µg/L), and high mineralization were detected. The mine waste dumps situated below the mine adit, together with a tailing impoundment, are also the main sources of contamination in this area. The highest concentrations of arsenic and antimony were observed in drainage water from the waste dumps. In spite of the fact that arsenic and antimony are attenuated by dilution and adsorption on ferric iron minerals in stream sediment, elevated concentrations of arsenic and antimony were also found in surface water in Borovský potok creek. Increased amounts of some monitored chemical elements were found in stream sediments of Murgaš adit outflow with a high proportion of Fe oxyhydroxides. Extremely high levels of arsenic (10,250 mg/kg) and antimony were detected in a soil sample in close proximity to the Murgaš mine adit. Significantly elevated contents of monitored elements in stream sediments were found in inflow from Murgaš mine adit where a high portion of Fe oxyhydroxides is present.

The Medzibrod deposit is bunched antimony veins lying in phyllites and black shales, which are sometimes impregnated with pyrite–arsenopyrite ores containing gold, 1–4 ppm. The main mineral ore is antimonite; berthierite, jamesonite, and pyrite are relatively abundant. Between 1938 and 1944, the deposit produced 57 kt of ore containing 2.8 wt% Sb and 4.48 ppm Au. The bearing is considered to be loaded [14, 15].

The Čučma deposit is a quartz vein with antimonite, accompanied by siderite, Fe dolomite, calcite, carbonates, tourmaline, albit, pyrite, arsenopyrite, markazine, pyrothine, chalcopyrite, sphalerite, tetrahedrite, gallate, gold, bismuth, antimony, berthierite, boulangerit, chalkostibit, bournonite, zinckenite, and jamesonite. In the middle ages, gold, silver, and copper ores, later iron, antimony, and manganese ores, were used. The Čučma deposit was obtained from 1918 to 1944 by 204 kt and after 1945 68 kt of antimony ore. Since 1952, mining has been stopped.

In surface water and groundwater, there could be the trace amounts of antimony and arsenic. There are very strict limits for these toxic elements in drinking water (5 µg/L for antimony and 10 µg/L for arsenic). The results of study of arsenic and antimony contamination at the Čučma abandoned deposit are presented in the article [16]. This mining area belongs to the important ore deposits in the south part of Slovenské Rudohorie Mts. The mine water from adits and tailing ponds represent the most important sources of contamination at this area. The maximum value of antimony (7,130 µg/L) was detected in mine water from the Jozef mine adit. The highest content of arsenic (1,350 µg/L) and also high concentrations of Fe and Mn were measured in mining water from the Gabriela mine adit. High concentration of antimony (86 µg/L) was registered in-house wells, as well. Most of the local inhabitants use the contaminated water for drinking purposes. Soil and stream sediments are also contaminated by As, Sb, Cu, Pb, Zn, Ni, Co, Fe, Mn, and Al in this area [16, 17].

The Poproč deposit consists of six cores with ore mineralization, Anna–Agnes, Borovičná hôrka, Barbora, Lazy, Ferdinand, and Libórius, which are located along steep tectonic surfaces. Mining work in modern history began in 1938. The operation was stopped in 1965 and gradually liquidated. The total amount of extracted and processed ore for the years 1939–1965 was about 259 kt. The yield of the flotation treatment plant during the last years of operation was 90.6–93.3% at the concentration of the concentrate 48.35–52.49% Sb and the Au content 3–6 g/t.

Abandoned Sb-deposit Poproč is located in the Gemeric tectonic unit, and hydrothermal mineralization occurs here in the form of veins mainly in phyllites. Stibnite is the most abundant ore mineral; pyrite, arsenopyrite, and few other Pb–Sb–Zn–Cu sulfides are also common. Natural water, soil, stream sediments, and plants in investigated areas of abandoned Sb-deposit Poproč are primarily affected by point sources of contamination (drainage from old mine, tailing impoundments, waste dumps). Weathering of open adits, dumps, and non-isolated tailing impoundments cause many problems such as water, soil, and stream sediment contamination mainly by arsenic and antimony in the area. Extremely high concentrations of Sb and As were observed in natural constituent in the catchments of Olšava river (waters, As max 2,400 µg/L, Sb max 410 µg/L; soils, As max 1,714 mg/kg, Sb max 6,786 mg/kg; stream sediments, As max 5,560 mg/kg, Sb max 1,360 mg/kg), but relatively high values of Fe, Pb, Zn, Mn, Al, and SO_4^{2-} were monitored.

A portion of water extractable fraction of Sb in soil ranges from 0.5 to 3.06% and in the stream sediments from 0.08 to 7.15%. This, however, points to low mobility of Sb, but due to a very high total content, leaching of soils and stream sediments may cause water pollution [18, 19].

Today, it is challenging to distinguish between anthropogenic and natural pollution caused by antimony. The enrichment of antimony occurs by contact of water with rocks, minerals, and soil. In the vicinity of ore deposits, water can be enriched with higher concentrations of antimony. In groundwater the threshold limit value (Sb = 0.005 mg/L) was exceeded, for instance, in the locality of Košice and surroundings (Zlatá Idka, Bukovec water reservoir), Low Tatras in the locality of Dúbrava (Liptovský Mikuláš), Spišsko-Gemerské Rudohorie Mts. (Čučma, Poproč), and in the Little Carpathians (Pernek, Pezinok) [20].

2 Antimony: Effects on the Environment

Concentrations in natural waters not polluted are generally very low (0.1–0.2 µg/L). The increased concentration of antimony was monitored in locality of mining activities. Adverse concentration of antimony comes from the mine tailing piles and sludge lagoon where the rocks rich in antimony were continually washed by the rainwater infiltrating into the groundwater resources or flowing into the surface water.

The chemistry of antimony and its natural occurrence in some water resources combines to create a strong, widespread human health risk, requiring management and removal from drinking water.

The chemical behavior of antimony is as complicated as that of arsenic, its neighbor in the periodic table. It is speculated that antimony could be a natural contaminant with arsenic in some drinking waters. Soluble forms of antimony (and arsenic) tend to be quite mobile in water, whereas less soluble species are adsorbed into clay or soil particles and sediments, where they are bound to extractable iron and aluminum [21, 22].

Antimony is a toxic heavy metal with effects similar to arsenic and lead. Intoxication with antimony is not as severe as in the case of arsenic because the compounds of antimony are absorbed slowly. Antimony is an inhibitor for some enzymes, has an effect on the metabolism of proteins and carbohydrates, and causes a failure of glycogen production in kidneys. Its ability to accumulate in bodies of organisms is low. While there is evidence that some antimony compounds are carcinogenic by inhalation, no such evidence exists for antimony in water. Known health risks by the oral route include an increase in blood cholesterol and a decrease in blood sugar. Findings on health aspects related to the occurrence of some heavy metals in drinking water are summarized in publication [6, 21].

Thus far, the World Health Organization (WHO) and institutes dealing with the monitoring of carcinogenic effects have not classified antimony as a carcinogen.

The limit concentration of antimony in drinking water in Slovakia is 5 µg/L [23]. This limit value is in accordance with the WHO recommendations [2] and the EU directive [3].

3   Antimony: Properties and Dissolution Chemistry

Antimony in its elemental form is a silvery white, brittle, fusible, crystalline solid that exhibits poor electrical and heat conductivity properties and vaporizes at low temperatures [24]. Antimony resembles a metal in its appearance and in many of its physical properties, but does not chemically react as a metal. It is also attacked by oxidizing acids and halogens. Antimony and some of its alloys are unusual in that they expand on cooling. Metallic antimony is too brittle to be used alone and, in most cases, has to be incorporated into an alloy or compound.

Antimony and its compounds are industrially crucial because of their usefulness in the manufacture of alloys, paints, paper, plastics, textiles, glass, clay products, and rubber. In recent years, high purity antimony has been used in the production of the semiconductor compound indium antimonide and in the formulation of bismuth telluride-type compound used for thermoelectric applications. Antimony trioxide (Sb_2O_3), the most important antimony compound, is used in halogen-compound flame-retarding formulations for plastics, paints, textiles, and rubbers. Lead–antimony alloys are used in starting–lighting–ignition batteries, ammunition, corrosion-resistant pumps and pipes, tank linings, roofing sheets, solder, cable sheaths, and antifriction bearings [24].

Antimony is present in water as Sb^{3-}, Sb^0, Sb^{3+}, and Sb^{5+} (Sb^{3+} is ten times more toxic than Sb^{5+}), depending on the pH of the water, the oxidation–reduction potential (Sb^{3+}/Sb^{5+} ratio), and the oxygen content. The most common form is antimonate–oxyanion (H_2SbO_4)$^-$ and ($HSbO_4$)$^{2-}$, or it can be present as antimonite (H_3SbO_3) [21, 25].

Both Sb (III) and Sb (V) ions hydrolyze efficiently in aqueous solution, thus making it difficult to keep antimony ions stable in solution except in highly acidic media [26]. Sb (V) is present as SbO_2^+ or $Sb(OH)_5$ under very acidic conditions, and $[Sb(OH)_6]^-$ or SbO_3^- exists in mildly acidic, neutral, and alkaline conditions (Fig. 2). The antimony pentoxide, Sb_2O_5, is hardly soluble in water and generates the antimonate anion upon dissolution, while the antimony trioxide, Sb_2O_3, also has a relatively low solubility in water [26, 27]. In the pH range 2–12, the solubility of Sb_2O_3 is independent of pH, thus indicating the formation of an undissociated substance and antimony hydroxide $Sb(OH)_3$. Sb (III) exists as SbO^+ or $Sb(OH)_2^+$ in acidic media and as $Sb(OH)_4^-$ or hydrated SbO_2^- in basic media (Fig. 2). At very reducing conditions in the presence of dissolved sulfide, Sb (III) sulfide species will be dominated, e.g., $HSb_2S_4^-$ and $Sb_2S_4^-$, at pH values less than and higher than 11.5, respectively. At low temperatures, antimony chloride complexes such as $SbCl_2^+$, $SbCl^{2+}$, $SbCl_3(aq)$, and $SbCl_4^-$ will dominate in chloride-rich acidic aqueous solutions.

Antimony is geochemically categorized as a chalcophile, occurring with sulfur and the heavy metals – lead, copper, and silver. Apart from stibnite (Sb_2S_3) and kermesite (Sb_2S_2O), which are the most common antimony-containing minerals found in hydrothermal deposits, antimony often occurs in minerals in solid solution with arsenic, for example, lead and copper minerals, such as guettardite Pb(Sb, As)$_2S_4^-$, jamesonite $FePb_4Sb_6S_{14}$, and tetrahedrite $Cu_{12}Sb_4S_{13}$ [27]. Tetrahedrite has basically the sphalerite crystallographic arrangement with one-fourth of the Cu replaced by Sb in its structure. As a result of the substantial amount of copper in the

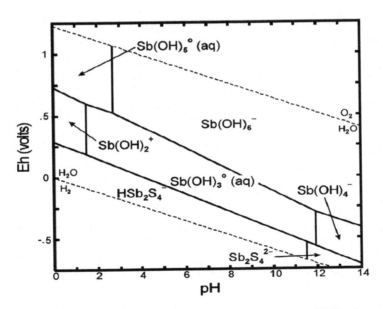

Fig. 2 Eh–pH diagram showing dominant aqueous species of antimony at 25°C and a concentration of 2.90×10^{-10} mg/L [27]

mineral, tetrahedrite is now becoming a potential source of copper, but the antimony content of the mineral is of great concern during the pyro-processing of the ore.

4   Methods of Heavy Metal Removal from Water

There are several technological methods for the removal of heavy metals in water treatment: precipitation (clarification), ion exchange, membrane technologies, adsorption, electrochemical processes, and recently also biological methods [28–33].

The most common method for the removal of heavy metals is water clarification – the *precipitation* of metal hydroxides and carbonates. This process is based on the dosing of appropriate agents (lime, iron salt, sodium carbonate, sodium hydroxide, and sulfates) to obtain the optimum pH value of a solution in which an insoluble solid phase of precipitated heavy metal hydroxides or carbonates is formed. The effectiveness of the precipitation depends on the type of contaminant, its concentration, water composition, and the type of agent.

Precipitation seems to be an ideal solution for the treatment of water containing heavy metals, provided that the process is not limited by specific effects that reduce the effectiveness of coagulation. For example, the efficiency of precipitation is lower at a higher concentration of metals in water. If the solution is too diluted, the precipitation will be too slow. The precipitation is also influenced by the pH value. Hydroxides are especially very sensitive to this parameter, and they are not competent enough in acid areas. In addition, the presence of other salts (ions) in water has an adverse effect on the precipitation process. The disadvantages of precipitation are the addition of other chemicals to the treatment process and the high production of sludge that should be processed and stored under specific conditions.

The advantage of precipitation is its relatively low cost compared to other metal removal methods. The coagulants used in this process are easily available. Precipitation can be used for a wide range of metals, and an acceptable level of effectiveness is achieved through its proper operation.

Ion exchange is based on the mutual exchange of ions with the same charge between an ion exchanger (an exchangeable ion) and the treated water (captured ion). The ion exchanger is a material capable of the reverse stoichiometric exchange of cations or anions in a condition of electroneutrality.

The advantage of the ion exchange process is the relatively low cost compared to other methods. The method is tried and tested, and all the components required for its operation are commercially available. It is possible to remove undesirable metals from water using the cation exchanger in a wide range up to the µg/L level.

The disadvantage of ion exchangers is that they disrupt ion exchange due to the high competitiveness of some ions (selenium, fluorine, nitrates, and sulfates) to finding a place in the ion exchanger. In addition, these ions reduce the efficiency due to suspended and organic substances, which may cause fouling of the ion exchanger filter. It is not possible to use the ion exchange method in the treatment of water with a high concentration of metals. Moreover, this method is sensitive to the pH value of the treated water and water quality (alkalinity, concentration of competing ions). The need to dispose of the regenerative agent used and the ion exchanger is also among the disadvantages of this material.

Adsorption processes are based on the adsorption of contaminants on the surface of an adsorption material. The molecules of the contaminant pass from the water environment to the solid adsorbent. It is possible to use activated alumina, iron-activated alumina (Fe-AA), activated carbon, iron-activated carbon, iron oxides, oxyhydroxides, or ferric hydroxide (GEH, CFH12, CFH18, Bayoxide E33, Everzit As), media containing TiO_2, CeO_2, ZrO_2, or MnO_2 layers on their surface, sand covered by iron hydroxide, low-cost materials (zeolites, carbonates, clay, peat, moss, ash, chitosan, sawdust, coconut husk, living or nonliving biomass, etc.), for removal of heavy metals.

Efficiency of heavy metal removal by adsorption material depends on the pH of water; oxidation–reduction potential of a given metal in water; concentration of substances in water that have a potential to affect (interfere with) adsorption or modify adsorbent surface loading; concentration of substances and colloid particles that can physically block the entry into the particle and the access to grains of adsorption media, respectively, specific surface area and distribution of pores of adsorption material; and hydraulic properties of filtration media in treatment (filtration rate, the Empty Bed Contact Time (EBCT), the filter medium height).

Membrane methods belong to a group of diffuse processes in which the selective properties of membranes are used (thin semipermeable films, the thickness of whose walls range from 0.05 to 2.0 mm) to eliminate contaminants from water. Depending on the type of membrane (structure and driving force), it is possible to divide these processes into microfiltration, ultrafiltration, nanofiltration, and reverse osmosis.

Today, *electrochemical methods* are not commonly used in the treatment of water and wastewater. These methods are still in the process of development, but it is important to note that they may become very useful for the removal of metals from water in the future.

Electrodialysis (ED) is a membrane process, by which ions are transported through semipermeable membrane, under the influence of an electric potential. The membranes are cation- or anion-selective, which basically means that either positive ions or negative ions will flow through. Cation-selective membranes are polyelectrolytes with negatively charged matter, which rejects negatively charged ions and allows positively charged ions to flow through.

Biological methods are based on the production of a special microbial culture capable of using heavy metals dissolved in water as a substratum for further microbial growth.

5   Materials Used for Sorption in Removal of Heavy Metals

The literature mostly describes the use of iron oxides, oxyhydroxides, and iron hydroxides, also known as GEH, Bayoxide E33, CFH12, CFH18, Everzit As, etc., for arsenic removal from water. They were manufactured and tested in particular for the removal of arsenic from water. A number of experiments and model studies on the adsorption of arsenic and other heavy metals are described in various publications [34–48]. These studies describe sorption processes at different pH values, initial heavy metal ion concentrations in water, the solid/liquid ratio, the particle size of a sorption material, and the temperature and composition of the water to be treated (concentration of iron, manganese, phosphorus, silicon, fluorides, sulfates, organic matter, etc.).

Bayoxide® E33 is a dry, granular amber-colored iron oxide composite medium, consisting primarily of α-FeOOH. It was developed by Severn Trent in cooperation with Bayer AG for the removal of arsenic and other contaminants (antimony, cadmium, chromate, molybdenum, selenium, and vanadium) from water. Bayoxide® E33 prefers to adsorb arsenic from these other ions. The advantage of this material is its ability to remove As^{3+} and As^{5+} too. Bayoxide® E33 has a capacity to treat water with As concentration of $11 \div 5{,}000$ µg/L [49–51].

CFH12 and CFH18 are granular sorption materials based on iron hydroxide (FeOOH). They were developed by Kemira Finland as efficient products for the removal of arsenic and other contaminants from water by adsorption. The advantage is their high adsorption capacity and higher efficiency at a lower cost, provided that the adsorption capacity is fully used (optimum filtration, backwash, and pH). CFH 12 and CFH18 differ from each other by their grain size and chemical composition (Table 1) [52–54].

GEH was obtained from the supplier (GEH Wasserchemie GmbH, Germany). GEH is a high-performance adsorbent developed by the Department of Water

Table 1 Physical and chemical properties of selected sorption materials

Parameter	Bayoxide E33	CFH12 a CFH18	GEH
Matrix/active agent	$Fe_2O_3 > 70\%$ and 90.1% α-FeOOH	FeOOH, Fe^{3+} $> 40\%$	52–57% $Fe(OH)_3$ and β-FeOOH
Material description	Dry granular media	Dry granular media	Moist granular media
Color	Amber	Brown red	Dark brown
Bulk density (g cm^{-3})	0.45	1.12–1.2	1.22–1.29
Specific surface area (m^2 g^{-1})	120–200	120	250–300
Grain size (mm)	0.5–2.0	0.32–2.5 or 0.5–1.8	0.32–2.0
Grain porosity (%)	85	72–80	72–77
pH	6.0–8.0	6.5–7.5	5.5–9.0

Quality Monitoring of the University of Berlin for the purpose of removing arsenic from water. GEH consists of ferric hydroxide and oxyhydroxide with a dry solid content of 57% ($\pm$10%) by mass and 43–48% by mass moisture content. Its iron content is 610 g/kg ($\pm$10%) relative to the dry solids [55].

The properties of GEH do not vary significantly from study to study. The density of water-saturated GEH (shipped conditions) has been noted as 1.32 g/cm^3 [56] and 1.25 g/cm^3 [57]. The surface areas of GEH range from 250 to 300 m^2/g, while porosity has been observed at 72–77% [56] and 75–80% [57]. GEH is delivered and provided in a water-saturated, granular form. The grain size of the GEH obtained from the manufacturer ranges from 0.2 to 2 mm.

GEH is highly selective toward arsenate; therefore, it requires an initial oxidation step in the presence of arsenite. In paper [58], the adsorption of arsenate occurred much more rapidly at lower pH values, while in higher pH waters, the adsorption rates were comparable for both arsenate and arsenite. GEH is slightly affected by the presence of sulfate but only when the influent pH is below 7. Increasing phosphate concentrations in influent water dramatically reduces arsenic removal [56].

Chemical composition was determined by the Institute of Inorganic Chemistry of the Faculty of Chemical and Food Technology of the Slovak University of Technology using the methods of X-ray microanalysis, SEM, and X-ray phase analysis; the values are listed in Table 2.

The shape and the external surface of sorption materials GEH, CFH12, and Bayoxide E33 were taken by scanning electron microscope. Figures 3 and 4 illustrate differences in the character of surfaces [59].

Table 2 Chemical composition of selected sorption materials

Material	Compound in mass (%)								
	MgO	Al$_2$O$_3$	SiO$_2$	P$_2$O$_3$	SO$_x$	K$_2$O	CaO	TiO$_2$	Fe$_2$O$_3$ (FeOOH)
E33	0.97	6.59	12.75	0.34	0.31	0.37	2.01	0.91	75.28
CFH12	3.75	0.45	1.18	–	8.49	0.27	2.72	0.50	82.65
CFH18	5.19	0.48	1.47	0.28	4.58	–	1.41	0.30	86.29
GEH	–	1.74	3.05	0.21	0.54	0.08	0.18	–	91.92

Fig. 3 The microstructure of GEH, Bayoxide E33, and CFH12 (40× magnification)

Fig. 4 The microstructure of GEH, Bayoxide E33, and CFH12 (5,000× magnification)

5.1 Activated Carbon

The most widely used adsorbent for water treatment is activated carbon. It is a well-known adsorbent due to its extended surface area, microporous structure, high adsorption capacity, and high degree of surface reactivity.

The structure consists of a distorted three-dimensional array of aromatic sheets and strips of primary hexagonal graphic crystallites. This structure creates angular pores between the sheets of molecular dimensions. Pore size ranges from 1 to 1,000 nm, and the extensive porosity is responsible for the high surface area of the material usually 500–1,500 m^2/g. Commercial activated carbon is manufactured from only a few carbon sources: wood, peat, coal, oil products, and nut shells.

The final pore structure depends on the nature of the starting material and the activation process. Macro- and mesopores can generally be regarded as the highways into the carbon particle and are crucial for kinetics. The micropores usually constitute the largest proportion of the internal surface of the activated carbon and contribute most to the total pore volume. Activated carbon has both chemical and physical effects on the substance where it is used as a treatment agent [60].

Adsorption is the most studied of these properties in activated carbon. Heavy metal removal by adsorption using commercial activated carbon has been widely used [61–64]. However, high costs of activated carbon and 10–15% loss during regeneration limit its use. This has led to a search for cheaper carbonaceous substitutes [60].

5.2 Activated Alumina

Activated alumina is a commercial filter media made by treating aluminum ore so that it becomes porous and highly adsorptive. It can also be described as a granulated form of aluminum oxide. Activated alumina is used for removing a variety of contaminants (fluoride, heavy metals, etc.) from water [65]. The medium requires periodic cleaning with an appropriate regenerator such as alum or acid in order to remain effective. Activated alumina has been used as an effective adsorbent especially for point of use applications.

There is more literature on the use of activated alumina for arsenic removal. The principle is that the soluble arsenic (AsO_4^{3-} and AsO_3^{3-}) in the water can be adsorbed on the surface of the AA[am-Al(OH)$_3$] and occupies the aluminous octahedron crystal lattice sites [66]. The maximum adsorptive capacity of AA is 5–24 (mg As adsorbed/g media) at equilibrium arsenic concentrations of 0.05–0.2 ppm [67, 68].

5.3   Low-Cost Adsorbents

Consequently, low-cost adsorbents have drawn attention to many researchers, and characteristics as well as application of many such adsorbents are reported. Some of the reported low-cost sorbents include bark-/tannin-rich materials, lignin, chitin/chitosan, eggshell, dead biomass, seaweed/algae/alginate, xanthate, zeolite, clay, ash, peat moss, bone gelatin beads, leaf mold, moss, iron-oxide-coated sand, sawdust, modified wool, modified cotton, coconut husk, rice husk, tea waste, agricultural waste (fly ash powder, bagasse, waste straw dust, sawdust, and coconut coir), eucalyptus and neem leaves, cast-iron filings (wastes from mechanical workshops, lathes), steel wool (commercially available, used for cleaning of wood surfaces prior to polishing), etc. [60, 69–71].

5.4   Adsorbent Properties

To be suitable for commercial applications, a sorbent should have high selectivity to enable sharp separations; high capacity to minimize the amount of sorbent needed; favorable kinetic and transport properties for rapid sorption; chemical and thermal stability, including extremely low solubility in the contacting fluid, to preserve the amount of sorbent and its properties; hardness and mechanical strength for long life; no tendency to promote undesirable chemical reactions; and the capability of being regenerated when used with commercial feedstocks.

When choosing the right filter, filtering-and-sorption or sorption material, it is always necessary to follow the given application and properties of different types of filter beds. Today there is a large number of publications available, dealing with arsenic or antimony removal from water using different sorption materials [37, 57, 72–80]. The published procedures are thus often adopted and adapted to the specific conditions. Where there is lack sufficient experience (knowledge) in the choice of sorption materials, it must be obtained, experimentally, best through long-term testing – pilot operation experiments.

Important parameters in the choice of sorption materials are [75]:

1. The concentration of the contaminant in the water.
2. The concentration of the contaminant after treatment.
3. The amount of treated water expressed as filtration rate, whereby filtration rate (m/h) = flow rate (m^3/h)/filter area (cross-section) (m^2).
4. Time of contact of water with material, expressed as EBCT (Empty Bed Contact Time); to calculate, we use the formula: contact time [min] = bed volume (m^3) * 6/flow rate (m^3/h).
5. Particle size (grain size) is important for the proper draft of operational flow rates due to the pressure drop and the contact time of the treated water with filtration material and backwash rates.
6. Density (kg/m^3). In the literature we encounter several densities, e.g., apparent density, expressing the max. vibration tapped density, bed density defined as the ratio of mass of a particulate material, and the total volume taken up by it (sum of the volume of the particles, the volume of the interparticle space, and the internal pore volume). Specific weight is used for the calculation of the volume and the weight of the sorption material.
7. The total surface area (BET) in m^2/g expresses the sorptive capacity of the given material, determined by the volumetric method (e.g., by physical adsorption of nitrogen at liquid nitrogen temperature). It is mainly used in the sorption of gases, having limited predicative value for water treatment, as it does not describe the content of micropores and transport pores in the sorbent material, while micropores are responsible for the adsorption. Transport pores serve for the supply of pollutant molecules to the micropores.

Sorption efficiency is reflected in the following parameters:

1. Adsorption capacity [µg/g] is the ratio of the mass of captured (adsorbed) contaminant in the bed [µg] and the weight of the bed in the filter [g], while the mass of adsorbed contaminant need to be determined experimentally.
2. "Bed volume" (BV) is a term often used to compare the efficacy of the technological process or the sorption material, representing the volume of water that flows through the filter bed V divided by the bed filter volume V_0 (the ratio V/V_0). Manufacturers of sorbents report this value together with adsorption capacity as data to characterize the effectiveness of the sorption process.
3. Filter length, L_F, is given in meter or in m^3/m^2 and represents the volume of water that flows through the filter unit area from the beginning of the filtration cycle; the higher the filter length, L_F, the higher the sludge capacity of the filter bed. In the literature for the removal of heavy metals, there is little data with this parameter; however, it needs to be used in characterizing the efficiency of sorption materials.

The following has an impact on the efficiency of removal of metals (As, Sb) from the water through sorption:

1. Water pH (lower pH increased sorptive capacity and lifetime of the medium).

2. The oxidation–reduction potential of the As and Sb (i.e., the ratio of As^{III}/As^{V}, Sb^{III}/Sb^{V}); it is well known that the pentavalent form of As and Sb is more easily removed from the water.
3. The concentration of substances present in the water that may affect (interfere with) the adsorption of As or modify the surface load of the sorption material.
4. Concentration of the substance and the colloidal particles in water that can physically block access of As to the interior of the particles or to the grains of adsorbent media.
5. Specific surface area and pore size distribution of the sorption material.
6. Hydraulic properties of the filter media during treatment (bed volume, filtration rate, the water retention time in the bed).

The first four factors are linked to the chemical equilibrium between the different substances present in the water and the filter material; the fourth and the last two factors are influenced primarily by the physical processes of mass transfer and properties of the used material. The substances whose presence in water can affect the sorption of arsenic and antimony include, for example, other heavy metals (vanadium), iron, manganese, silicate, sulfate, phosphate, fluoride, organics, etc. [81, 82].

The disadvantages of the use of sorption materials in the removal of heavy metals may be the costs associated with purchase, recovery, or disposal. It is therefore necessary to evaluate and compare this method of treatment with the methods used thus far.

6   Removal of Antimony by Adsorption at WR Dúbrava

The Dúbrava water supply resource is situated in the western part of the Low Tatras mountain range. Geological and hydrogeological conditions of this region are very complex where the water of crystalline and Mesozoic basements is interconnected. Higher antimony concentration in the sources of water for the water supply occurs mainly due to existence of the antimony deposit in Dúbrava and its higher content in granitoids of this part of the Low Tatras region. In the middle of the eighteenth century, antimony ore mining started in this site. Until the end of antimony mining in 1993, deposit Dúbrava was one of the most important producers of Sb in Czecho-slovakia. Since 1753 were mined 1,046 kt antimony ore, the total mined ore contained 27 kt of antimony [8].

Moreover, the concentration of antimony in mining water was considerably increased at relatively high capacities of wells. Adverse effect comes from the mine tailing piles and sludge lagoon where the rocks rich in antimony were continually washed by the rainwater infiltrating into the groundwater resources or flowing into to the surface stream of Križianky. Contaminated water of the Križianka River and water of its alluvial deposits have deteriorated water quality in the springs of Močidlo and Brdáre. In the past, three springs of the Dúbrava water resource

(Brdáre, Močidlo, Škripeň) were used for supplying population with drinking water (capacity of about 40 L/s), but today only one spring is used for this purpose (spring Škripeň that does not contain antimony). Two other springs are contaminated with antimony [8].

Water quality monitoring data provided by the Water Company of the Region of Liptov indicates the water quality parameters for the separate springs of the Dúbrava water resource and is shown in Table 3. The highest contamination from antimony was observed in water from the Brdáre spring, where the concentrations ranged from 80.3 to 91.3 µg/L. The concentration of antimony in water from the Močidlo spring was 70.6–82.0 µg/L. Apparently, the best water quality was monitored in the Škripeň spring, where the concentration of Sb was lower than 1 µg/L in every sample taken during the monitoring period. No other heavy metals were present in the Dúbrava water resource. The groundwater analysis in locality Dúbrava is shown in Table 4.

The pilot tests for removing antimony were carried out at the Dúbrava chlorination plant (Fig. 5). At present, only water from the Škripeň well is conveyed into the storage tank of the chlorination plant. After its disinfection, the water is gravitationally distributed to the point of consumption. For the purpose of these simulation tests, there was a need to convey the water from the Brdáre well to the chlorination plant through a separate pipe in order to avoid mixing it with the water from the Škripeň well [8].

Table 3 Water quality of the Dúbrava water resource according to selected parameters for the period 2000–2005

Parameter	Dúbrava – spring		
	Močidlo	Škripeň	Brdáre
pH	7.65–7.90	7.55–7.95	7.75–7.95
Alkalinity (mmol/L)	1.7–3.8	1.8–3.8	1.7–2.2
Conductivity (mS/m)	23.1–38.6	23.0–42.6	22.5–28.7
Ca^{2+} (mg/L)	30–54	48–52	28–32
Mg^{2+} (mg/L)	8.5–28.0	15.8–24.3	9.7–15.8
Sb (µg/L)	70.6–82.0	<1.0	80.3–91.3

Table 4 Filtration conditions [84]

Parameter	GEH	CFH12	Bayoxide E33
Grain size (mm)	0.32–2.0	1.0–2.0	0.5–2.0
Medium height (cm)	60	60	60
Mass of sorption material (g)	1,324	1,416	998
Average flow through column (mL/min)	147.3	147.8	140.0
Average filtration rate (m/hod)	4.50	4.51	4.27
Total filtration time (hod)	1,174	1,174	1,174
EBTC (min)	8.0	7.97	8.41

Raw water (the Brdáre spring) passed through the filtration system (Fig. 5), and the concentration of antimony was monitored in raw and treated water at the outlets of the filtration columns. Simultaneously, the flow rates were measured at the outlet of each column. A system of several valves was used for feeding the water for the filtration system (from top to bottom) and for the filter backwash (from bottom to top) as well as for regulating the filtration rates.

The aim of first pilot-scale experiment was to verify the sorption properties of granular iron-based filter materials (GEH, CFH12, Bayoxide E33) in the Dúbrava water resource during the process of antimony removal from water.

In order to verify the effectiveness of the antimony elimination process, tree adsorption columns filled with the sorption material were used. The adsorption column was made of glass material with a diameter of 5.0 cm and medium height of 60.0 cm. The adsorption column with a volume of 1,178.1 cm³ covered an area of 19.635 cm². Water flowed through the column from the top to the bottom. The water discharge was measured continually, and the filtration rate achieved approximately 4.5 m/h [84]. The amount of water flowing through the column was monitored using a water meter placed in front of the column inlet. The filtration conditions are shown in Table 5.

Antimony samples after passing through columns were collected into plastic bottles and immediately acidified with highly pure nitric acid (Merck). All bottles

Fig. 5 Dúbrava chlorination plant and model filtration columns

Table 5 Analysis of groundwater in the area of Dúbrava [84]

Parameter	Unit	RW	Parameter	Unit	RW
pH		7.53	NH_4^+	mg/L	0
Conductivity	mS/m	21	Fe total	mg/L	0.02
Color	mg/L Pt	2	Mn	mg/L	0.001
Turbidity	ZF	0	Cl^-	mg/L	8.23
$ANC_{4.5}$	mmol/L	2.962	NO_3^-	mg/L	5.12
$BNC_{8.3}$	mmol/L	0	SO_4^{2-}	mg/L	21.85
Ca + Mg	mmol/L	1.175	F^-	mg/L	0.18
TDS (105°C)	mg/L	100	COD_{Mn}	mg/L	0.42

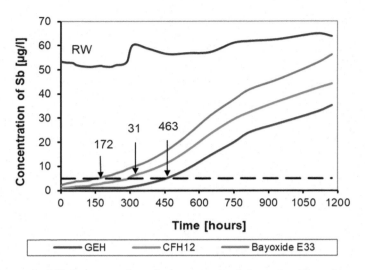

Fig. 6 Removal of Sb from water depending on operational time at breakthrough concentration 5 μg Sb/L (raw water concentration) [84]

were submerged in 10% nitric acid solution over 3 days and triple rinsed with deionized water. Agilent 7500CE ICP-MS (ORS technology) was used to determine antimony concentration in solution. The detection limit for Sb by ICP-MS was 1 μg/L [84, 85].

The results of the technological process are shown in Figs. 6 and 7; there is demonstrated relationship between antimony concentration and operational time or bed volumes treated (volume of the water passed through filtration column to volume of the adsorption column). Figure 6 shows the breakthrough curves of antimony as a function of water volumes treated in for each sorption material when reaching the limit concentration of antimony (5 μg/L). The effectiveness of the monitored sorption materials in the antimony removal process can be seen.

On the basis of the results obtained, it can be stated that all materials used are suitable for removal of antimony from water, although it is recommended in the literature to remove arsenic from water. The antimony removal efficiency is shown in Table 6.

For GEH sorption material has exceeded the value of 5 mg/L Sb after 463 h of operation of the filter device. The amount of water that has passed through this filtering device during this period is 4.088 m^3, i.e., 3,470 times the volume of the medium of GEH. For the sorption material CFH12 (Kemira), the limit value was exceeded after 312 h of operation, with the amount of water that exceeded the filter device during this time period of 2.85 m^3, i.e., 2,421 times the volume of the filter material CFH12. In the Bayoxide E33 sorption material, the limit value was exceeded after 172 h of operation, with the amount of water that exceeded the filter device during this time period of 1.50 m^3, i.e., 1,273 times the volume of the filter medium Bayoxide E33.

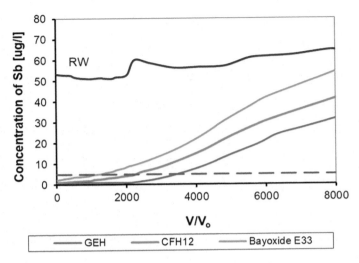

Fig. 7 Removal of Sb from water depending on the bed volume (volume of filtered water V to the volume of the filtration media (V_0) at breakthrough concentration 5 μg Sb/L (raw water concentration)) [84]

Table 6 Antimony removal efficiency of the water during filtration–adsorption

Parameter	GEH	CFH12	Bayoxide E33
Total time of filtration (h)	1,174	1,174	1,174
Filtration time (hod) at breakthrough concentration 5 μg Sb/L	463	312	172
Total amount of water passed through filtration column (m³)	10.11	10.08	9.74
Amount of water passed through filtration column at breakthrough concentration 5 μg Sb/L (m³)	4.088	2.852	1.460
Bed volume (V/V_0)	3,470	2,421	1,274

The adsorption capacity of the individual adsorbents was calculated based on the condition of not exceeding the antimony limit on the effluent from the filters (Fig. 8). Under the given operating conditions (average antimony concentration in raw water 55.6 μg/L, filtration rate 4.5 m/h) and weight of 1,324 g of GEH in column was adsorbed 222.16 mg of antimony. In the column with CFH12 (weight of 1,416 g) was adsorbed 149.71 mg of antimony, and in the column with 998 g of Bayoxide E33 was adsorbed 90.77 mg of antimony. From these results, the adsorption capacity of the GEH filter material was 167.8 μg/g, CFH12 105.7 μg/g, and Bayoxide E33 90.9 μg/g.

Considering the minimum differences in the filtration rates and based on the results presented in Figs. 6, 7, and 8, it can be concluded that GEH is the most suitable material for antimony removal compared to the other sorbents used in the test.

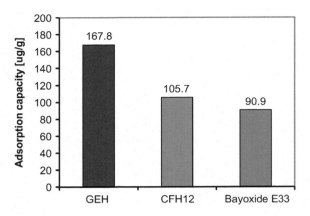

Fig. 8 Adsorption capacity of adsorbents used (v μg/g) [84]

Table 7 The values of bed volume and adsorption capacity of GEH material for different concentration of Sb in raw water, filtration rates, heights media, and EBCT

Material	Concentration of Sb (μg/L) in RW	Average filtration rate (m/h)	Height media (mm)	EBCT (min)	Bed volume (V/V_0)	Adsorption capacity (μg/g)
GEH	55.6	4.5	60	8.0	3,470	167.8
GEH	58.3	5.0	50	6.0	1,700	83.6
GEH	58.3	5.5	53	6.2	2,260	103.7
GEH	72.6	5.5	52	5.7	1,610	81.4
GEH	81.4	3.4	51	9.1	2,030	145
GEH	81.4	5.6	51	5.4	1,342	96.9

In Table 7 are summarized the results of other experiments conducted in the Dúbrava with material GEH [8, 20, 59, 83–85]. The table contains the bed volume and the adsorption capacity for the various antimony concentrations in the treated water and the filtration rate used, the filter media height, and the contact time treated water with the filter media GEH at the breakthrough concentration of 5 μg Sb/L.

If the need to water for the Liptovský Mikuláš region is increased there, it will be possible to use water from the Dúbrava water source. For removal of antimony from the water, we recommend using closed filters with a GEH filling and a filtration speed of 4.5 m/h. After further studies, it will be possible to carry out an economic assessment of the whole technological process and arrive at clear conclusions about the use of GEH materials in water treatment processes.

The aim of the second pilot-scale experiment was to compare the efficacy of antimony removal from water at the Dúbrava water resource using three different heights (50, 70, 90 cm) of filter beds with GEH material.

The effectiveness of antimony elimination from water was studied in a model facility, where raw water passed through three adsorption columns filled with GEH material in a direction from top to bottom. The adsorption column was made of glass, the column diameter was 5.0 cm, and the column height was 80 and 100 cm [75].

Without undergoing any pretreatment, the raw water passed through filtration equipment, while the concentration of antimony was monitored in raw and treated water at the outlet from individual filter columns. At the same time, the water flow at the outlet of each column was also monitored. Technological tests were aimed at verifying the possibilities of using GEH sorption material for water treatment–removal of Sb.

The results of the model tests were used to evaluate the courses of antimony concentration at the outlet from the columns from the time of the model facility operation, depending on the filter length, L_F (expressed in m^3/m^2, or in meters), and bed volume (BV). Based on the material balance of antimony in model facilities, we calculated the amounts of adsorbed antimony; from these data we calculated the adsorption capacities in μg/g. All published results are related to the concentrations of 5 μg/L of Sb at the outlet from the column, i.e., for the limit concentration of Sb in drinking water [75].

Within the given model tests, the concentration of antimony in raw water ranged from 90 to 108 μg/L Sb (average 90.3 μg/L Sb). The filtration rate in the case of a column with a bed height of 50 cm ranged at 5.3–5.6 m/h; at 70 cm bed height, it ranged from 5.1 to 5.5 m/h; and at 90 cm bed height, it ranged from 5.0 to 5.5 m/h. Filtration conditions are shown in Table 8.

Figure 9 shows the course of the concentration of antimony depending on the operational time of the model facility. The figure also shows the limit value of antimony in drinking water according to the Decree of the Ministry of Health of the Slovak Republic No. 247/2017 for drinking water (5 μg/L) [86]. Given that the experiments have been completed prior to reaching a concentration of 5 μg/L of Sb at the outlet from the columns for a medium height of 70 and 90 cm, the remaining value of the Sb concentration was additionally calculated through extrapolation.

Based on the achieved results, Table 9 summarizes the measured and calculated values for the removal of antimony from water using the GEH material and three adsorption bed heights, and the results are related to the value of 5 μg/L Sb at the outlet from the filter bed.

For mathematical processing and generalization of data in Table 4, we used the linear regression method. Figures 10, 11, and 12 show the equations of lines for GEH adsorption capacities, the V/V_0 ratio (bed volume), the contact time of water with the

Table 8 The conditions of filtration (the average values) [75]

Parameter	GEH		
Grain size (mm)	0.32–2.0	0.32–2.0	0.32–2.0
Medium height (cm)	50	70	90
Medium volume (cm^3)	0.982	1.364	1.751
Medium weight (g)	1,227.2	1,705.8	2,189.3
Average flow through column (mL/min)	178.0	176.4	173.4
Average filtration rate (m/h)	5.44	5.39	5.30
Total filtration time (h)	423	423	423
Filtration time (h) at breakthrough concentration 5 μg Sb/L	147	483	784

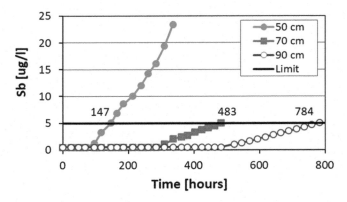

Fig. 9 Compare of efficiency of the materials GEH in the removal of Sb from water

Table 9 Measured and calculated values the sorption of antimony from water [75]

Height media (cm)	Volume media (cm³)	Average filtration rate (m/h)	EBCT (min)	Bed volume (V/V_0)	Filtration length L_F (m)	Amount of adsorbed Sb at filter bed (μg)	Adsorption capacity (μg/g)
50	981.75	5.44	5.5	1,537	768.1	138,341	112.7
70	1,364.63	5.39	7.7	3,736	2,596.9	405,987	238.0
90	1,751.44	5.30	10.1	4,659	4,155.7	727,326	332.6

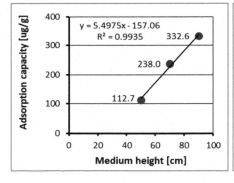
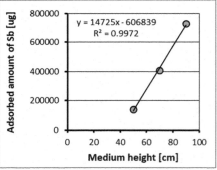

Fig. 10 Adsorption capacities and the amount of adsorbed antimony at three different filter bed heights set for 5 μg/L of Sb at the outlet from the column [75]

filter bed material, and the value of the filter length, L_F, for 5 μg/L Sb at the outlet of the individual columns for 50, 70, and 90 cm bed height.

Figures 10, 11, and 12 show that the monitored parameters have a linear relationship, except the V/V_0 parameter (bed volume) which does not have a linear relationship, as can be seen not only visually but also based on the standard deviation R^2. Therefore, it is appropriate to supplement this parameter with the filter length

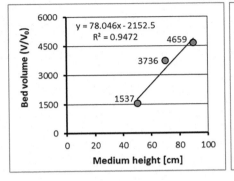
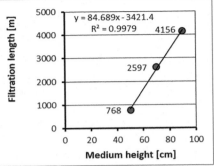

Fig. 11 Bed volume (V/V_0 ratio) and the filter length (L_F) for three different filter bed heights (for 5 µg/L of Sb at the outlet from the column) [75]

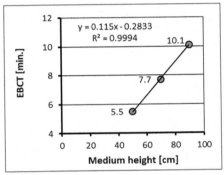
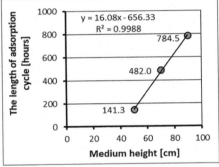

Fig. 12 The Empty Bed Contact Time (EBCT) and the length of the adsorption cycle for individual filter bed heights set for 5 µg/L Sb at the outlet from the column [75]

indicator, which is used for filter materials, but in the sorption materials, this figure is usually not given in the literature [75].

7 Conclusion

The conducted technological tests with underground spring water in the Dúbrava location showed that with the help of the GEH sorption material, we can reduce the antimony content in water to the value 5 µg/L determined by the Decree of the Ministry of Health of the Slovak Republic No. 247/2017, which lays down details on drinking water quality, drinking water quality control, monitoring, and risk management of drinking water supply [82].

Model tests were intended to monitor the effectiveness of antimony removal from water source Dúbrava with three different sorption materials (GEH, CFH12, and Bayoxide E33). From this results show that GEH is the most suitable material for antimony removal compared to the other sorbents used in the test. Therefore, the effectiveness of antimony elimination from water was studied in a model facility, where raw water passed through three adsorption columns filled with GEH material using different height (50, 70, or 90 cm) of filter bed and to determine the most frequently used parameters indicating the effectiveness of sorption (adsorption capacity and bed volume) on the basis of the measured values through linear regression [75].

For the known filtration rate (flow) and concentration of antimony in water, we can propose the volume (height) of the adsorption column bed and determine the efficiency of antimony removal from the water, expressed either as bed volume (the V/V_0 ratio) or as a filter length, L_F, using the linear regression equation. It is also possible to calculate (estimate) the amounts of adsorbed antimony in the filter bed and the adsorption capacities of the materials used for the given technological process of water treatment [75]. If the water contact time (EBCT) with the sorption material in filter column is longer, the higher the antimony removal efficiency from the water will be achieved.

Assuming that the linear relationship will also apply to other filter bed heights (e.g., 120 cm, 150 cm, etc.), we can determine the length of the colon's adsorption cycle (in hours) after which the concentration of Sb at the outlet will achieve just 5 µg/L. For 120 cm bed height, it would be 1,273 h, and for 150 cm it would be about 1,756 h. If we compare it with real results, the increase in bed height from 90 to 150 cm, i.e., about 60 cm, would extend the length of the work cycle to about two times (from 784.5 to 1,756 h) [75]. To increase the efficiency of antimony removal from water source Dúbrava, it is possible to use two filter columns connected in series.

Our results also showed that in addition to the adsorption capacity and the V/V_0 ratio (bed volume), it is necessary to express the efficiency of the used procedure also by the filter length parameter (although this figure is not used for the sorption materials in literature). This is due to the fact that the bed volume parameter did not have a linear dependency for the used heights of adsorption column beds during our experiments [75].

Obtained results confirmed the findings published by foreign authors who consider these sorption materials to be more efficient in arsenic than antimony removal. Therefore, obtained results in this phase of works provide a certain background for usage of monitored filtration (sorption) materials also in antimony removal from water.

This method of water treatment is suitable mainly in localities where water treatment does not include coagulation, sedimentation, and filtration as well as in emergency situations. The advantage of this technology is total reliability, promptness, and simplicity of the operation. The disadvantage can be the cost of sorption materials (6 to 8€/kg) and pH of treated water (lower pH increases the sorption capacity and operational life of the medium). The presence of salts, colloid particles, organic substances, and other heavy metals in treated water can affect antimony adsorption or block the access of antimony to grains of adsorption medium. The disadvantage can be the higher concentrations of metals adsorbed in sorption material (after exhaustion of sorption capacity and replacing of material) and the necessity to dispose of used material on waste dump.

8 Recommendations

Our recommendations are as follows:

1. The results of experiments obtained have proved that the GEH material is more effective for the removal of antimony from the water compared to the Bayoxide E33 and CFH12.
2. The effectiveness of antimony removal from water is significantly lower compared to arsenic removal, the materials used in this work are developed to remove arsenic from water, so new materials need to be found and tested under operating conditions.
3. In general, the effectiveness of heavy metal removal depends on the filtration rate, height (volume) of the filter media (i.e., contact time with sorption materials), and heavy metal concentrations in raw water.
4. To increase the efficiency of heavy metal removal from water source using adsorption, it is necessary to decrease pH value of raw water (lower pH increased sorption capacity and lifetime of the sorption material).
5. It is necessary to increase oxidation potential of the heavy metals (it is well known that the pentavalent form of As and Sb is more easily removed from the water compared to trivalent form of As and Sb) by oxidation (aeration, adding disinfectant agent) of raw water.
6. When choosing the right filter, or sorption material, it is always necessary to follow the given operational conditions and have knowledge of different types of sorption materials (sorption capacity, bed volume, grain size, etc.); the efficiency of sorption materials needs to be experimentally verified by long-term testing – pilot plan experiments.
7. The higher efficiency of heavy metal removal from water will be achieved by increasing the water contact time (EBCT) with the sorption material in filter column (optimal filtration rate needs to be verified).
8. The sorption material used must be sufficiently washed prior to the start of the experiment; it is necessary to remove the dust particles and air bubbles in the filter column.
9. When choosing the right treatment technology for removing of heavy metals, it is necessary to know water quality, heavy metal concentrations in raw water, for what purpose the treated water is to be served, which concentration must be obtained by the selected water treatment, etc. Therefore, pilot tests to verify the effectiveness of heavy metal removal directly on a water source are needed.

References

1. Newcombe RL, Möller G (2006) Arsenic removal from drinking water: a review. Blue Water Technologies, Hayden. http://www.blueh2o.net/docs/asreview%20080305.pdf
2. WHO (2011) Guidelines for drinking-water quality, 4th edn. WHO, Geneva
3. Drinking Water Directive 80/778/EEC, COM(94) 612 Final. Quality of water intended for human consumption
4. Water Quality and Treatment (1990) A handbook of Community Water Suppliers, 4th edn. AWWA, McGraw-Hill, p 1194. ISBN: 9780070015401
5. Ďurža O, Khun M (2002) Environmental geochemistry of some heavy metals. Univerzita Komenského, Bratislava, p 115. ISBN: 8022316571 (in Slovak)
6. US EPA (1984) Antimony: an environmental and health effects assessment. Washington, US Environmental Protection Agency, Office of Drinking Water
7. Lexa J, Bačo P, Chovan M, Petro M, Rojkovič I, Tréger M (2004) Metalogenetická mapa Slovenskej republiky (Metallogenetic map of Slovakia), 1st edn. Bratislava: Ministerstvo životného prostredia SR. ISBN: 8088974550
8. Ilavský J, Barloková D, Munka K (2014) Influence of mining operations to presence of antimony in water source Dúbrava and possibilities of its removal. Acta Montan Slovaca 19(3):141–150
9. Fľaková R, Ženišová Z, Ondrejková I, Krčmár D, Sracek O (2017) Occurrence of antimony and arsenic at mining sites in Slovakia: implications for their mobility. Carpath J Earth Environ 12(1):41–48
10. Ondrejková I, Ženišová Z, Fľaková R, Krčmář D, Sracek O (2013) The distribution of antimony and arsenic in waters of the Dúbrava abandoned mine site, Slovak Republic. Mine Water Environ 32(3):207–221. https://doi.org/10.1007/s10230-013-0229-5
11. Ženišová Z, Fľaková R, Jašová I, Krčmář D (2010) Water contamination by antimony and arsenic in the vicinity of abandoned deposit Dúbrava. Podzemná voda 16(1):1–19. (in Slovak)
12. Jašová I, Ženišová Z, Fľaková R (2009) Contamination of surface and groundwater in abandoned Pernek mining area. Acta Geologica Slovaca 1(1):39–46. (in Slovak)
13. Fľaková R, Ženišová Z, Krčmár D, Ondrejková I, Fendeková M, Sracek O, Chovan M, Lalinská B (2012) The behavior of arsenic and antimony at Pezinok mining site, southwestern part of the Slovak Republic. Environ Earth Sci 66(4):1043–1057. https://doi.org/10.1007/s12665-011-1310-7
14. Fľaková R, Ženišová Z, Ondrejková I, Krčmár D, Petrák M, Matejkovič P (2011) Contamination of natural waters, soils and stream sediments in abandoned Sb deposit Medzibrod. Miner Slovaca 43:419–430. (in Slovak)
15. Chovan M, Lalinská B, Šottník P, Hovorič R, Petrák M, Klimko T (2010) Mineralogical and geochemical characterization of contamination sources at abandoned Sb-Au deposit Medzibrod. Miner Slovace 42:95–108. (in Slovak)
16. Fľaková R, Ženišová Z, Ondrejková I, Krčmář D, Galo I (2011) Water contamination by antimony and arsenic at the Čučma abandoned deposit (Slovenské rudohorie Mts.) Acta Geologica Slovaca 3(1):57–74. (in Slovak)
17. Ženišová Z, Fľaková R, Jašová I, Cicmanová S (2009) Antimony and arsenic in waters influenced by mining activities in selected regions of Slovakia. Podzemná voda 15(1):100–117. (in Slovak)
18. Fľaková R, Ženišová Z, Jašová I, Krčmář D (2009) Water contamination of arsenic and antimony in abandoned deposit Poproč. Podzemná voda 15(2):132–148. (in Slovak)
19. Jurkovič Ľ, Šottník P, Fľaková R, Jankulár M, Ženišová Ž, Vaculík M (2010) Abandoned Sb-deposit Poproč: source of contamination of natural constituents in Olšava river catchment. Mineral Slovace 42:109–120. (in Slovak)
20. Barloková D, Ilavský J (2009) Removal of arsenic and antimony from water. Vodní hospodářství 59(2):45–49. (in Slovak)

21. WHO (2003) Antimony in drinking water. Background document for the development of WHO guidelines for drinking water quality. WHO/SDE/WSH/03.04/74, 2003
22. Gray NF (2008) Drinking water quality: problems and solutions, 2nd edn. Cambridge University Press, p 538. ISBN: 9780521702539
23. The Decree of the Ministry of Health of the Slovak Republic No. 247/2017 on requirements for drinking water quality, drinking water quality control, monitoring and management risk of drinking water supply
24. Anderson CG (2001) Hydrometallurgically treating antimony-bearing industrial wastes. J Metals 53(1):18–20. https://doi.org/10.1007/s11837-001-0156-y
25. Pitter P (2009) Hydrochemie, 4th edn. Institute of Chemical Technology Press, Praha, p 568. ISBN: 978-80-7080-701-9. (in Czech)
26. Filella M, Belzile N, Chen Y (2002) Antimony in the environment: a review focused on natural waters II. Relevant solution chemistry. Earth-Sci Rev 59:265–285
27. Krupka KM, Serne RJ (2002) Geochemical factors affecting the behaviour of antimony, cobalt, europium, technetium and uranium in Vadose sediments. Pacific Northwest National Laboratory, Richland
28. Petrusevski B, Sharma S, Schippers JC, Shordt K (2007) Arsenic in drinking water. Thematic overview paper 17, IRC
29. Gannon K, Wilson DJ (1986) Removal of antimony from aqueous system. Separat. Sci Technol 21:475–493
30. Anjum S, Gautam D, Gupta B, Ikram S (2009) Arsenic removal from water: an overview of recent technologies. J Chem 2(3):7–52
31. Bellack E (1971) Arsenic removal from potable water. J Am Water Works Assoc 63(7): 454–458
32. MacPhee MJ, Charles GE, Cornwell DA (2001) Treatment of arsenic residual from drinking water removal processes. EPA/600/R-11/090
33. Jekel M, Seith R (2000) Comparison of conventional and new techniques for the removal of arsenic in a full scale water treatment plant. Water Supply 18:628–631
34. Aragon M, Kottenstette R, Dwyer B, Aragon A, Everett R, Holub W, Siegel M, Wright J (2007) Arsenic pilot plant operation and results – Anthony, New Mexico, SANDIA report, SAND2007–6059
35. EPA (2003) Arsenic treatment technology evaluation handbook for small systems. US EPA, Office of Water (4606M), EPA 816-R-03-014
36. Badruzzaman M, Westerhoff P, Knappe DRU (2004) Intraparticle diffusion and adsorption of arsenate onto granular ferric hydroxide. Water Res 38:4002–4012
37. Barloková D, Ilavský J (2012) The use of granular iron-based sorption materials for nickel removal from water. Pol J Environ Stud 21(5):1229–1236
38. Bathnagar A, Choi Y, Yoon Y, Shin Y, Jeon BH, Kang JW (2009) Bromate removal from water by granular ferric hydroxide (GFH). J Hazard Mater 170:134
39. Biela R, Kučera T, Pěkný M (2017) Monitoring the effectiveness of advanced sorption materials for removing selected metals from water. J Water Supply Res Technol 66(6):jws2017127
40. Cumming LJ, Wang L, Chen ASC (2009) Arsenic and antimony removal from drinking water by adsorptive media. U.S. EPA Demonstration Project at South Truckee Meadows General Improvement District (STMGID), NV final performance evaluation report, EPA/600/R-09/016
41. Deliyanni EA, Peleka EN, Matis KA (2009) Modeling the sorption of metal ions from aqueous solution by iron-based adsorbents. J Hazard Mater 172:550–558. https://doi.org/10.2166/aqua. 2017.127
42. Filella M, Belzile N, Chen YW (2002) Antimony in the environment: a review focused on natural waters. I. Occurrence. Earth Sci Rev 57:125–176
43. Leiviskäa T, Khalida MK, Sarpolab A, Tanskanena J (2017) Removal of vanadium from industrial wastewater using iron sorbents in batch and continuous flow pilot systems. J Environ Manage 190:231–242. https://doi.org/10.1016/j.jenvman.2016.12.063
44. Paolucci AM, Chen ASC, Wang L (2011) Arsenic removal from drinking water by adsorptive media. Final performance evaluation report, EPA/600/R-11/074, U.S. EPA Demonstration Project at Geneseo Hills Subdivision in Geneseo, IL

45. Simeonidis K, Papadopoulou V, Tresintsi S, Kokkinos E, Katsoyiannis IA, Zouboulis AI, Mitrakas M (2017) Efficiency of iron-based oxy-hydroxides in removing antimony from groundwater to levels below the drinking water regulation limits. Sustainability 9(2):238. https://doi.org/10.3390/su9020238

46. Szlachta M, Wójtowicz P (2016) Treatment of arsenic-rich waters using granular iron hydroxides. Desalin Water Treat 57(54):26376–22638. https://doi.org/10.1080/19443994.2016.1204947

47. Westerhoff P, Benn T, Chen A, Wang L, Cumming L (2008) Assessing arsenic removal by metal (hydr)oxide adsorptive media using rapid small Scale column test. EPA/600/R-08/051

48. Ilavský J, Barloková D, Munka K (2015) Antimony removal from water by adsorption to iron-based sorption materials. Water Air Soil Pollut 226:2238. https://doi.org/10.1007/s11270-014-2238-9

49. Severn Trent Services (2017) http://www.severntrentservices.com/en_us/LiteratureDownloads/Documents/565_0200.pdf

50. Naeem A, Westerhoff P, Mustafa S (2007) Vanadium removal by metal (hydr)oxide adsorbents. Water Res 41:1596–1602

51. Sazakli E, Zouvelou SV, Kalavrouziotis I, Leotsinidis M (2015) Arsenic and antimony removal from drinking water by adsorption on granular ferric oxide. Water Sci Technol 71(4):622–629

52. Backman B, Kettunen V, Ruskeeniemi T, Luoma S, Karttunen V (2007) Arsenic removal from groundwater and surface water – field tests in the Pirkanmaa region, Finland. Geological Survey of Finland, Espoo, pp 1–40

53. Thirunavukkarasu OS, Viraraghavan T, Subramanian V (2003) Arsenic removal from drinking water using granular ferric hydroxide. Water SA 29:161–170

54. Kemwater ProChemie (2016) http://www.prochemie.cz/chem/tech-list-hydroxid-železity-kemira-cfh.pdf

55. GEH-Wassechemie (2017) http://www.geh-wasserchemie.de/files/datenblatt_geh101_en_web.pdf

56. Driehaus W, Jekel M, Hildebrandt U (1998) Granular ferric hydroxide – a new adsorbent for the removal of arsenic from natural water. J Water Supply Res Technol 47:30–35

57. Westerhoff P, Highfield D, Badruzzaman M, Yoon Y (2005) Rapid small scale column tests for arsenate removal in iron oxide packed bed columns. J Environ Eng 131(2):262–271. https://doi.org/10.1061/(ASCE)0733-9372(2005)131:2(262)

58. Bissen M, Frimmel FH (2003) Arsenic – a review. Part II: Oxidation of arsenic and its removal in water treatment. Acta Hydrochim Hydrobiol 31:97–107

59. Ilavský J, Barloková D, Hudec P, Munka K (2014) Iron-based sorption materials for the removal of antimony from water. J Water Supply Res Technol 63(6):518–524

60. Ghaedi M, Mosallanejad N (2013) Removal of heavy metal ions from polluted waters by using of low cost adsorbents: review. J Chem Health Risks 3(1):07–22

61. Gupta SK, Chen KY (1978) Arsenic removal by adsorption. J Water Pollut Control Fed 50:493–506

62. Huang CP, Fu PLK (1984) Treatment of As(V) containing water by activated carbon. J Water Pollut Control Fed 61:233–242

63. Gu Z, Fang J, Deng B (2005) Preparation and evaluation of GAC-based iron-containing adsorbents for arsenic removal. Environ Sci Technol 39(10):3833–3843

64. Chang Q, Lin W, Ying WC (2010) Preparation of iron-impregnated granular activated carbon for arsenic removal from drinking water. J Hazard Mater 184(1–3):515–522. https://doi.org/10.1016/j.jhazmat.2010.08.066

65. Ghorai S, Pant KK (2005) Equilibrium, kinetics and breakthrough studies for adsorption of fluoride on activated alumina. Sep Purif Technol 42:265–271

66. Xiao TF, Hong B, Yang ZH (2001) Hydrogeochemistry of arsenic and its environmental effects. Geol Sci Technol Inf 20(1):71–76

67. Ghosh MM, Yuan JR (1987) Adsorption of inorganic arsenic and organo-arsenicals on hydrous oxides. Environ Prog 6(3):150–157

68. Jekel MR (1994) Removal of arsenic in drinking water treatment. In: Nriagu JO (ed) Arsenic in the environment, Part I: Cycling and characterization 119–132. Wiley, New York, p 448. ISBN: 978-0-471-57929-8
69. Bailey SE, Estes TJ, Bricka RM, Adrian DD (1999) A review of potentially low–cost sorbents for heavy metals. Water Res 33(11):2469–2479
70. Chiban M, Zerbet M, Carja G, Sinan F (2012) Application of low-cost adsorbents for arsenic removal: a review. J Environ Chem Ecotoxicol 4(5):91–102. https://doi.org/10.5897/JECE11.013
71. Singh N, Gupta SK (2016) Adsorption of heavy metals: a review. Int J Innov Res Sci Eng Technol 5(2):2267–2281. https://doi.org/10.15680/IJIRSET.2016.050146
72. Mohan D, Pittman Jr CU (2007) Arsenic removal from water/wastewater using adsorbents – a critical review. J Hazard Mater 142(1–2):1–53
73. Ilavský J, Barloková D, Hudec P, Munka K (2015) Read-As and GEH sorption materials for the removal of antimony from water. Water Sci Technol Water Supply 15(3):525–532
74. Rubel PE (2003) Design manual: removal of arsenic from drinking water by adsorptive media. EPA/600/R-03/019, U.S. Environmental Protection Agency, National Risk Management Research Laboratory, Cincinnati
75. Barloková D, Ilavský J, Munka K (2017) Removal of antimony from water using GEH sorption material at different filter bed volumes. In: Environmental engineering, 10th international conference, Vilnius Gediminas Technical University, Vilnius, pp 1–7. https://doi.org/10.3846/enviro.2017.069
76. Sperlich A, Werner A, Genz A, Amy G, Worch E, Jekel M (2005) Breakthrough behavior of granular ferric hydroxide (GFH) fixed-bed adsorption filters: modeling and experimental approaches. Water Res 39(6):1190–1198. https://doi.org/10.1016/j.watres.2004.12.032
77. Saha B, Bains R, Greenwood F (2005) Physicochemical characterization of granular ferric hydroxide (GFH) for arsenic (V) sorption from water. Sep Sci Technol 40(14):2909–2932. https://doi.org/10.1080/01496390500333202
78. Guan XH, Wang J, Chussuei CC (2008) Removal of arsenic from water using granular ferric hydroxide: macroscopic and microscopic studies. J Hazard Mater 156(1–3):178–185. https://doi.org/10.1016/j.jhazmat.2007.12.012
79. Biela R, Kučera T (2016) Arsenic removal from water by using sorption materials. Advances and trends in engineering sciences and technologies. In: Proceedings of the international conference on engineering sciences and technologies, Tatranská Štrba, High Tatras Mountains, 27–29 May 2015, pp 245–250
80. Filella M, Williams PA, Belzile N (2009) Antimony in the environment: knowns and unknowns. Environ Chem 6(2):95–105. https://doi.org/10.1071/EN09007
81. Nguyen VL, Chen WH, Young T, Darby J (2011) Effect of interferences on the breakthrough of arsenic: rapid small scale column tests. Water Res 45:4069–4080. https://doi.org/10.1016/j.watres.2011.04.037
82. Zeng H, Arashiro M, Giammar D (2008) Effect of water chemistry and flow rate on arsenate removal by adsorption to an iron-based sorbent. Water Res 42:4629–4636. https://doi.org/10.1016/j.watres.2008.08.014
83. Ilavský J, Barloková D (2007) Removal of heavy metals from water by sorption materials. Vodní hospodářství 57(8):302–304
84. Ilavský J (2008) Removal of antimony from water by sorption materials. Slovak J Civil Eng 16(2):1–6
85. Ilavský J, Barloková D, Munka K (2015) The use of iron-based sorption materials and magnetic fields for the removal of antimony from water. Pol J Environ Stud 24(5):1983–1992
86. The Decree of the Ministry of Health of the Slovak Republic No. 247/2017, which laying down details on drinking water quality, drinking water quality control, monitoring and risk management of drinking water supply

Wastewater Management and Water Resources in Slovakia

Štefan Stanko and Ivona Škultétyová

Contents

Abstract Slovakia is a small country typical with high mountains on the north part, two lowlands reach the south border, and the middle mountains between them. This character of Slovakia defines the two river drainage basins orientated to the north Baltic Sea by the river Poprad and the next two directed to the south to the Black Sea. The Slovak Republic territory is 49,014 km^2, with a population of 5.4 million, and is located in the temperate climate zone of the northern hemisphere with regularly alternating seasons. About 38% of the country is forested. Based on longitudinal measurements, the average annual air temperature is 7°C. The longitudinal average amount of precipitation is around 760 mm (Kriš et al., Sustainability of Slovak water resources, presentation in project SWAN – towards sustainable water resources management in central Asia. In: TEMPUS IV programme. www.wrmc.uz, 2013).

The capacity of natural surface water source amounts is about 90.3 m^3 s^{-1}. Ecological discharges are 36.5 m^3 s^{-1}. Water reservoirs across Slovakia enable

Š. Stanko (✉) and I. Škultétyová
Department of Sanitary and Environmental Engineering, Faculty of Civil Engineering,
Slovak University of Technology in Bratislava, Bratislava, Slovakia
e-mail: stefan.stanko@stuba.sk

A. M. Negm and M. Zeleňáková (eds.), *Water Resources in Slovakia:*
Part I - Assessment and Development, Hdb Env Chem (2019) 69: 335–354,
DOI 10.1007/698_2018_285, © Springer International Publishing AG 2018,
Published online: 31 August 2018

335

increasing the discharges in dry periods in 53.8 m^3 s^{-1}. Reservoirs can provide approximately 4,000 l s^{-1} of the high quality of water used for drinking purposes. Water off-take, currently amounting to 39 m^3 s^{-1}, is equal to about 29% of the discharges during dry periods and to 10% of the longitudinal mean discharge (Kriš et al., Sustainability of Slovak water resources, presentation in project SWAN – towards sustainable water resources management in central Asia. In: TEMPUS IV programme. www.wrmc.uz, 2013).

The water consumption is significant for agricultural, for industry, and for drinking purposes, such as water supply system and production of bottled water. The very significance of Slovakia is the huge amount of mineral waters across the country, which reaches the high quality not only for drinking but for health purposes too. The connection to the public water supply system of the population is over 95%.

The worse situation is with the public connection to the public sewer system, which reaches over 66% in the year 2017. This situation is continuously increased in the last 30 years, which is conditional by the investments. There are many people who work in the water sector, which is covered by the Ministry of Environment, governmental and public institutions, and a lot of private companies, which have a goal to improve and protect the public health.

Keywords Drinking water, Wastewater management, Water resources, WWTP

1   Introduction

"Water is not only a constituent part of the human, animal and plant organism, but it is also a functional element without which it is impossible to imagine the origin, development, existence, activity, and health of living creatures in nature." "However, water is not every time available in desirable quantities and, in addition, it does not always have the quality as required." These above two definitions from the book *Fight for Water*, which were defined by the Slovak academic Štefan Vladimír Bella in 1956, generally describe the role of water management, which typically characterized the water management in Slovakia in the last 50 years, and these ideas are still valid [1, 2].

The increasing water demand in agriculture and industry has the direct influence on extending the Slovakia GDP (see Fig. 1). The last but not the least is that the drinking water and following wastewater discharge had a significant influence on public water supply and water management development in Slovakia too (see Fig. 2). Especially, the press on water quality changed the approach for water supply when the required limits on drinking water and the limit of pollution in the rivers and the lakes have a strong influence on water and wastewater management in the last years [3–7].

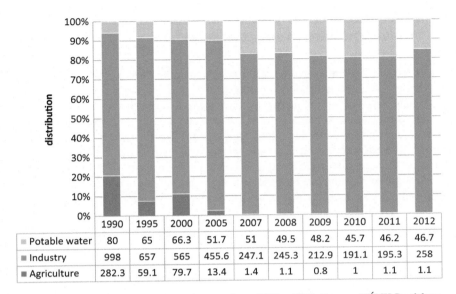

Fig. 1 Charged water in Slovakia, development from 1990 to 2012. Source: VÚVH Bratislava

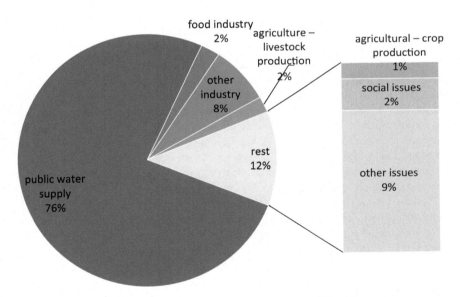

Fig. 2 Water consumption distribution in Slovakia. Source: SHMÚ Bratislava

2   Water Resources in Slovakia, Ground and Surface Waters, Drinking Waters, Wastewaters

Water resources are managed by water management in Slovakia which are directed under the definition that this is a complex of legislative, organizational, technical, ecological, and economic activities and measures related to systematic water resources protection of the country [8]. The water sources in Slovakia are divided into surface and groundwater. The huge amount of surface water is used as utility waters, hydropower waters, irrigation waters, waterways, and fisheries water. The not huge but important are potable waters. All these types of water exploitation have a significant influence on general water management in Slovakia and a significant influence on investments, reconstructions, and operational costs.

2.1   Surface Waters

The utility water increased in the last years and depended on the industry. This amount of water represents over 400,000 m^3/year. The big role of water demand represents the "water law," which defines that a small water consumption of 1,250 m^3/month or 15,000/year is free of charge together with the irrigation water.

The hydropower water becomes very interesting in the last year, especially with the small hydropower plant building and installation. It is the renewable energy which can save the environment pollution and decrease global warming. Together with Gabčík dam, the hydropower plants share about 40% of the available power performance in Slovakia. The hydropower potential becomes very important, and this is collaborated in the government document: "Concept of utilization of the hydropower potential of the Slovak watercourses by 2030." The strategic goal of the concept is to meet the strategic objectives of renewable electricity production set by European and national legislation while taking into account the environmental aspects and the principles of sustainable development [9].

Figure 1 shows the changed distribution of charged water demand from the year 1990. The development was marked by agriculture decreasing in the 1990s. The consequence was the decreasing of the food security and increasing of the food dependency of importing foods. The closest last years are not evaluated, but the tendency is to increase the food security and quality which has a direct impact on again irrigation development and changed the charged surface water distribution.

The irrigation systems have been used in recent years only by cost-intensive grower crops that are priced in the market at a rate that is in the making of the price where the irrigation water supply can also be included in the market. The irrigation technical service units were used in the last years in the form of rental directly to the agribusinesses or to the organizations that agribusinesses ensured their operation.

Another way of water source exploitation is waterways. The responsibility belongs to the institution Slovak Water Management Company, administered by the governmental institution. It's mainly concerned to the Danube rivers – international waterway and river Váh.

The Danube navigation carried out by the Slovak Republic is in accordance with international conventions and applicable laws in close cooperation with the State Navigation Authority of Bratislava and the navigation authorities of Austria and Hungary, in particular the setting up of a fairway and all related facilities, the regular measurements of rock thresholds, the continuous operation of the navigation yard on the Gabčíkovo VD, as well as the crossing (comps) operation in its supply channel.

2.2 Groundwater

Groundwater is primarily intended to supply the drinking population's water. This is the definition according to § 3 par. 4 of Act no. 364/2004 Coll. of Law on Water and on amendment to the Act of SNR No. 372/1990 Coll. of Law.

Data on groundwater abstraction are registered in the SHMÚ collection register in Bratislava (Slovak Hydrometeorological Institute). They are provided by their users on the basis of the obligation arising from Act no. 384/2009 Coll. of Law on water and the Implementing Decree of the Ministry of Agriculture and Forestry SR no. 418/2010 Coll. of Law on the implementation of some provisions of the Water Act.

The groundwater abstraction in Slovakia is divided into the following groups: public water supply, food industry, other industry, agriculture – livestock production, agricultural – crop production, social issues, other issues.

2.3 Geothermal Waters

Geothermal water is mainly used as a source of energy and also in agriculture and tourism. The use of geothermal energy is not only economic but also ecological in economic terms (see Fig. 3).

In the period 1971–2011, 141 geothermal boreholes were drilled in 27 geothermal areas and $2,084\ 1\ s^{-1}$ geothermal waters. On the basis of the reported data at Slovak Hydrometeorological Institute in Bratislava, the users of the geothermal wells located in 35 localities used them in the period 2000–2012. These are boreholes, which are not in the records of the SPA and Salt Inspectorate.

The largest use of geothermal energy in Slovakia is currently for recreational purposes (87% of the number of resources used). It is used in seasonal summer swimming pools (11 boreholes in 10 localities) as well as year-round thermal baths (29 boreholes in 23 localities). For the heating of buildings, thermal energy is used from more than 22 wells (48% of the number of resources used). It's about heating, e.g., hospitals and interiors of aqua park buildings mainly.

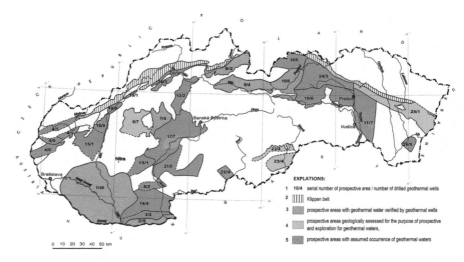

Fig. 3 Map of the use of geothermal waters. Source: MoE Bratislava

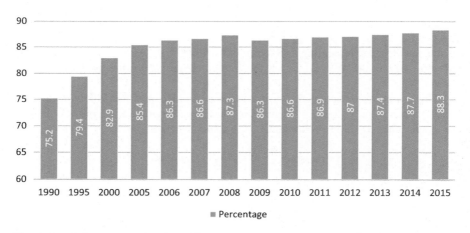

Fig. 4 Population connected to the public water supply system. Source: VÚVH Bratislava

2.4 Drinking Water Supply

The population's drinking water supply from public water mains reached 88.3% in 2015, an increase of 5.4% compared to 2000 (see Fig. 4). Despite the increase in the number of inhabitants connected to public water mains, the abstraction of drinking water from water management facilities is decreasing [10]. This decline is also reflected in the specific consumption of households, which declines year-on-year, and its value in 2015 (77.3 l/capita/day) is alarmingly close to the hygienic minimum (80 l/capita/day).

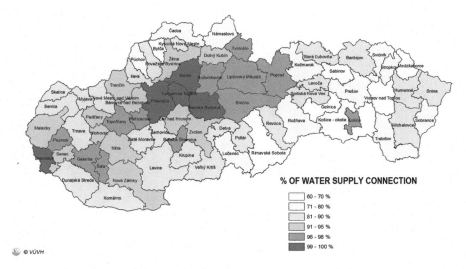

Fig. 5 Distribution of public water supply connections in Slovakia. Source: VUVH Bratislava, 2015

The level of development of public water mains is uneven across the region, with one of the decisive factors in this condition being the lack of groundwater resources in passive areas, in the south of central Slovakia and in most eastern Slovakia (see Fig. 5).

2.5 Renewable Water Resources (RWRs)

RWRs are defined by the sum of the amount of rainfall (after deduction of the amount of water consumed by evapotranspiration) and the amount of water coming from the territory of the Slovak Republic and the water coming into the country from impurities from the neighboring countries. In terms of green growth, it is necessary to ensure the most efficient use of water resources, which are important not only for the economic activities of the landscape but also for the quality of life and health of the population [11, 12]. The important indicator is the *intensity of surface water resources uses* (see Fig. 6). The indicator is expressed as a percentage of total surface water abstraction to the total available surface water reserves (including inflows from neighboring countries).

2.6 Quality of Drinking Water

Evaluation of the quality of drinking water in public water supply is based on the results of control of water companies. Water quality is evaluated on the basis of

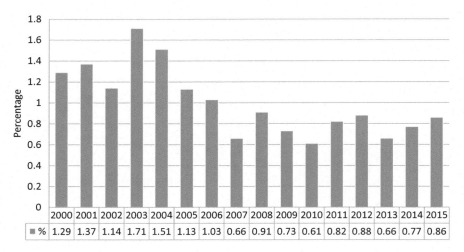

Fig. 6 Intensity of surface water resources uses. Source: VÚVH Bratislava, 2015

the number or, respectively, the share of determination of individual indicators of drinking water quality exceeding the relevant hygiene limits.

Drinking water quality was evaluated by the Government Regulation No. 354/2006 Coll. of Law, "laying down requirements for water intended for human consumption and quality control of water intended for human consumption" [13], as amended by the Decree of the National Council of the Slovak Republic No. 496/2010 Coll. of Law and according to Decree of Ministry of economy SR No. 528/2007 Coll. of Law, laying down details on the requirements for limiting exposure from natural radiation.

The main performances of water quality are (a) microbiological and biological indicators (*Escherichia coli, coliform bacteria, enterococci, cultured microorganisms at 36°C, microscopically detectable microscopy, abiosestone*), (b) physical-chemical indicators (*nitrates, color, manganese, sulfate, turbidity, iron*), and (c) radiological indicators (*total bulk alpha activity, radon volume activity 222*).

Drinking water supplied to consumers by a public water supply system must be disinfected by health. Drinking water disinfection is predominantly carried out by chemical treatment by chlorination. Government Regulation No. 354/2006 Coll. of Law establishes a limit value of 0.3 mg l^{-1} for the active chlorine content in drinking water. If the water is disinfected with chlorine, the minimum value of active chlorine in the distribution network must be 0.05 mg l^{-1}. In the case of demonstrating the good quality of the drinking water source and the grid, the health authority may allow water to be supplied without hygienic safety (Table 1) [14].

The proportion of non-compliant analyses of 0.3 mg l^{-1} exceeded 1.62% in 2015. The minimum free chlorine content did not reach 10.92% of drinking water samples.

The requirements for the drinking water quality are systematically updated. The last update is the Decree of the Ministry of Health of the Slovak Republic, establishing details on drinking water quality—as are depicted in Tables 2, 3, 4, 5, 6, and 7, drinking water quality control, monitoring program, and risk management in the supply of drinking water under No. 247/2017 Collection of Laws [14]. This

Table 1 Microbiological and biological indicators

No.	Indicator	Symbol	Limit	Limit type	Unit
1	Escherichia coli	EC	0	HLM	KTJ/100 ml
			0	HLM	KTJ/10 ml
			0	HLM	KTJ/250 ml
2	Coliform bacteria	CB	0	LM	KTJ/100 ml
			0	LM	KTJ/10 ml
			0	LM	KTJ/250 ml
3	Enterococci	EC	0	HLM	KTJ/100 ml
			0	HLM	KTJ/10 ml
			0	HLM	KTJ/250 ml
4	Pseudomonas aeruginosa	PA	0	HLM	KTJ/250 ml
5	Cultivable microorganisms at 22°C	CM22	200	HLM	KTJ/ml
			500		KTJ/ml
			100		KTJ/ml
6	Cultivable microorganisms at 36°C	CM36	50	LM	KTJ/ml
			100	LM	KTJ/ml
			20	LM	KTJ/ml
7	Live organisms	LO	0	LM	Count/ml
			0	LM	Count/ml
8	Fibrous bacteria (excluding ferric and MN bacteria)	FB	0	LM	Count/ml
			0	LM	Count/ml
9	Micromycetes determined by microscopy	MM	0	LM	Count/ml
			0	LM	Count/ml
10	Dead organisms	DO	30	LM	Count/ml
			30	LM	Count/ml
11	Iron and manganese bacteria	FeMnBa	10	LM	% of cover field
			10	LM	% of cover field
12	Abioseston	AB	10	LM	% of cover field
			10	LM	% of cover field
13	Clostridium perfringens (including spores)	CP	0	LM	KTJ/100 ml
			0	LM	KTJ/100 ml

Source: Min. of Health, 2017

decree does not apply to spring water, spring water and natural mineral water suitable for the preparation of infant formulas, natural mineral water, and natural medicinal water. When selecting drinking water quality indicators, local conditions are taken into account in each supply system. The monitoring program will also include the indicators needed to assess the impact of water mains on the quality of drinking water.

Drinking water quality control can be done by sampling or analyzing spot samples of drinking water or continuous measurements [15].

The monitoring program contains (a) the scope and frequency of the water quality control in the drinking water source, during its treatment, accumulation, and distribution, (b) the extent and frequency of the quality control of drinking water quality at the sites examined, and (c) the sources of drinking water, the range, and the drinking water supply system (Table 2, 3, 4, 5, and 6).

Table 2 Physical and chemical indicators: inorganic indicators [14]

No.	Indicator	Symbol	Limit	Limit type	Unit
14	Antimony	Sb	5.0	HLM	mikrog/l
15	Arsenic	As	10.0	HLM	mikrog/l
16	Boron	B	1.0	HLM	mg/l
17	Nitrates	NO^{3-}	50.0	HLM	mg/l
18	Nitrites	NO^{2-}	0.50	HLM	mg/l
19	Fluoride	F^-	1.50	HLM	mg/l
20	Chrome		50.0	HLM	mikrog/l
21	Cadmium	Cd	5.0	HLM	mikrog/l
22	Cyanides	CN^-	50.0	HLM	mikrog/l
23	Copper	Cu	2.0	LM	mg/l
24	Nickel	Ni	20.0	HLM	mikrog/l
25	Lead	Pb	10.0	HLM	mikrog/l
26	Mercury	Hg	1.0	HLM	mikrog/l
27	Selenium	Se	10.0	HLM	mikrog/l

Source: Min. of Health, 2017

Table 3 Physical and chemical indicators: organic indicators [14]

No.	Indicator	Symbol	Limit	Limit type	Unit
28	Acrylamide	–	0.10	HLM	mikrog/l
29	Benzene	–	1.0	HLM	mikrog/l
30	Monochlorobenzene	MCB	10.0	LM	mikrog/l
31	Dichlorobenzene	DCB	0.30	LM	mikrog/l
32	1,2-Dichloroethane	DCA	3.0	HLM	mikrog/l
33	Total organic carbon	TOC	3.0	LM	mg/l
34	Pesticides	PL	0.10	HLM	mikrog/l
35	Pesticides together	PLs	0.50	HLM	mikrog/l
36	Polycyclic aromatic hydrocarbons	PAU	0.10	HLM	mikrog/l
37	Benzo (a) pyrene	B(a)P	0.010	HLM	mikrog/l
38	Epichlorohydrin	–	0.10	HLM	mikrog/l
39	Tetrachloroethylene and trichlorethylene	PCE + TCE	10.0	HLM	mikrog/l
40	Vinyl chloride	–	0.50	HLM	mikrog/l
41	Microcystin LR	LR	1.0	LM	mikrog/l

Source: Min. of Health, 2017

Table 4 Physical and chemical indicators: indicators investigated for disinfection and chemical treatment of drinking water [13]

No.	Indicator	Symbol	Limit	Limit type	Unit
42	Free chlorine	Cl2	0.30	LM	mg/l
43	Bromate	BrO^{3-}	10.0	HLM	mikrog/l
44	2,4-Dichlorophenol	DCP	2.0	LM	mikrog/l
45	2,4,6-Trichlorophenol	TCP	10.0	LM	mikrog/l
46	Chlorine dioxide	ClO^{2-}	0.20	LM	mg/l
47	Chlorite	ClO^{2-}	0.20	HLM	mg/l
48	Chlorate	ClO^{3-}	0.20	HLM	mg/l
49	Ozone	O^3	50.0	LM	mikrog/l
50	Trihalomethanes together	THMs	0.10	HLM	mg/l
51	Haloacetic acids	HAAs	60.0	HLM	mikrog/l
52	Silver	Ag	50.0	HLM	mikrog/l
53	Aluminum	Al	0.20	LM	mg/l

Source: Min. of Health, 2017

Table 5 Physical and chemical indicators: indicators which may adversely affect the properties of drinking water [14]

No.	Indicator	Symbol	Limit	Limit type	Unit
54	Absorbance (254 nm, 1 cm)	A^{254}	0.080	LM	
55	Ammonium ions	$NH4^+$	0.50	LM	mg/l
56	Color	–	20.0	LM	mg/l
57	Chemical consumption of oxygen by manganese	CHSKMn	3.0	LM	mg/l
58	Chlorides	ClMn	250	LM	mg/l
59	Manganese		50.0	LM	mikrog/l
60	Reaction to water	pH	6.5–9.5	LM	
61	Sulfates	$SO4^{2-}$	250	LM	mg/l
62	Taste	–	Acceptable to the consumer	LM	
63	Temperature	–	8–12	RV	°C
64	Turbidity	–	5.0	LM	FNU
65	Odor	–	Odorless		
66	Iron	Fe	0.20	LM	mg/l
67	Conductivity	EK	125.0	LM	mS/m of 20°C
68	Sodium	Na	200	LM	mg/l

Table 6 Physical and chemical indicators: substances whose presence in drinking water is desirable [14]

No.	Indicator	Symbol	Limit	Limit type	Unit
69	Magnesium	Mg	10.0–30.0	RV	mg/l
			125	LM	mg/l
70	Calcium	Ca	>30	RV	mg/l
71	Calcium and magnesium	Ca + Mg	1.1–5.0	RV	mmol/l

Table 7 Radiological indicators [14]

No.	Indicator	Symbol	Limit	Limit type	Unit
72	Tritium	3H	100.0	IV	Bq/l
73	Radon	222Rn	100.0	IV	Bq/l
74	Total alpha bulk activity	aVcalfa	0.10	IV	Bq/l
				IV	
75	Total beta activity	avcbeta	0.50	IV	Bq/l
76	Indication dose	ID	0.10	IV	mSv/y
77	Natural radionuclides	238U	3.0	LM	Bq/l
78		234U	2.80	LM	Bq/l
79		226Ra	0.50	LM	Bq/l
80		228Ra	0.20	LM	Bq/l
81		210Pb	0.20	LM	Bq/l
82		210Po	0.10	LM	Bq/l
83		222Rn	300.0	LM	Bq/l
84	Artificial radionuclides	14C	240.0	LM	Bq/l
85		90Sr	4.90	LM	Bq/l
86		239Pu/	0.60	LM	Bq/l
		240Pu			
87		241Am	0.70	LM	Bq/l
88		60Co	40.0	LM	Bq/l
89		134Cs	7.20	LM	Bq/l
90		137Cs	11.0	LM	Bq/l
91		131	6.20	LM	Bq/l

The important indicator for general drinking water use is *the minimum analysis* which is designed to check and obtain regular information on the stability of the drinking water source, the effectiveness of drinking water treatment (especially for disinfection control, if any), and the microbiological quality and sensory characteristics of the drinking water supplied. The minimum analysis is the set of 26 indicators (Table 8).

Explanations of table abbreviations:

Table 8 Twenty-six indicators of minimal drinking water quality analysis [14]

No.	Indicator
1	Escherichia coli
2	Coliform bacteria
3	Enterococci
4	Cultivable microorganisms at 22°C
5	Cultivable microorganisms at 36°C
6	Live organisms
7	Fibrous bacteria (excluding ferric and manganese bacteria)
8	Micromycotes determined by microscopy
9	Dead organisms
10	Iron and manganese bacteria
11	Abioseston
12	Clostridium perfringens including spores
13	Nitrates
14	Nitrites
15	Absorbance (254 nm, 1 cm)
16	Ammonium ions
17	Color
18	Chemical consumption of oxygen by manganese
19	Manganese
20	Reaction to water
21	Taste
22	Temperature
23	Turbidity
24	Odor
25	Iron
26	Conductivity

LM   *The limit value* is a value of the indicator of the quality of drinking water, beyond which loses drinking water of satisfactory quality in the variable whose value has been exceeded.

HLM   *The highest limit* is the value of a health indicator of the quality of drinking water, the excess of which excludes the use of water as drinking water.

RV   *The recommended value* is the value or range of drinking water quality indicator values that are desirable from the point of view of health and whose exceeding or non-compliance does not exclude the use of water as drinking water.

IV   *The indicative dose* is the value of the effective dose, on average, per the calendar year, of the intake of natural radionuclides or artificial radionuclides with potable water, except for 3H, 40 K, 222Rn and short-time semi-precursor transformation products 222Rn.

2.7 Wastewaters and Sewer Systems

The increasing of cities population, the territory industrialization, and the development of services have significantly outstripped the rate of development of the water infrastructure (sewerage networks and WWTPs) [16, 17]. The existence of water infrastructure is also a prerequisite for further social and economic development at local, regional, state, and global level [18–20].

The level of wastewater drainage in Slovakia reached about 66% of the total population in 2015–2016, which represents an absolute amount of 3.6 million residents. Of these, about 98% are connected to a public sewage system with a wastewater treatment plant. The length of sewer networks reached above 11,000 km. The growth rate of the connected population (see Fig. 7) and proportion distribution of sewer systems (see Fig. 8) are shown.

The growth of sewer systems also puts pressure on reconstruction and restoration of existing infrastructure. Particularly sewers built up to 1989 are not of a qualitative level, which corresponds to the present requirements of construction. The distribution of sewerage in the country also depends to a large extent on the living standard of the population and on GDP in the individual regions as corresponds with the proportion of population connected to the public sewer system showed in Fig. 8.

Despite the increase in public connection to public sewerage, the level of population drainage continues to lag behind the development of public water supply systems. In view of the objectives and requirements of Council Directive 91/271/EEC on urban wastewater treatment, the Slovak Republic has concentrated on the maximum attention and funds for the construction of public sewage systems and the improvement of effluent (WWTP) efficiency. The effects of this process are

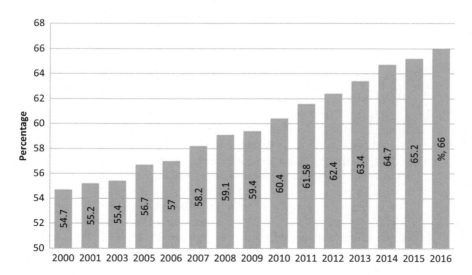

Fig. 7 Population connected to the public sewer system. Source: VÚVH Bratislava, 2016

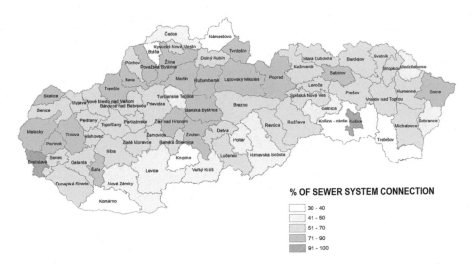

Fig. 8 Proportion of population connected to the public sewer system in Slovakia. Source: VÚVH Bratislava, 2016

manifested by the gradual increase of the inhabitants living in the houses connected to the public sewerage system but especially by the improvement of the parameters of discharged treated wastewater, respectively, by reducing the discharge into the aquatic environment.

Sewerage built over the last two decades is also marked by various technologies. Public procurement also had an impact on the quality of construction. The public procurement system largely preferred plastic pipes, initially corrugated. Later, PVC pipes were filled into the trenches. Only in exceptional cases were the sewers constructed from polypropylene or fiberglass tubes – mainly in larger cities and sections of highway drainage.

Storage of sewer pipes was very often performed not entirely according to the technologically correct procedure [21]. The problems with plastic pipes were mainly due to incorrect compaction of the backfill, which, after about 5–10 years of operation, has an emergency [22, 23].

Wastewater (WW) management approach in Slovakia is divided into two ways of WW disposal: (a) centralized and (b) decentralized [24, 25].

The most of WW disposal is a centralized system, which includes (a) combined, (b) separated, and (c) dry weather flow sewer system [26, 27].

The historical cities mostly exploit the combined sewer system which disposes off both storm and dry weather flow in one pipe. The advantage is that only one pipe was built up and only one pipe we had to operate. The disadvantage is a huge amount of wastewater, which is mixed water of storm water/rainwaters and dry weather flow (sewage or fecal waters). The behavior of this type of sewer is different in wet and dry weather. In wet weather, due to rain/storm events, the fecal waters are diluted with storm water [28]. This means that sewage water pollutes the storm water. We can see the impact especially if we have to overflow a huge amount of water to the receiving waters through the combined overflow structure (CSO). The systems of CSO chambers are very popular in Slovakia, but now it is defined as a point source of pollution [29–31]. There are a lot of computes, how to design the CSO chamber, with the aim to keep maximum discharge of overflow waters keeping the receiving waters protection. We use two ways of calculation [31, 32]: (1) the method of dilution ratio (1) and (2) method of boundary rain. First one uses the dilution ratio storm: sewage/fecal waters = 8:1 or earlier 4:1 (2). The second one uses the limit values of the rain multiplying the reduction catchment area.

$$Q_{\text{wwtp}} = q_{rl} \cdot S_r \tag{1}$$
$$Q_{\text{wwtp}} = \text{PE} \cdot Q_{24} \cdot 8 \tag{2}$$

q_{rl} specific reduction amount of storm waters from 7.5 to 25 l s ha^{-1}, S_r the reduction catchment area, Q_{wwtp} amount of waste water to WWTP, Q_{24} 24-h discharge of dry weather flow, PE number population equivalent.

3 Conclusion and Recommendations

Urban water and waste management is the sustainable activity in high development urban areas, which support the public health, comfortability, and active development of urban areas. A lot of water and wastewater management activities, which include expert planning, economy, and management of a wide range of activities joining with this practice, can ensure and improve the sustainable life level of people living in urban areas.

The works related to the development of drinking water supply and its subsequent drainage as well as the drainage of rainwater belongs to the basic idea of creating and engraving towns and municipalities. Currently, not only a quantitative but also a qualitative approach to this issue is a very important task.

Generally water quality, especially quality of drinking water, is now very important. In Slovakia, drinking water from the water supply is still of high quality and must fulfill the requirements of the law. This fact also has an important but indirect impact on water trade and pricing policy.

Increasingly, we have been confronted with the sale of water in plastic PET bottles for a long time. It should be remembered that it is a substitute source of water supply where such water is much more expensive in terms of its transport to the consumer, but its quality may be at a time endangered by long-term storage. Another significant factor is the dislocation of these PET bottles, which greatly affects the waste management system and disproportionately damages the environment.

Despite the availability of drinking water over more than 90% in Slovakia, bottled water is a huge food chain item. However, this influence is due to the general ignorance of the inhabitants of the country about the quality of drinking water from the water mains and the determined unjustified concern about the health of the population.

This approach of inhabitants of towns and cities in Slovakia is an example of insufficient education in the field of water management, especially in the area of massive population supply with drinking water from the water supply. Only permanent education in schools and public institutions, as well as promotional activity, can make this awareness change for the better.

Water resources belong to the wealth of Slovakia, and it is an important and necessary task to protect them. At present, we are witnessing the disproportionate development of housing construction in suburban areas of large cities and extensive development without the sound development of infrastructure. Buildings are often built in areas where there is a huge supply of groundwater. This activity, directly and indirectly, threatens groundwater supplies mainly qualitatively.

With regard to climate change, this concern is even more overwhelming, as not only today but long ago we know that water, soil, and air are the most precious basic assumptions of life and need to be protected at all costs. Unfortunately, very often we are witnessing the opposite.

In the present highly sophisticated times when economic interests exceed the environment, this is especially important to see as a threat and the best possible activity to reverse the native influence of the company's behavior and to focus the general interest in water protection. This is also highly dependent on the correct system of water supply in the country.

References

1. Bella Š, Bella V (1956) Boj s vodou a o vodu. SVTL, Bratislava
2. Kriš J, Škultétyová I, Stanko Š (2013) Sustainability of Slovak water resources, presentation in project SWAN – towards sustainable water resources management in central Asia. In: TEMPUS IV programme. www.wrmc.uz
3. Regulation of the Government of the Slovak Republic no. 296/2005 laying down quality and quality objectives for surface water and the limit values for indicators of waste water and special water pollution
4. Škultétyová I, Stanko Š, Kriš J, Božíková J, Rusnák D (2011) Riziko kontaminácie zo skládok odpadov. Acta Hydrol Slovaca 12:96–104 ISSN 1335-6291
5. Škultétyová I, Stanko Š, Mahríková I (2010) Environmental engineering: selected papers, vol 1. STU v Bratislave, Bratislava, p 162 ISBN 978-80-227-3289-5
6. Stanko Š, Rusnák D, Škultétyová I, Božíková J, Kriš J (2011) Klimatické zmeny a ich možné vplyvy na tvorbu povrchového odtoku. Acta Hydrol Slovaca 12:112–118 ISSN 1335-6291
7. Volodina GB, Garelik C, Lysenko IO, Mahríková I, Okrut SV, Popov NS, Škultétyová I, Stanko Š (2011) Vodnaja ekologija i vlijanije dejateľnosti čeloveka na sostojanije vodnych resursov. Izd-vo IP Česnokova AV, Tambov, p 230 ISBN 978-5-903435-82-1A4
8. Water management in the Slovak Republic (2012) Ministry of Environment of the Slovak Republic, Bratislava
9. Ministry of Environment SR, Bratislava (2011) Concept of utilization of the hydropower potential of the Slovak watercourses by 2030

10. Hlavínek P, Raclavský J, Mičin J, Baur R, Shilling W (2004) Strategic rehabilitation of water distribution and wastewater collection systems. In: Proceedings of 22nd international NO-DIG conference and exhibition, Hamburg, 11/2004

11. Krejčí V et al (2002) Odvodnění urbanizovaných území – Koncepční přístup. NOEL 2000 s.r.o. Brno, Česká Republika. ISBN 80-86020-39-8

12. Stanko Š, Mahríková I, Gibala T (2006) Development of a complex system for pipeline design in Slovakia. Security of water supply systems: from source to tap. Springer, Dordrecht, pp 137–147 ISBN 1-4020-4563-8

13. Government Regulation no. 354/2006 Coll. of the Slovak Republic, Laying down requirements for water intended for human consumption and quality control of water intended for human consumption

14. Decree of the Ministry of Health of the Slovak Republic No. 247/2017 Collection of Laws, dated 9 October 2017, laying down details of drinking water quality, drinking water quality control, monitoring and risk management of drinking water supply

15. Čerešňák D, Bujnová A, Stanko Š, Rusnák D, Belica P (2012) Súčasnosť a výhľad stavu v odkanalizovaní. Vodní hospodářství 62(1):7–12 ISSN 0862-5549

16. Rusnák D, Urcikán P, Stanko Š (2008) Stokovanie a čistenie odpadových vôd: Stokovanie III. Kanalizačné rúry. Stavba, prevádzka a obnova stôk. STU v Bratislave, Bratislava, p 186 ISBN 978-80-227-2889-8

17. Škultétyová I, Dubcova M, Galbova K, Stanko S, Holubec M (2016) Life cycle assessment applied to wastewater treatment plant. In: International multidisciplinary scientific geoconference surveying geology and mining ecology management, SGEM. Ecology, economics, education and legislation conference proceedings, vol 1, pp 399–405. ISBN 978-619-7105-65-0

18. Chabal L, Stanko S (2014) Sewerage pumping station optimization under real conditions. GeoSci Eng 60(4):19–28 ISSN 1802-5420

19. Grič J (2008) Pokyny pre občanov k vypúšťaniu odpadových vôd do podtlakových šácht a tým do podtlakovej kanalizácie Vajnory, Vajnorské noviny, ročník XIV 9/2008

20. Hlavínek P, Prax P, Raclavský J, Šulcová V, Tuhovčák L (2004) Využití expertního systému pro plánování rekonstrukcí stokových sítí. In: Konference s mezinárodní účastí "Odpadové vody 2004", AČE SR, Tatranské Zruby, Slovensko, pp 119–123. ISBN 80-89088-33-3

21. Montero C, Villaneuva A, Hlavinek P, Hafskjold L (2004) Wastewater rehabilitation technology survey. CARE-S report D12

22. Stanko Š, Mahríková I (2007) Sewer system condition, type of sewers and their impact on environmental management, In: ARW on natural disasters and water security: risk assessment, emergency response and environmental management, Yerevan, Armenia, 18–22 Oct 2007: threats to global water security. Springer, Dordrecht, pp 359–364. ISBN 978-90-481-2343-8 (PB), ISBN 978-90-481-2336-0 (HB), ISBN 978-90-481-2344-5 (e-book)

23. Stanko Š (2009) Kapacitné posúdenie stokovej siete racionálnou metódou. In: Rekonštrukcie stokových sietí a ČOV: 6. bienálna konferencia s medzinárodnou účasťou. Podbanské, 21–23 Oct 2009, pp 229–232. ISBN 978-80-89602-64-50

24. STN EN 752 – 2 stokové siete a systémy kanalizačných potrubí mimo budov. časť 2: funkčné požiadavky

25. Urcikán P (2011) Stokovanie a čistenie odpadových vôd II. STU v Bratislave, Bratislava ISBN 978-80-227-3434-9

26. Holubec M, Gregusova V, Stanko S, Škultétyová I, Galbova K (2016) Impact of various inlet configurations on overall efficiency in waste water settling tanks. In: International multidisciplinary scientific geoconference SGEM. In: Water, resources, forest, marine and ocean ecosystems conference proceedings, vol 1, pp 453–458. ISBN 978-619-7105-61-2, WOS: 000391653400059

27. Stanko Š, Mahríková I (2006) Implementation of fiber optic data cables in sewage system. Integrated urban water resources management. Springer, Dordrecht, pp 171–180 ISBN 1-4020-4684-7

28. Rusnák D, Stanko Š, Škultétyová I (2015) Rain model – boundary condition in sewer network appraisal. Pollack Periodica 11(2):105–112
29. Stanko Š, Hrudka J, Škultétyová I, Holubec M, Galbová K, Gregušová V, Mackuľak T (2017) CFD analysis of experimental adjustments on wastewater treatment sedimentation tank inflow zone. Monatsh Chem 148(3):585–591. https://doi.org/10.1007/s00706-017-1927-7 ISSN 0026-9247
30. Stanko S, Molnarova L, Holubec M, Galbova K, Škultétyová I (2016) Flow measurements infiltrated water in the sewer network. In: International multidisciplinary scientific geo-conference surveying geology and mining ecology management. Water, resources, forest, marine and ocean ecosystems conference proceedings, vol 1, pp 343–350. ISBN 978-619-7105-61-2; WOS: 000391653400045
31. Stanko Š, Škultétyová I, Hrudka J, Holubec M, Galbová K (2016) Numerické matematické modelovanie objektov stokových sietí. Vodní hospodářství 66(12):10–13 ISSN 1211-0760
32. Urcikán P, Rusnák D, Stanko Š (2001) Stanovenie bilančných veličín na posudzovanie činnosti odľahčovacích komôr. Vodní hospodářství 51(5):119–121 ISSN: 0322-886X

Possibilities of Alternative Water Sources in Slovakia

Gabriel Markovič

Contents

Abstract The world is facing severe challenges in the management of water in various urban and regional locations nowadays. What we have to admit are significant gaps in our knowledge about the existing alternative sources of water for potable as well as non-potable use. So the biggest challenge for developers, engineers, and architects is increasing water supply through alternative water sources.

This chapter contains results of measurements and evaluation in the field of alternative water resources. It also contains an evaluation of water consumption in a family house's secondary source of water – water from well where was made. Mains water supply for the activities as flushing of toilets and clothes washing were entirely replaced by the water from well. The chapter also provides an analysis of the potential use of another water resource (rainwater) for the second object – school building in TUKE campus – according to measured data of rainfall and flowed volumes of runoff for this actual school building in TUKE campus.

Keywords Drainage, Infiltration, Rainwater harvesting, Savings

G. Markovič (✉)
Faculty of Civil Engineering, Institute of Architectural Engineering, Institute of Environmental Engineering, Technical University of Košice, Košice, Slovakia
e-mail: gabriel.markovic@tuke.sk

A. M. Negm and M. Zeleňáková (eds.), *Water Resources in Slovakia:*
Part I - Assessment and Development, Hdb Env Chem (2019) 69: 355–372,
DOI 10.1007/698_2017_191, © Springer International Publishing AG 2018,
Published online: 10 April 2018

355

1 Introduction

Stormwater management is a relatively new issue in Slovakia. Percolation of rainwater as a part of stormwater management as well as rainwater harvesting is becoming more and more critical as a drainage solution in Slovakia.

Majority of new properties have limitations in a stormwater discharge into the combined sewer. This problem is usually solved by stormwater sewer and its discharge into receiving waters or by infiltration systems.

Prefabricated, easy-to-install plastic infiltration chambers are commonly used. Rainwater harvesting is not a usual method in the Slovak industrial or commercial sphere yet.

Rainwater management is so important, especially in the new development project. It has to be taken into account from the initial phase of every new project. For developers, planners, designers, as well as realization company, the priority should be to find the most sustainable and effective way of handling with rainwater. Rainwater harvesting and infiltration of rainwater provide effective and sustainable combination with this source of water.

On the other hand, stormwater infiltration can cause groundwater pollution. This is the fact that cannot be neglected. We should ensure sufficient stormwater treatment that naturally depends on stormwater runoff quality. As Stahre [1] mentioned, sustainability in SWM deals with quantity and so with quality issues at the same time, and this should be the primary target in stormwater management in our conditions too.

Urban drainage systems can be divided into two most commonly used systems: combined sewer system and separate sewer system. Combined sewer systems convey stormwater and wastewater away in one pipe. Where there are combined systems, there is a risk of combined sewer overflows which represents transfers of untreated wastewater to receiving waters [2]. Whereas, separate sewer system carries stormwater and wastewater in separate pipes, usually laid side by side [3].

2 Water Resource Management

Water on Earth takes up about 2/3 of the total area. Most of the water is contained in seas and oceans – up to 96%. The rest are rivers, lakes, artificial reservoirs, underground water, water vapor, water in living organisms, and glaciers. It means that only 2.5% is fresh and only 0.007% is available for people via rivers, lakes, and reservoirs. Freshwater is the most important source in the world for every sphere of our life, and handling with this source appears as the most significant challenge for mankind in our future [4].

Peoples currently confiscate more than half of accessible freshwater runoff, and this amount is expected to increase significantly in the coming decades. A significant amount, 70%, of the freshwater currently withdrawn from all freshwater resources is used for agriculture. Of course, we expected that the world's population would be increased significantly by 2050 that represents still higher and higher demand for water consumption and higher pressure on freshwater resources. The recent global water assessments suppose that about around 70% of the future world population will face water shortages and 16% will have insufficient water to grow their essential food requirement by 2050 [4]. The climate change, increased frequency, and intensity of precipitation as well as drought events – all of this represent complication for the future freshwater management and the handling with this resource.

A cycle of water circulation in nature consists of the following phases: precipitation, infiltration, runoff, and evaporation.

The natural terrain represents the most suitable and effective type of landscape for rainwater infiltration where rainwater naturally infiltrates into the soil and becomes a part of subsoil water. Only about 20% of rainfall water is naturally drained to recipients.

Impervious surfaces result from the new and further urbanization and lead to increased volume of stormwater runoff within existing areas. This not results only to the increased volume of runoff but also to the increased speed of runoff. It also results in peak flows that in many cases represent volumes of rainwater which are not possibly drained with existing drainage system [1, 5, 6].

In densely urban areas, natural terrain is replaced by paved surfaces. When rainwater reaches these surfaces, almost 80% of this water flows to wastewater disposal system or rivers, and only 20% infiltrates into the soil. This leads to the abovementioned ecological damages [7].

As it was written in introduction, combination of rainwater harvesting and infiltration system should provide sustainability in stormwater management. However, we have to remember that the rainwater harvesting measure as a part of source control in the SWM is not only one alternative possibility for building objects. There are several possibilities of an alternative source of water in buildings. Mainly in every new building, respectively, every new building object has to consider the possibility of alternative water source for this building in this times.

Every building has water sources available on building sites that can supplement the traditional water source – water main. These water sources are usually for non-potable usage and vary significantly in quality. Most of the water resources have limited application, and especially water from reuse solutions is never suitable for human consumption [8, 9].

Alternative water is often treated to non-potable standards. Typical uses of alternative water include toilet flushing, washing clothes and cleaning, or specific function as fountain filling, cooling tower makeup, or as a firewater.

In general, when we considered about implementing an alternative water project, the following should be taken into account [10, 11]:

- Available alternative water source and its potential for usage
- Treatment of alternative water source and the water quality requirements of the application for specific solution

- Design of infrastructure requirements such as piping, storage, and pumps
- Overflow and backflow prevention requirements
- Cost-effectiveness of alternative water use
- Required permits from local or state government entities and the timing to secure them

In general, the following sources can be used when implementing an alternative water project (Table 1):

- Rainwater
- Water well
- Gray water
- Other waters (reclaimed wastewater, air handling condensate, water purification system discharge water)

Table 1 Considerations for alternative water sources

Alternative water source	Water quality concerns	Potential treatment	Potential use	Considerations
Well water	Pathogens, organics dissolved, dissolved solids	Filtration	Irrigation, toilet flushing, washing clothes	Minimal treatment is needed for irrigation and flushing toilets
Rainwater	Suspended solids, pathogens	Filtration, possible sedimentation, and disinfection	Irrigation, toilet flushing, washing clothes	Minimal treatment is needed for irrigation and flushing toilets
Gray water	Pathogens, sediments, organics dissolved solids, hardness	Possible sedimentation and biological treatment	Toilet and urinal flushing, irrigation	Subsurface irrigation is most appropriate unless water is disinfected
Other waters (reclaimed wastewater, air handling condensate, water purification system discharge water)	Heavy metals, bacterial growth, dissolved solids, hardness, pathogens, sediments, organics	Filtration, disinfection, biological treatment	Cooling tower makeup, industrial uses, irrigation	Condensed water can be corrosive to metals because condensate can be slightly acidic; water may absorb copper from cooling coils; highly dissolved solids can pose issues for cooling towers and landscape

2.1 Measurements of Water Savings by Use of Water Well as an Alternative Water Source in the Family House

We started our research and own measurements of water well quantity at the selected family house. The family house that was tested was located in the north part of Slovakia. There are three inhabitants of this family house.

System for water well supply consists of pump station equipment, horizontal and vertical plastic pipes connected well with the family house, a primary filter for treatment and fittings, and fixture plumbing (Figs. 1 and 2). The appliances located on both the first and second floors are washing machine and toilet.

The in situ measurements will show us the exact volume of used water well as a savings of potable water by replacing activities (flushing toilets and washing clothes) in the house.

According to the user behavior and their water habits, we observed the real substitution and savings of potable water by water well. The measurements were recorded for a 1-year period and will always continue. The methodology was based on monitoring of the week amount of water used for toilet flushing, washing machines, and total water consumption.

The volume of water used for toilets flushing and washing machines was recorded via water meter, which was set on the water supply pipelines on the basement floor. The water meter was installed earlier, so we need to set the starting values. The water meter placed on supply pipe from well had the starting amount of $53.174 \, m^3$, and the water meter placed on water main pipe had the starting amount of $114.0185 m^3$ [9].

The graph shown in Fig. 3 depicts total consumption of potable water as well as water well for research period 22 May 2015 to 22 April 2016 from family house.

The red line depicts the values of the potable water used for every activity except for flushing toilets and for washing machines in this family house. Total consumption of potable water for this research period was $39.7 \, m^3$ (Fig. 3). Blue line depicts the values of the water well used for flushing toilets and for washing machines (water well entirely replaced potable water for these activities). Total consumption of water well for this research period was $33.6 \, m^3$ (Fig. 3).

The graph shown in Fig. 4 depicts a comparison of the total consumption of potable water as well as water well for research period 22 May 2015 to 22 April 2016 from family house. The comparison shows that a savings of potable water by replacing water well is more than 50% during research period (Figs. 4, 5, and 6).

2.2 Measurements of Quantity of Rainwater Runoff in TUKE Campus

Rainwater harvesting is not the only part of source control measure in the SWM; it is also the way on how to control water consumption and how to support qualitative and reasonable water use for different purposes. One of the objectives of the WFD

Fig. 1 Water stress and scarcity will increase globally due to population increases and climate changes [4]. Source: World Meteorological Organisation (WMO), Geneva, 1996; Global Environment Outlook 2000 (GEO), UNEP, Earthscan, London, 1999

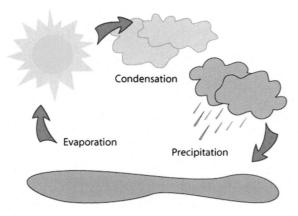

Fig. 2 Water cycle [12]

Fig. 3 Pump station for water well [9]

[13, 14] is to promote sustainable water use, based on long-term protection of available water resources, and we can say that RWH contributes to this objective.

The research of quantity of rainwater as a potential alternative source of water as well as the research of percolation facility efficiency takes place at the Faculty of Civil Engineering in Kosice City. The percolation shafts which are tested are located in the premises of TUKE (Technical University of Kosice). The equipment that provides us information about the quality and quantity of water from runoff are rain gauge located on the roof of the University library, real school building PK6,

Fig. 4 Connecting pipes, filter, and water meter

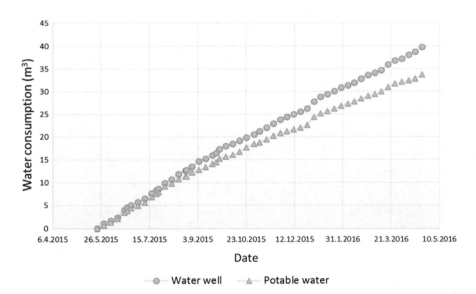

Fig. 5 Total consumption of water well and potable water during research period [9]

and two infiltration shafts for rainwater runoff drainage (Fig. 7). Roof area of the PK6 building is 548.55 m^2 [7, 15–18].

For the recording of rainfall intensity, a heated rain gauge was used for all year-round measuring. This type of rain gauge was a necessity for our measurement with respect to weather conditions in Slovak Republic, and we need to perform the

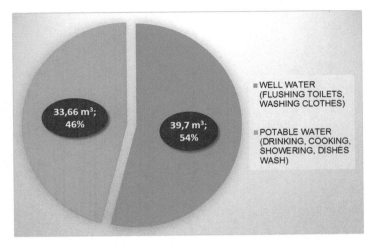

Fig. 6 Comparison of the total consumption of water well and potable water during research period [9]

Fig. 7 Infiltration shafts located near building PK6 and rain gauge located on the roof of the University library

measuring of liquid precipitation (rain) as well as solid precipitation (snow). The rain gauge is made of a stainless material with round catchment area of 200 cm^2 and measuring based on tipping bucket mechanism. Tipping bucket is located inside the rain gauge body right under the funnel outlet. Liquid precipitation, as well as solid precipitation, falls down the funnel outlet into the divided bucket. The bucket is calibrated on 0.2 mm amount of water, then it tips, and second half of bucket can be filled with rainwater. Rainwater after bucket tipping is drained into the drainage hole. The material of tipping bucket is plastic with a very thin layer of titanium, and it is hanged on stainless steel axial holder. Tipping continues during the time of

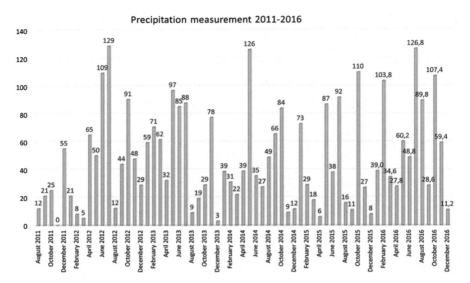

Fig. 8 Measured values of rainfall during our research August 2011 to December 2016

rainfall [19]. Figure 8 represents the measured monthly rainfall totals during our research. Data are presented for the period August 2011 to December 2016.

Table 2 summarizes the real measured monthly rainfall totals against the theoretical volumes of collected rainwater. Data is presented for the period April 2012 to December 2016 because from April 2012 we began measuring the rainwater volumes from all roof area of the building PK6 and precipitation measurements simultaneously (Notice: August 2012 is without data due to equipment failure) [20].

Table 3 summarizes the measured monthly rainfall totals with corresponding theoretical volumes of collected rainwater and comparison with real volumes of collected rainwater from our measurements. Data are presented for the period April 2012 to December 2016 (data is present from April 2012 because at that time we began measuring the flow from all roof area of the building PK6 and precipitation measurements simultaneously) (Notice: August 2012 is without data due to equipment failure) [20].

The next step of our research was to compare the current total consumption of potable water in a PK6 building against the volume of rainwater runoff from the roof that represents a potential source of water in the PK6 building.

Table 4 represents real volumes of rainwater from 548 m^2 roof of the PK6 building against the total consumption of potable water used for every activity in this building measured from January 2014 to December 2016 (Note: September and October 2014 are without data due to equipment failure. Note 2: data of total consumption of potable water are available only from the year 2014).

The graph shown in Fig. 9 depicts the total consumption of potable water used for every activity in this building as well as flowed volumes of rainwater as a

Table 2 Theoretical volume of rainwater from PK6 building (548 m^2) according to the measured values of precipitation from April 2012 to December [20]

Month	Rainfall (mm)	Theoretical volume from 548.55 m^2 (m^3)
April 2012	65	35.6
May 2012	50	27.4
June 2012	109	60.0
July 2012	129	70.6
August 2012	12	6.7
September 2012	44	24.0
October 2012	91	49.6
November 2012	48	26.1
December 2012	29	15.8
January 2013	59	32.6
February 2013	71	38.8
March 2013	62	33.8
April 2013	32	17.6
May 2013	97	53.2
June 2013	85	46.8
July 2013	88	48.2
August 2013	9	4.9
September 2013	19	10.5
October 2013	29	15.9
November 2013	78	42.5
December 2013	3	1.6
January 2014	39	21.2
February 2014	31	17.0
March 2014	22	12.1
April 2014	39	21.3
May 2014	126	69.2
June 2014	35	19.4
July 2014	27	15.0
August 2014	49	26.7
September 2014	66	35.9
October 2014	84	46.1
November 2014	9	5.0
December 2014	12	6.4
January 2015	73	40.0
February 2015	29	15.9
March 2015	18	9.9
April 2015	6	3.3
May 2015	87	47.7
June 2015	38	20.8
July 2015	92	50.5
August 2015	16	8.8
September 2015	11	6.0

(continued)

Table 2 (continued)

Month	Rainfall (mm)	Theoretical volume from 548.55 m^2 (m^3)
October 2015	110	60.3
November 2015	27	14.8
December 2015	8	4.4
January 2016	39	21.4
February 2016	104	56.9
March 2016	35	19.0
April 2016	28	15.2
May 2016	60	33.0
June 2016	49	26.8
July 2016	127	69.6
August 2016	90	49.3
September 2016	29	15.7
October 2016	107	58.9
November 2016	59	32.6
December 2016	11	6.1

potential alternative source of water for this building measured during January 2014 to December 2016.

The red line depicts the values of the potable water used for every activity in the PK6 building during research period (Fig. 9). Blue line depicts the values of the real rainwater volumes from roof construction of the PK6 building as a potential alternative source of water for this building (Fig. 9).

The comparison showed that there are some months with a higher volume of rainwater drained from roof construction than total consumption of potable water for every activity in this building which represents a significant potential for savings of potable water by replacing it with rainwater in the PK6 building (Fig. 9).

3 Conclusion and Recommendations

Diversifying our existing water sources helps secure our water supply system against water scarcity, droughts, and floods. Using alternative water sources to meet demand makes better use of all the water available to us and saves drinking water for specific purposes where high-quality water is needed. So it is a fundamental challenge for developers, engineers, and architects to find any source of alternative kind of water for saving so precious drinking water.

In identifying alternative sources of water, the first consideration is what those sources will be used for. Potable water, which we can use for drinking, cooking, and bathing, among other uses, must meet a high level of purity and safety. Non-potable water is less pure, but, when handled correctly, it can be excellent for landscape irrigation, makeup water for cooling towers, and toilet flushing. Many alternative

Table 3 Measured monthly rainfall totals with corresponding theoretical volumes of collected rainwater and real amount of rainwater from the roof of PK6 building (548 m^2)

Month	Rainfall (mm)	Theoretical volume from 548.55 m^2 (m^3)	Real volume from 548.55 m^2 (m^3)
April 2012	65	35.6	26.7
May 2012	50	27.4	18.9
June 2012	109	60.0	40.8
July 2012	129	70.6	49.6
August 2012	12	6.7	–
September 2012	44	24.0	17.9
October 2012	91	49.6	36.5
November 2012	48	26.1	16.9
December 2012	29	15.8	12.1
January 2013	59	32.6	19.9
February 2013	71	38.8	23.5
March 2013	62	33.8	22.8
April 2013	32	17.6	11.8
May 2013	97	53.2	30.6
June 2013	85	46.8	30.2
July 2013	88	48.2	36.6
August 2013	9	4.9	3.8
September 2013	19	10.5	8.9
October 2013	29	15.9	13.7
November 2013	78	42.5	38.4
December 2013	3	1.6	1.3
January 2014	39	21.2	10.9
February 2014	31	17.0	12.4
March 2014	22	12.1	8.3
April 2014	39	21.3	13.3
May 2014	126	69.2	44.9
June 2014	35	19.4	12.6
July 2014	27	15.0	13.9
August 2014	49	26.7	20.8
September 2014	66	35.9	–
October 2014	84	46.1	–
November 2014	9	5.0	4.1
December 2014	12	6.4	4.7
January 2015	73	40.0	22.9
February 2015	29	15.9	8.9
March 2015	18	9.9	4.8
April 2015	6	3.3	2.1
May 2015	87	47.7	19.9
June 2015	38	20.8	11.0
July 2015	92	50.5	23.3
August 2015	16	8.8	3.9

(continued)

Table 3 (continued)

Month	Rainfall (mm)	Theoretical volume from 548.55 m² (m³)	Real volume from 548.55 m² (m³)
September 2015	11	6.0	4.1
October 2015	110	60.3	35.6
November 2015	27	14.8	7.9
December 2015	8	4.4	2.2
January 2016	39	21.4	9.3
February 2016	104	56.9	28.4
March 2016	35	19.0	7.6
April 2016	28	15.2	8.1
May 2016	60	33.0	17.4
June 2016	49	26.8	12.2
July 2016	127	69.6	33.4
August 2016	90	49.3	24.6
September 2016	29	15.7	7.6
October 2016	107	58.9	31.3
November 2016	59	32.6	19.6
December 2016	11	6.1	2.8

water sources are best suited to non-potable uses, though some can be made potable with additional treatment.

If we can provide separate plumbing in and around buildings for potable and non-potable water, it opens up significant new options for water supply. Installing separate supply piping for landscape irrigation and cooling tower makeup water is fairly easy while installing separate non-potable supply plumbing for toilet flushing, which requires dual piping throughout a building, is more difficult.

Using any source of alternative water as another source of water in the buildings provides a lot of advantages for users. This source of water is free of charge and an independent source of water contrary to the water supply from the water company. Of course, these systems also have some disadvantages which we must take into account, i.e., unpredictability of water volume of an alternative source, increased demand for maintenance of this system is given by required water quality, etc.

The essential conclusion from our research is that both the theoretical and real volumes from our measurements show a significant potential for savings of potable water by the use of alternative water source in both measured places. In the case of school-type buildings, potential of water savings replaced by rainwater is significantly higher which is caused by the absence of purposes such as showering, bathing, laundry, etc. The most volume of potable water at school buildings in TUKE campus is consumed by flushing toilets, apparently the most suitable purpose for the use of rainwater. In the case of water from well, the results from research show savings more than 50% of potable water by replacing with water from well.

Table 4 Measured values of rainwater from PK6 building with comparison of total consumption of potable water in this building

Month	Real volume from 548.55 m^2 (m^3)	Total consumption of potable water in PK6 building (m^3)	Excess/lack of water (m^3)
January 2014	10.9	–	10.9
February 2014	12.4	6	6.4
March 2014	8.3	18	−9.7
April 2014	13.3	14	−0.7
May 2014	44.9	14	30.9
June 2014	12.6	15	−2.4
July 2014	13.9	9	4.9
August 2014	20.8	6	14.8
September 2014	–	4	–
October 2014	–	11	–
November 2014	4.1	23	−18.9
December 2014	4.7	17	−12.3
January 2015	22.9	8	14.9
February 2015	8.9	11	−2.1
March 2015	4.8	13	−8.2
April 2015	2.1	10	−7.9
May 2015	19.9	23	−3.1
June 2015	11	12	−1
July 2015	23.3	9	14.3
August 2015	3.9	3	0.9
September 2015	4.1	9	−4.9
October 2015	35.6	7	28.6
November 2015	7.9	8	−0.1
December 2015	2.2	31	−28.8
January 2016	9.3	12	−2.72
February 2016	28.4	6	22.38
March 2016	7.6	12	−4.44
April 2016	8.1	14	−5.9
May 2016	17.4	12	5.35
June 2016	12.2	11	1.16
July 2016	33.4	8	25.37
August 2016	24.6	0	24.55
September 2016	7.6	3	4.58
October 2016	31.3	13	18.3
November 2016	19.6	10	9.6
December 2016	2.8	20	−17.2

Water is the most precious resource. People still allow so precious commodity flow away, and we still want to shed water as quickly as possible. Many engineers still apply conventional drainage at first as a drainage concept for their projects.

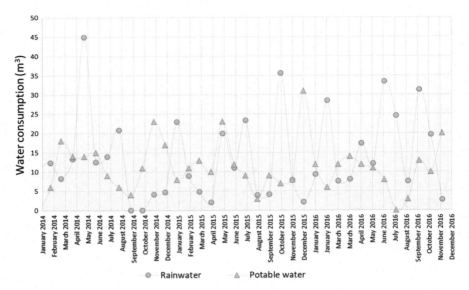

Fig. 9 Total consumption of potable water and measured rainwater volumes from roof construction of PK6 building as a potential alternative source of water for this building

Continuous growth of population and consequent growing need for potable water are global problems which lead to a search for new ways of effective use of water resources. So the time of cheap, easily accessible water has passed, and it is time to change our approach to this life-sustaining resource.

References

1. Stahre P (2006) Sustainability in urban storm drainage, planning and examples. Svenskt Vatten, Stockholm. ISBN 91-85159-20-4
2. Semadeni-Davies A, Bengtsson L (2000) Theoretical background. Urban drainage in specific climates. Urban drainage in cold climate, IHP-V technical documents in hydrology, no. 40, vol II. UNESCO, Paris
3. Butler D, Davies JW (2011) Urban drainage.3rd edn. Spon Press, London. ISBN 978-0-415-45526-8
4. Petiteville I, Lecomte P, Ward S, Dyke G, Steventon M, Harry J (2017) The earth observation handbook. http://www.eohandbook.com
5. Słyś D, Stec A (2014) The analysis of variants of water supply systems in multi-family residential. Ecol Chem Eng S 21(4):625–635
6. Słyś D, Stec A, Zelenakova M (2014) A LCC analysis of rainwater management variants. Ecol Chem Eng S 19(3):359–372
7. Markovič G, Vranayová Z (2013) Infiltration as a means of surface water drainage. Košice, TU
8. Markovič G (2016) A case study of the use of alternative water source in building. In: Cassotherm 2016: no conference proceedings of scientific papers. TU, Košice, pp 81–86. ISBN 978-80-553-3041-9

9. Markovič G, Vranayová Z, Káposztasová D (2016) Alternative water resources as a way for reducing dependence on mains water supply. In: IIAHS world congress on housing. Institute for Research and Technological Development in Construction Sciences, Coimbra, pp 1–8. ISBN 978-989-98949-4-5

10. Dolnicar S, Hurlimann A (2009) Drinking water from alternative water sources: differences in beliefs, social norms and factors of perceived behavioural control across eight Australian locations. Water Sci Technol 60(6):1433–1444

11. Pollicea A, Lopeza A, Laeraa G, Rubinob P, Lonigrob A (2003) Tertiary filtered municipal wastewater as alternative water source in agriculture: a field investigation in Southern Italy. Sci Total Environ 324(1–3):201–210

12. Káposztásová D (2016) Environmental and economic evaluation of small scale water-energy cycle in the cities of the future. In: INFRAEKO 2016. Oficyna Wydawnicza Politechniki Rzeszowskej, Rzeszow, pp 155–164. ISBN 978-83-7934-076-7

13. The Water Framework Directive 2000/60/EC

14. The Water Framework Directive (2002) Tap into it! Office for Official Publications of the European Communities, Luxembourg, 12 pp, ISBN 92-894-1946-6

15. Markovič G, Zeleňáková M (2014) Measurements of quality and quantity of rainwater runoff from roof in experimental conditions. In: ICITSEM 2014: international conference on innovative trends in science, engineering and management 2014, Dubaj, 12 and 13 Feb 2014. Mudranik Technologies, Bangaluru, pp 145–151. ISBN 978-93-83303-19-9

16. Markovič G, Vranayová Z, Káposztasová D (2014) The analysis of the possible use of harvested rainwater in real conditions at the university campus of Kosice. In: Proceedings of the 2014 international conference on environmental science and geoscience (ESG'14), Venice, 15–17 Mar 2014, p 46

17. Markovič G, Vranayová Z (2015) Measurements of the quality of rainwater run-off and its potential for savings of potable water in TUKE campus. WSEAS Trans Environ Dev 11:302–311. ISSN 1790-5079

18. Zeleňáková M, Markovič G, Káposztásová D, Vranayová Z (2014) Rainwater management in compliance with sustainable design of buildings. In: Procedia engineering: 16th conference on water distribution system analysis, WDSA2014 – urban water hydroinformatics and strategic planning, No. 89, Bari, 14–17 July 2014, pp 1515–1521. ISSN 1877-7058

19. Markovič G, Vranayová Z, Káposztasová D (2016) Rainwater as an important alternative water supply system in buildings. In: INFRAEKO 2016. Oficyna Wydawnicza Politechniki Rzeszowskej, Rzeszow, pp 217–226. ISBN 978-83-7934-076-7

20. Markovič G, Zeleňáková M, Kaposztasova D, Hudakova G (2014) Rainwater infiltration in the urban areas. In: Environmental impact 2, vol 181. WITT Press, Southampton, pp 313–320. ISBN 978-1-84564-762-9. ISSN 1743-3541

Part VI
Conclusion

Update, Conclusions, and Recommendations for Water Resources in Slovakia: Assessment and Development

Martina Zeleňáková and Abdelazim M. Negm

Contents

Abstract This chapter presents an update regarding the water quality assessment and development in Slovakia. Also, the main conclusions and recommendations of the chapters presented in this volume are summarized. Therefore, this chapter presents a summary of the most important findings presented by the contributors of the volume. Topics which are covered include water resources in Slovakia, their assessment and development, water supply and demand, irrigation water, groundwater, water and sediment quality with emphasizing to mining water, wastewater management in Slovakia, and rainwater management. Also, a set of recommendations for future research work is pointed out to direct the future research toward the development of water resources which is one of the strategic themes of the Slovak Republic.

Keywords Assessment, Development, Groundwater, Irrigation, Mining water, Rainwater, Slovakia, Supply and demand, Water resources

M. Zeleňáková
Department of Environmental Engineering, Faculty of Civil Engineering, Technical University of Košice, Košice, Slovakia
e-mail: martina.zelenakova@tuke.sk

A. M. Negm (✉)
Water and Water Structures Engineering Department, Faculty of Engineering, Zagazig University, Zagazig, 44519, Egypt
e-mail: Amnegm@zu.edu.eg; Amnegm85@yahoo.com

A. M. Negm and M. Zeleňáková (eds.), *Water Resources in Slovakia:*
Part I - Assessment and Development, Hdb Env Chem (2019) 69: 375–386,
DOI 10.1007/698_2018_326, © Springer International Publishing AG, part of Springer Nature 2018,
Published online: 20 June 2018

1 Introduction

Water is an essential medium regarding the transport, decomposition, and accumulation of pollutants, whether of natural or anthropogenic origin, which in excessive amounts represent considerable risks for all kinds of living organisms, thus also for human beings. The step toward adequate protection of water resources is to know their quality. Systematic investigation and evaluation of the occurrence of surface water and groundwater within the country is a fundamental responsibility of the state, as an indispensable requirement for ensuring the preconditions for permanently sustainable development as well as for maintaining standards of public administration and information. The primary requirement in this context is to optimize water quality monitoring and assessment and the implementation of necessary environmental measures.

The chapter presents a brief of the essential findings of the studies on the assessment and development of water resources in Slovakia and then the main conclusions and recommendations of the volume chapters in addition to few recommendations for researchers and decision-makers.

2 Update

In the following the national studies regarding the water resources, assessment, and development in Slovakia are presented. The most studies were done concerning water quality. The brief results of the studies are introduced.

Hiller et al. [1] studied the concentrations and fractionation of metals (antimony, arsenic, cadmium, chromium, cobalt, copper, lead, mercury, molybdenum, nickel, vanadium, and zinc) in the surface sediments of the two water reservoirs Ruzin and Velke Kozmalovce. When the risk assessment code was applied to the fractionation study, cadmium and cobalt came under the high-risk and the very high-risk category for the environment and therefore might cause an adverse effect to aquatic life. Heavy metal concentrations were investigated in perches (*Perca fluviatilis*) in the Ruzin water reservoir and in two of its most common parasites by Brázová et al. [2]. Samples of fish and both parasites were analyzed for As, Cd, Cr, Cu, Hg, Mn, Ni, Pb, and Zn. Zinc was found to be the dominant element, and its antagonistic interaction with copper was confirmed. Hiller et al. [3] investigated polycyclic aromatic hydrocarbon (PAH) distribution and predicted their possible sources in three water reservoirs from the Slovak Republic. "The results showed that the highest total PAH concentrations were associated with sediments from the Velke Kozmalovce. Evidently increased environmental pollution as a consequence of the 25-year manufacture of polychlorinated biphenyls (PCBs) in eastern Slovakia was observed" by Kočan et al. [4]. The manufacturer's effluent canal causes the contamination of the Laborec river and large Zemplinska Sirava reservoir since PCB levels in the canal sediment are still high.

The impact of forestry, agriculture, and urban activities on the quality of surface water was analyzed in the study by Pekárová and Pekár [5]. "It is shown that the nitrate concentrations in surface water have decreased in Slovakia since 1989 as a result of decreased use of inorganic nitrogen fertilisers (lower intensity of agricultural production in Slovakia). Numerous recent laboratory studies have shown that vegetation can influence soil water flow by inducing very low levels of water repellency" [6]. In the study, Lichner et al. [6] extended on this previous research by developing a field-based test using a miniature infiltrometer to assess low levels of water repellency from physically based measurements of liquid flow in soil. Pekárová et al. [7] summarized investigations of air and water temperature in the Bela River in Slovakia. While the air temperature within 50 years increased significantly by 1.5°C, in the case of water temperature, this increase was merely by 0.12°C. In the second part of the study, the impact of the riparian vegetation growing along the riverbanks was evaluated.

Gulisa et al. [8] conducted an ecologic study to determine whether nitrate levels in drinking water were correlated with non-Hodgkin lymphoma and cancers of the digestive and urinary tracts in an agricultural district in Trnava District. "These ecologic data support the hypothesis that there is a positive association between nitrate in drinking water and non-Hodgkin lymphoma and colorectal cancer" [8].

Combined sewer systems in Slovakia were evaluated in the study of Sztruhár et al. [9]. Over 300 combined sewer overflows (CSOs) were visited and their structural condition evaluated. Samples of overflowing water were taken from eight overflows and analyzed for common constituents. The first flush of organic pollution was not confirmed in any of the events.

3   Conclusions

The following conclusions have mainly extracted from the chapters[1] presented in this volume of the Handbook of Environmental Chemistry. Surface water and groundwater resources of Slovakia are rich enough to ensure current and prospective water needs. The surface waters are formed by surface water inflow to Slovakia and by surface water runoff rising at the Slovak territory. Groundwater resource formation is dependent mainly on geological-tectonic conditions, hydrogeological parameters of the rock environment, and climatic conditions. Surface water resources in Slovakia are bound to two different European river basins. The Danube River Basin covers 96% of the Slovak territory; Danube River flows toward the Black Sea. The Poprad and Dunajec river basins cover 4% of the territory; both streams are tributaries of the Vistula River flowing toward the Baltic Sea. Surface water and groundwater bodies were delineated on the Slovak territory according to the Water

---

[1]Therefore, whenever the words "chapter titled" appears followed by the title of the chapter, it means that the chapter is contained in this volume.

Framework Directive requirements. In total, 1,487 surface water bodies are on the list at present, 84 of them in the Vistula River Basin District and 1,413 in the Danube River Basin District. The largest rivers of Slovakia besides of Danube and Morava rivers, which have their springs outside the Slovak territory, are Váh, Nitra, Hron, Ipeľ, Slaná, Hornád, Bodva, Bodrog, and Poprad with their tributaries. Groundwater bodies are divided into three levels – there are 16 quaternary, 59 pre-quaternary, and 27 geothermal structures.

The water supply in Slovakia is mostly assured from the public water supply systems based either on the surface water or on groundwater sources. Surface water sources are represented by water reservoirs or by direct water takeoff from the surface streams. As groundwater sources, either wells or springs can be utilized. There are 295 water reservoirs in Slovakia, 8 water reservoirs were constructed until now for drinking water supply purposes, and more than 200 small water reservoirs serve mostly for irrigation. The amount of water abstraction from the groundwater sources is generally higher than from the surface water. The number of inhabitants supplied with water from the public water supply sources has been increasing steadily, reaching the number of 4.7853 million (88.3%) of inhabitants of Slovakia. The number of supplied municipalities reached 2,380 with the share of 82.4% on the total number of municipalities of Slovakia. However, the trend of the water consumption, both total and specific for a private household, is declining in the long-term scale, declining from 195.5 L capita^{-1} day^{-1} in 1990 to 77.3 L capita^{-1} day^{-1} in 2015. The quality of drinking water from the public water supply systems has been showing a high level in the long-term period. The development of public sewerage systems lags behind that of public water supplies. In 2015, about 1,044 municipalities had the public sewerage system in place. This makes only 36.2% of the total number of 2,890 municipalities in Slovakia. The main economic sectors using the surface water are industry, public drinking water supply, and irrigation of arable lands.

Agriculture had been intensified in Europe as well as in Slovakia after the World War I, alongside with industry. In Czechoslovakia and Slovakia, conditions for designing and preparing implementation and operation of irrigation were well created. It was a long-term process mainly after 1960. Organizations from designing to irrigation's operation were established. After 1990, the situation has changed, and nowadays, we are resolving the need for irrigation again. The chapter titled "Irrigation of Arable Land in Slovakia - History, and Perspective" in this volume analyzes development of organizations and management of irrigation in the past and looks at the future development. In Slovakia, the manager and organizer of irrigation constructions have become, and still is, the state. Irrigation water can cause damage to irrigated crops and human and animal health. Therefore, it is essential to monitor the irrigation water regularly. There is a long-term tradition of irrigation water quality monitoring in Slovakia. The number of monitoring stations varied during years from more than 200 to 11 in recent years. The irrigation water quality increased over the years. An advantage of the irrigation water monitoring is in increasing quality of agriculture production, in a reduction of a risk of bacterial and virus infection of humans and animals. Another risk mainly in the agricultural sector is drought. Drought by itself cannot be considered a disaster. However, if its impacts on local people, economies, and the environment are severe, and their ability to cope with and recover from it is difficult, it should be considered as a disaster. Droughts and floods are a recognizable category of natural risk. Hydrological assessments of drought impacts require detailed characteristics. The chapter titled "Small Water Reservoirs - Source of Water for Irrigation" proposed a new conceptual framework for drought identification in landscape with agricultural use. They are described hydrological drought characteristics with impacts at the agricultural landscape and food security and the issues related to drought water management. In the past, the Slovak Republic was not considered a country immediately threatened with drought. The situation had changed at the turn of the millennium, especially after the extreme weather conditions in 2014 and also in 2015, when, for example, the historical minima were recorded. Drought could affect the whole hydrological cycle.

Surface water-groundwater interaction is a dynamic process which can be influenced by many factors most associated with the hydrological cycle. Besides the fluctuation of surface water and groundwater levels and their gradient, this interaction is also influenced by the parameters of the aquifer (regional and local geology and its physical properties). The next significant factors are precipitation, the water level regime of rivers or reservoirs based on the area of interest, and last but not least the properties of the riverbed itself. The investigation of the interaction between the surface water and groundwater was applied utilizing modern numerical simulations on the Gabčíkovo-Topoľníky channel. On the other hand, the channel Network at Žitný Ostrov Area is one of the main channels of irrigation and drainage channel network at Žitný Ostrov. Žitný Ostrov area is situated in the southwestern part of Slovakia, and it is known as the biggest source of groundwater in this country. For this reason, experts give it heightened attention from different points of view. The channel network was built up in this region for drainage and safeguarding of irrigation water. The water level in the whole channel network system affects the groundwater level and vice versa. With regard to the mutual interaction between channel network and groundwater, it has been necessary to judge the impact of channel network silting up by alluvials and the rate of their permeability to this interaction. The results of simulation of real and theoretical scenarios of interaction between groundwater and surface water along the Gabčíkovo-Topoľníky channel produced valuable information about how the clogging of the riverbed in the channel network influences the groundwater level regime in the area. The lower boundary of unsaturated soil zone is formed by groundwater level. At this level, water from unsaturated soil zone flows to groundwater and vice versa. Groundwater penetrates the unsaturated zone. By capillary rise, groundwater can supply water storage in the root zone and thus influence on actual evaporation in this soil layer. The degree to which this occurs depends on given soil texture and the groundwater level position with regard to the position of lower root zone boundary. The chapter titled "Impact of Soil Texture and Position of Groundwater Level on Evaporation from the Soil Root Zone" quantifies the impact of soil texture on the involvement of groundwater in the evaporation process. The results were obtained by numerical experiment on

GLOBAL model. The measurements used for model verification and numerical simulation were gained in East Slovakia Lowland.

Water is a necessary component of the human environment, as well as all vegetal and animal ecosystems. Unfortunately, water quality not just in Slovakia, but also in other countries of the world, worsened in the course of the twentieth century, and this trend has not been stopped even at present. Current legislation evaluating the quality of water bodies in Slovakia is based on the implementation of the Water Framework Directive (2000/60/ES). The Directive requires eco-morphological monitoring of water bodies, which is based on an evaluation of the rate of anthropogenic impact. This does not refer only to riverbeds but also to the state of the environs of each stream. While in the past point sources of pollution were considered as the most significant source of pollution in surface streams, after the installation of treatment plants for urban and industrial wastewater, nonpoint sources of pollution emerged as the critical sources of pollution in river basins. The chapter titled "Assessment of Water Pollutant Sources and Hydrodynamics of Pollution Spreading in Rivers" deals with the distribution and quantity assessment of pollutant sources in Slovakia during the period 2006–2015. The primary point sources evaluated are the ones representing higher values than the 90 percentile of the empirical distribution of total mass and also the mass of applied manures and fertilizers as nonpoint pollutant sources.

The development of computer technologies enables us to solve ecological problems in water management practice very efficiently. Mathematical and numerical modeling allows us to evaluate various situations of spreading of contaminants in rivers without immediate destructive impact on the environment. However, the reliability of models is closely connected with the availability and validity of input data. Hydrodynamic models simulating pollutant transport in open channels require large amounts of input data and computational time, but on the other hand, these kinds of models simulate dispersion in surface water in more detail. As input data, they require digitization of the hydro-morphology of a stream, velocity profiles along the simulated part of the stream, calculation of the dispersion coefficients, and also the locations of pollutant sources and their quantity. The greatest extent of uncertainty is linked with the determination of dispersion coefficient values. These coefficients can be accurately obtained by way of field measurements, directly reflecting conditions in the existing part of an open channel. It is not always possible to obtain these coefficients in the field, however, because of financial or time constraints. The results and obtained knowledge about values of longitudinal dispersion coefficients and dispersion processes that are presented in the chapter titled "Assessment of Water Pollutant Sources and Hydrodynamics of Pollution Spreading in Rivers" can be applied in numerical simulations of pollutant spreading in a natural stream.

The chapter titled "Assessment of Heavy Metal Pollution of Water Resources in Eastern Slovakia" presents sediment quality monitoring which is among the highest priorities of environmental protection policy. The chapter introduces ways to control and minimize the incidence of pollutant-oriented problems and to provide for water of appropriate quality to serve various purposes such as drinking water supply, irrigation water, etc. The quality of sediments is identified in terms of their physical, chemical, and biological parameters. The particular problem regarding sediment quality monitoring is the complexity associated with analyzing a large number of measured variables. The research was realized to determine and analyze selected heavy metals present in sediment samples from six river basins on east of Slovakia, representing by the rivers Hornad, Laborec, Torysa, Ondava, Topla, and Poprad. Sampling points were selected based on the current surface water quality monitoring network. The investigation was focused on heavy metals (Zn, Cu, Pb, Cd, Ni, Hg, As, Fe, Mn). The content of heavy metals reflected the scale of industrial and mining activities in a particular locality. The degree of sediment contamination in the rivers has been evaluated using an enrichment factor, pollution load index, geo-accumulation index, and potential environmental risk index. Acid mine drainage (AMD) has been a detrimental by-product of sulfidic ore mining for many years. In most cases, this acid comes primarily from oxidation of iron sulfide, which is often found in conjunction with valuable metals. AMD is a worldwide problem, leading to ecological destruction in watersheds and the contamination of human water sources by sulfuric acid and heavy metals, including arsenic, copper, and lead. The Slovak Republic belongs to the countries with long mining tradition, especially in connection with the mining of iron, copper, gold, silver, and another polymetallic ore. The abandoned mine Smolnik is one of these mines where AMD is produced. Acid mine drainage from an abandoned sulfide mine in Smolnik, with the flow rates of 5–10 L s^{-1} and a pH of 3.7–4.1, flows into Smolnik creek and adversely affects the stream's water quality and ecology. High rainfall events increase the flow of Smolnik creek, which ranges from 0.3 to 2.0 m^3 s^{-1} (monitored 2006–2016). Increased flow is also associated with a pH increase and precipitation of metals (Fe, Al, Cu, and Zn) and their accumulation in sediment. The dependence of pH on flow in Smolnik creek was evaluated using regression analysis. The study also deals with the metal distribution between water and sediment in the Smolnik creek depending on pH and the metal concentrations. Acid mine drainage is the product of the natural oxidation of sulfide minerals. The simultaneous influence of water, oxygen, and indigenous microorganisms represents the necessary conditions for AMD formation. The occurrence of AMD is associated mainly with the presence of sulfide minerals in the polymetallic, coal, and lignite deposits. AMD contaminates the groundwaters and soils because it contains mainly sulfuric acid, heavy metals, and metalloids. During the exploitation, and mostly after the mine closure, the produced AMD pollutes the environment. The continuance of AMD generation is difficult to halt. Self-improvement situation is not possible. It is necessary to monitor the quality of AMD and develop the methods of their treatment. Slovakia belongs to the countries with significant mining tradition, especially with regard to the exploitation of iron, copper, gold, and silver. Currently, only one deposit is being exploited, namely, Au ore deposit in Hodruša. The other deposits are mostly flooded. They present the suitable conditions for creation and intensification of chemical and biological-chemical oxidation of the sulfide minerals, i.e., formation of AMD. In Slovakia, Smolnik and Pezinok deposits, as well as the Šobov dump, are typical examples of the old mining loads with the production of AMD.

For the supply of drinking water in Slovakia, as was mentioned, groundwater resources are mainly used with 87.3% of inhabitants which are supplied with drinking water from underground resources; approximately 22% of this amount has to be treated. Water treatment is mostly needed for the removal of iron and/or manganese. Concentrations of dissolved iron and manganese are evaluated every year within the groundwater monitoring done by the Slovak Hydrometeorological Institute (SHMI) for the whole territory of Slovakia. "Iron and manganese compounds in water give rise to technological problems, failures of water supply systems, and deterioration of water quality with respect to sensory properties. If these waters are slightly over-oxidized [10], unfavorable incrustations are formed." The objective of the pilot plant tests in the water treatment plant Kúty which is presented in the chapter titled "Influence of Mining Activities on Quality of Ground Water" was to verify the efficiency of manganese and iron removal from water with using of different filtration materials with MnO_2 layer on the surface – Klinopur-Mn, Greensand, and Cullsorb M. "The increased pollution of water resources leads to a deterioration in the quality of surface water, and groundwater and it initiates the application of various methods for water treatment. The Slovak Technical Standards – STN 75 7111 drinking water and the enactment the Decree of the Ministry of Health of the Slovak Republic No. 151/2004 on requirements for drinking water and monitoring of the quality of drinking water quality has resulted in the reduction of heavy metal concentrations or, for the first time, in defining the limit concentrations for some heavy metals (As, Sb), respectively. Based on this fact, some water resources in Slovakia have become unsuitable for further use, and they require appropriate treatment" [11]. The chapter titled "Wastewater Management and Water Resources" introduced a method to verify the sorption properties of some new sorption materials for the removal of antimony (Bayoxide E33, GEH, CFH12). Technological tests were carried out at the facility of the Slovak Water Company Liptovský Mikuláš in the locality of Dúbrava. Technological tests have proved that the new sorption materials can be used for reduction of antimony concentration in water to meet the values set under the Decree of the Ministry of Health of the Slovak Republic No. 247/2017 on requirements for drinking water – 5 µg/L.

Slovakia is the small country typical with high mountains on the north part; two lowlands reach the south border and with the middle mountains between them. This character of Slovakia, as was already mentioned, defines the two river drainage basins orientated to the north - Baltic Sea by the river Poprad and the next two directed to the south to the Black Sea. The Slovak Republic territory is 49,014 km^2; its population is 5.4 million and is located in the temperate climate zone of the Northern Hemisphere with regularly alternating seasons. About 38% of the country is forested. Based on longitudinal measurements, the average annual air temperature is 7°C. The longitudinal average amount of precipitation is around 760 mm. The capacity of natural surface water sources amounts is about 90.3 $m^3 s^{-1}$. Ecological discharge is 36.5 $m^3 s^{-1}$. Water reservoirs across Slovakia enable increasing the discharges in dry periods in 53.8 $m^3 s^{-1}$. Reservoirs can provide approximately 4,000 L s^{-1} of the high quality of water used for drinking purposes. Water offtake, currently amounting to 39 m^3 s^{-1}, is equal to about 29% of the discharges during dry periods and to 10% of the longitudinal mean discharge. The water consumption is significant for agricultural, for industry, and for drinking purposes, such as water supply system and production of bottled water. The very significant for Slovakia is the huge amount of mineral waters across the country, which reaches the high quality not only for drinking but for health purposes too. The connection to the public water supply system of the population is over 95%. The worse situation is with the public connection to the public sewer system, which reaches over 66% in the year 2017. This situation is continuously increased in the last 30 years, which is conditional by the investments. There are many people who work in the water sector, which is covered by the Ministry of Environment, governmental and public institutions, and a lot of private companies, which have a goal to improve and protect the public health. The world is facing severe challenges in the management of water in various urban and regional locations nowadays. What we have to admit are significant gaps in our knowledge about existing alternative sources of water for potable as well as non-potable use. So the biggest challenge for developers, engineers, and architects is increasing water supply through alternative water sources. The chapter titled "Urban Rainwater Drainage" contains results of measurements and evaluation in the field of alternative water resources. It is also an evaluation of water consumption in a family house secondary source of water – water from well where it was made. Main water supply for the activities as flushing of toilets and clothes washing was entirely replaced by the water from well. This chapter also provides analyses of the potential use of another water resource (rainwater) for the second object – school building in the university campus – according to measured data of rainfall and flowed volumes of runoff for this actual school building.

4 Recommendations

The assessment of water resources of a country is a national responsibility, and relevant activities should be proposed so that the specific needs of a country are met. Many of its component activities may be done at the local and regional levels. This national responsibility should be divided among neighboring countries in the case of cross-border water resources, and international programs and projects may provide valuable help.

With respect to the importance of the assessed information on sustainable development and the maintaining of the integrity of ecosystems, all countries are urgently called upon to achieve a level of assessment of water resources corresponding to needs as soon as possible.

The policy should be such that all national and international activities of assessing water resources are fully coordinated and financed over the long term. The approach to achieving this goal may differ in individual countries but will typically include the mandating of regulations and administrative decisions, especially in terms of allocating financial resources.

The assessment of water resources requires significant financial resources if support of sustainable social-economic development is raised with this. These resources, however, represent only a small portion (for example, 0.2–1.0%) of financial resources expended on investments and activities in the water sector as a whole. Governments are urgently called on to allocate national and international funds for priority assessment of activities in the area of water resource management.

Reliable information about the state and trends of a country's water resources (surface water, waters in an unsaturated zone, and groundwater) both quantity and quality, are assessed for several purposes. These purposes include evaluating the potential sources and potential for storage of present and foreseeable demand and the protection of people and property against dangerous associations with water; planning, designing, and operating water projects; and monitoring the off-take of water units for anthropogenic impacts, variability, and climate change and for other environmental factors.

Integrated monitoring and information systems should be established, and data should be collected and preserved on all aspects of water sources which are necessary for complete understanding of the nature of these sources and for their sustainable development. Information includes not only hydrological data but also associated geological, climatological, hydrobiological, and topographical data and data on types of soil, the use of soil, and desertification and deforestation, as well as information about subjects such as the using and reusing of water, wastewater treatment, point and exceptional sources of pollution, and runoff into the seas and oceans. This includes the installation of observation networks and other mechanisms for gathering data determined for the monitoring of various climatic and topographical regimes and for the development of tools for storing data. In places where the national, regional, and international level information related to water with the number of information systems is managed, it is essential that these systems be coordinated.

Recommended actions and/or activities for water resource management in Slovakia include the following:

1. Defining informational needs of users and the creation of internal state policy, an elective framework, effective institutional structures, and economic instruments suitable for the assessment of water resources.
2. Introducing and maintaining of active and effective cooperation in the area of assessment of water resources and activities of hydrological prognosticating among national agencies within a country and between countries with respect to cross-border water resources.
3. Encouraging those who are responsible for gathering and storing data to apply processes which were elaborated and approved on the international level when assessing their activities in the area of water resources.
4. Developing and distributing information on resources for determining the benefits and costs of activities of water resource assessment and helping with internal state services so that benefits from water resources assessments are demonstrated.

5. Elaborating of practical and legislative provisions for long-term sustainability of activities in the field of the use and prognosticating of water resources and allocating the necessary financial resources, especially in the case of developing countries.
6. Installing of monitoring systems designed for the provision of valid and comparable data associated with water.
7. Ensuring the continuous functioning of such systems for the support of studies which require long-term data, such as, for example, data related to climate change.
8. Modernizing of equipment and processes for preserving, confirming, and securing such data.
9. Implementing technology for processing such data and assimilation of associated information.
10. Comparing, selecting, and applying of hydrological technology corresponding to the needs of each country and ensuring the transmission of suitable technology, especially between hydrological services.

References

1. Hiller E, Jurkovič L, Šutriepka M (2010) Metals in the surface sediments of selected water reservoirs, Slovakia. Bull Environ Contam Toxicol 84:635. https://doi.org/10.1007/s00128-010-0008-y
2. Brázová T, Torres J, Eira C, Hanzelová V, Miklisová D, Šalamún P (2012) Perch and its parasites as heavy metal biomonitors in a freshwater environment: the case study of the Ružín water reservoir, Slovakia. Sensors 12(3):3068–3081. https://doi.org/10.3390/s120303068
3. Hiller E, Sirotiak M, Jurkovič L, Zemanová L (2009) Polycyclic aromatic hydrocarbons in bottom sediments from three water reservoirs, Slovakia. Bull Environ Contam Toxicol 83(3):444–448
4. Kočan A, Petrik J, Jursa S, Chovancová J, Drobná B (2001) Environmental contamination with polychlorinated biphenyls in the area of their former manufacture in Slovakia. Chemosphere 43(4–7):595–600
5. Pekárová P, Pekár J (1996) The impact of land use on stream water quality in Slovakia. J Hydrol 180(1–4):333–350
6. Lichner L, Hallett P, Feeney D, Ďugová O, Šír M, Tesař M (2017) Field measurement of soil water repellency and its impact on water flow under different vegetation. Biologia 62:537–541
7. Pekárová P, Miklánek P, Halmová D, Onderka M, Pekár J, Kučárová K, Liová S, Škoda P (2011) Long-term trend and multi-annual variability of water temperature in the pristine Bela River basin (Slovakia). J Hydrol 400(3–4):333–340
8. Gulisa G, Czompolyova M, Cerhan JR (2002) An ecologic study of nitrate in municipal drinking water and cancer incidence in Trnava District, Slovakia. Environ Res 88(3):182–187
9. Sztruhár D, Sokáč M, Holienčin A, Markovič A (2002) Comprehensive assessment of combined sewer overflows in Slovakia. Urban Water 4(3):237–243
10. Barloková D (2008) Natural zeolites in the water treatment process. Slovak J Civ Eng 16(2):8–12. https://www.svf.stuba.sk/buxus/docs/sjce/2008/2008_2/file1.pdf
11. Ilavský J (2008) Removal of antimony from water by sorption materials. Slovak J Civ Eng 2:1–6. https://www.svf.stuba.sk/buxus/docs/sjce/2008/2008_2/file3.pdf

Index

A. M. Negm and M. Zeleňáková (eds.), *Water Resources in Slovakia:*
Part I - Assessment and Development, Hdb Env Chem (2019) 69: 387–392,
https://doi.org/10.1007/978-3-319-92853-1,
© Springer International Publishing AG, part of Springer Nature 2019

Printed by Printforce, the Netherlands